194
Advances in Polymer Science

Editorial Board:
A. Abe · A.-C. Albertsson · R. Duncan · K. Dušek · W. H. de Jeu
J.-F. Joanny · H.-H. Kausch · S. Kobayashi · K.-S. Lee · L. Leibler
T. E. Long · I. Manners · M. Möller · O. Nuyken · E. M. Terentjev
B. Voit · G. Wegner · U. Wiesner

Advances in Polymer Science

Recently Published and Forthcoming Volumes

Surface-Initiated Polymerization II
Volume Editor: Jordan, R.
Vol. 198, 2006

Surface-Initiated Polymerization I
Volume Editor: Jordan, R.
Vol. 197, 2006

Conformation-Dependent Design of Sequences in Copolymers II
Volume Editor: Khokhlov, A. R.
Vol. 196, 2006

Conformation-Dependent Design of Sequences in Copolymers I
Volume Editor: Khokhlov, A. R.
Vol. 195, 2006

Enzyme-Catalyzed Synthesis of Polymers
Volume Editors: Kobayashi, S., Ritter, H., Kaplan, D.
Vol. 194, 2006

Polymer Therapeutics II
Polymers as Drugs, Conjugates and Gene Delivery Systems
Volume Editors: Satchi-Fainaro, R., Duncan, R.
Vol. 193, 2006

Polymer Therapeutics I
Polymers as Drugs, Conjugates and Gene Delivery Systems
Volume Editors: Satchi-Fainaro, R., Duncan, R.
Vol. 192, 2006

Interphases and Mesophases in Polymer Crystallization III
Volume Editor: Allegra, G.
Vol. 191, 2005

Block Copolymers II
Volume Editor: Abetz, V.
Vol. 190, 2005

Block Copolymers I
Volume Editor: Abetz, V.
Vol. 189, 2005

Intrinsic Molecular Mobility and Toughness of Polymers II
Volume Editor: Kausch, H.-H.
Vol. 188, 2005

Intrinsic Molecular Mobility and Toughness of Polymers I
Volume Editor: Kausch, H.-H.
Vol. 187, 2005

Polysaccharides I
Structure, Characterization and Use
Volume Editor: Heinze, T.
Vol. 186, 2005

Advanced Computer Simulation Approaches for Soft Matter Sciences II
Volume Editors: Holm, C., Kremer, K.
Vol. 185, 2005

Crosslinking in Materials Science
Vol. 184, 2005

Phase Behavior of Polymer Blends
Volume Editor: Freed, K.
Vol. 183, 2005

Polymer Analysis/Polymer Theory
Vol. 182, 2005

Interphases and Mesophases in Polymer Crystallization II
Volume Editor: Allegra, G.
Vol. 181, 2005

Interphases and Mesophases in Polymer Crystallization I
Volume Editor: Allegra, G.
Vol. 180, 2005

Enzyme-Catalyzed Synthesis of Polymers

Volume Editors:
Shiro Kobayashi · Helmut Ritter · David Kaplan

With contributions by
D. L. Kaplan · S. Kobayashi · S. Matsumura · M. Ohmae
M. Reihmann · H. Ritter · A. Singh · H. Uyama · P. Xu

 Springer

The series *Advances in Polymer Science* presents critical reviews of the present and future trends in polymer and biopolymer science including chemistry, physical chemistry, physics and material science. It is adressed to all scientists at universities and in industry who wish to keep abreast of advances in the topics covered.
As a rule, contributions are specially commissioned. The editors and publishers will, however, always be pleased to receive suggestions and supplementary information. Papers are accepted for *Advances in Polymer Science* in English.
In references *Advances in Polymer Science* is abbreviated *Adv Polym Sci* and is cited as a journal.

Springer WWW home page: http://www.springer.com
Visit the APS content at http://www.springerlink.com/

Library of Congress Control Number: 2005933611

ISSN 0065-3195
ISBN-10 3-540-29212-8 Springer Berlin Heidelberg New York
ISBN-13 978-3-540-29212-8 Springer Berlin Heidelberg New York
DOI 10.1007/11549307

This work is subject to copyright. All rights are reserved, whether the whole or part of the material is concerned, specifically the rights of translation, reprinting, reuse of illustrations, recitation, broadcasting, reproduction on microfilm or in any other way, and storage in data banks. Duplication of this publication or parts thereof is permitted only under the provisions of the German Copyright Law of September 9, 1965, in its current version, and permission for use must always be obtained from Springer. Violations are liable for prosecution under the German Copyright Law.

Springer is a part of Springer Science+Business Media

springer.com

© Springer-Verlag Berlin Heidelberg 2006
Printed in Germany

The use of registered names, trademarks, etc. in this publication does not imply, even in the absence of a specific statement, that such names are exempt from the relevant protective laws and regulations and therefore free for general use.

Cover design: *Design & Production* GmbH, Heidelberg
Typesetting and Production: LE-TEX Jelonek, Schmidt & Vöckler GbR, Leipzig

Printed on acid-free paper   02/3141 YL – 5 4 3 2 1 0

Volume Editors

Prof. Shiro Kobayashi
R & D Center for Bio-based Materials
Kyoto Institute of Technology
Matsugasaki, Sakyo-ku
Kyoto 606-8585, Japan
kobayash@kit.ac.jp

Prof. Helmut Ritter
Institut für Organische Chemie und
Makromolekulare Chemie
Lehrstuhl II, Gebäude 26.33, Ebene 00
Raum 40
Universitätsstraße 1
40225 Düsseldorf, Germany
H.Ritter@uni-duesseldorf.de

Prof. David Kaplan
Department of Biomedical Engineering
Science and Technology Center
Medford, MA 02155, USA
david.kaplan@tufts.edu

Editorial Board

Prof. Akihiro Abe
Department of Industrial Chemistry
Tokyo Institute of Polytechnics
1583 Iiyama, Atsugi-shi 243-02, Japan
aabe@chem.t-kougei.ac.jp

Prof. A.-C. Albertsson
Department of Polymer Technology
The Royal Institute of Technology
10044 Stockholm, Sweden
aila@polymer.kth.se

Prof. Ruth Duncan
Welsh School of Pharmacy
Cardiff University
Redwood Building
King Edward VII Avenue
Cardiff CF 10 3XF, UK
DuncanR@cf.ac.uk

Prof. Karel Dušek
Institute of Macromolecular Chemistry,
Czech
Academy of Sciences of the Czech Republic
Heyrovský Sq. 2
16206 Prague 6, Czech Republic
dusek@imc.cas.cz

Prof. W. H. de Jeu
FOM-Institute AMOLF
Kruislaan 407
1098 SJ Amsterdam, The Netherlands
dejeu@amolf.nl
and Dutch Polymer Institute
Eindhoven University of Technology
PO Box 513
5600 MB Eindhoven, The Netherlands

Prof. Jean-François Joanny
Physicochimie Curie
Institut Curie section recherche
26 rue d'Ulm
75248 Paris cedex 05, France
jean-francois.joanny@curie.fr

Prof. Hans-Henning Kausch
Ecole Polytechnique Fédérale de Lausanne
Science de Base
Station 6
1015 Lausanne, Switzerland
kausch.cully@bluewin.ch

Prof. Shiro Kobayashi
R & D Center for Bio-based Materials
Kyoto Institute of Technology
Matsugasaki, Sakyo-ku
Kyoto 606-8585, Japan
kobayash@kit.ac.jp

Prof. Kwang-Sup Lee
Department of Polymer Science &
Engineering
Hannam University
133 Ojung-Dong Daejeon,
306-791, Korea
kslee@hannam.ac.kr

Prof. L. Leibler
Matière Molle et Chimie
Ecole Supérieure de Physique
et Chimie Industrielles (ESPCI)
10 rue Vauquelin
75231 Paris Cedex 05, France
ludwik.leibler@espci.fr

Prof. Timothy E. Long
Department of Chemistry
and Research Institute
Virginia Tech
2110 Hahn Hall (0344)
Blacksburg, VA 24061, USA
telong@vt.edu

Prof. Ian Manners
School of Chemistry
University of Bristol
Cantock's Close
BS8 1TS Bristol, UK
Ian.Manners@bristol.ac.uk

Prof. Martin Möller
Deutsches Wollforschungsinstitut
an der RWTH Aachen e.V.
Pauwelsstraße 8
52056 Aachen, Germany
moeller@dwi.rwth-aachen.de

Prof. Oskar Nuyken
Lehrstuhl für Makromolekulare Stoffe
TU München
Lichtenbergstr. 4
85747 Garching, Germany
oskar.nuyken@ch.tum.de

Prof. E. M. Terentjev
Cavendish Laboratory
Madingley Road
Cambridge CB 3 OHE, UK
emt1000@cam.ac.uk

Prof. Brigitte Voit
Institut für Polymerforschung Dresden
Hohe Straße 6
01069 Dresden, Germany
voit@ipfdd.de

Prof. Gerhard Wegner
Max-Planck-Institut
für Polymerforschung
Ackermannweg 10
Postfach 3148
55128 Mainz, Germany
wegner@mpip-mainz.mpg.de

Prof. Ulrich Wiesner
Materials Science & Engineering
Cornell University
329 Bard Hall
Ithaca, NY 14853, USA
ubw1@cornell.edu

Advances in Polymer Science
Also Available Electronically

For all customers who have a standing order to Advances in Polymer Science, we offer the electronic version via SpringerLink free of charge. Please contact your librarian who can receive a password or free access to the full articles by registering at:

springerlink.com

If you do not have a subscription, you can still view the tables of contents of the volumes and the abstract of each article by going to the SpringerLink Homepage, clicking on "Browse by Online Libraries", then "Chemical Sciences", and finally choose Advances in Polymer Science.

You will find information about the

- Editorial Board
- Aims and Scope
- Instructions for Authors
- Sample Contribution

at springeronline.com using the search function.

Preface

This special volume on the enzyme-catalyzed synthesis of polymers focuses on various methods of polymer synthesis using enzymes as catalysts. There are three cases for such synthetic processes: (1) In living cells (in vivo) enzymes catalyze the synthesis of all biopolymers besides other biological substances via biosynthetic (metabolic) pathways. In test tubes (in vitro) enzymatic catalysis is achieved for the synthesis of polymers via (2) biosynthetic pathways or (3) non-biosynthetic pathways. The present volume is concerned with case (3). Therefore, studies such as the synthesis of polyesters via fermentation using micro-organisms and synthesis of proteins using *E. coli* are not included.

All enzymes are categorized into six groups: (1) oxidoreductases, (2) transferases, (3) hydrolases, (4) lyases, (5) isomerases, and (6) ligases. They show respective catalytic functions specific to the substrate group in vivo. In this volume, however, each chapter is organized on the basis not of the enzyme category but on the class of polymers synthesized. Enzymatic synthesis of polymers (often referred to as enzymatic polymerization) is operated by in vitro enzymatic catalysis, which is close to or very similar to in vivo enzymatic catalysis depending on the reaction.

In vivo enzymatic catalysis has advantageous characteristics: high catalytic activity (high turnover), reactions under mild conditions with respect to temperature, pressure, solvent, neutral pH, etc., and excellent reaction control of stereo-, regio-, chemo-, enantio- and choro-selectivities. These characteristics can also be realized in vitro depending upon the catalyst. Therefore, the reactions contribute to the alleviation of environmental problems due to their low loading processes (energy saving), clean processes (without forming by-products), and the biodegradable nature of product polymers in many cases. These are often difficult to realize by conventional methods.

The first chapter by Reihmann and Ritter reviews the recent developments of peroxidase-catalyzed oxidative polymerization of phenol and derivatives with a phenolic OH group. The importance of enzymatic polymerization in general is emphasized. Properties of product polyphenols, characteristics of the enzyme catalysis, and significance of the process and the product are discussed. The second chapter by Uyama and Kobayashi is concerned with the oxidative polymerization of polyphenols, which are compounds containing more than two phenolic OH groups. These compounds include catechols and flavonoids such as urushiol and catechin. The anti-oxidant properties of flavonoids were greatly enhanced by the polymerization. In the next chapter, Xu, Singh and Kaplan described the synthesis of polyaniline and their derivatives by using oxidoreductase enzymes and biomimetic catalysts. In the presence of templates, the polymer structure was controlled to produce processable materials, which can be used for electrical contact applications, such as synthetic–biological interfaces. The dip-pen nanolithography technique was used for the processing.

The fourth chapter by Matsumura overviews the lipase-catalyzed polyester synthesis via ring-opening polymerization of lactones. A hydrolysis enzyme of lipase, catalyzing the hydrolysis of esters to break the ester-bond in vivo, catalyzed the ester-bond formation to produce polyesters. In comparison with conventional methods, characteristics of the ring-opening polymerizability by lipase catalysis was pointed out, and the reaction mechanism was extensively discussed. A chemical recycling process using a polymer-oligomer equilibrium is proposed. The following chapter by Uyama and Kobayashi describes various polycondensations for the polyester synthesis catalyzed by lipase. Divinyl esters were found to be very effective monomers for the reactions. Dehydration was accomplished in water with lipase catalyst.

In the sixth chapter, Kobayashi and Ohmae reviewed polysaccharide synthesis by hydrolase-catalyzed polymerizations. The general characteristics of enzymatic polymerization and the importance of a new concept for monomer design called transition-state analogue substrates are mentioned. New enzyme-recognizable substrates were designed as two families, sugar fluoride monomers and sugar oxazoline monomers. The polycondensation of the former and the ring-opening polyaddition of the latter enabled the first successful chemical synthesis of natural polysaccharides like cellulose, xylan, chitin, hyaluronan, chondroitin and their derivatives as well as unnatural polysaccharides. Reaction mechanisms and high-order molecular structure formation from the product polymers are also discussed.

In the last chapter by Singh and Kaplan, vinyl polymerizations induced by oxidoreductase enzymes are described, where a mediator is normally used. The reaction is of radical-type to form a $C-C$ bond main chain. Vitamin C-functionalized vinyl monomers and others were polymerized. Such in vitro reactions are unique because nature does not utilize the vinyl-type $C-C$ bond formation reaction except for natural rubber synthesis via polycondensation.

The interdisciplinary topics presented are expected to bridge the fields of materials science (including polymer science) and life science.

Kyoto, Düsseldorf, Medford   *Shiro Kobayashi, Helmut Ritter, David Kaplan*
October 2005

Contents

Synthesis of Phenol Polymers Using Peroxidases
M. Reihmann · H. Ritter   1

Enzymatic Synthesis and Properties of Polymers from Polyphenols
H. Uyama · S. Kobayashi 51

Enzymatic Catalysis in the Synthesis of Polyanilines and Derivatives
of Polyanilines
P. Xu · A. Singh · D. L. Kaplan 69

Enzymatic Synthesis of Polyesters via Ring-Opening Polymerization
S. Matsumura 95

Enzymatic Synthesis of Polyesters via Polycondensation
H. Uyama · S. Kobayashi 133

Enzymatic Polymerization to Polysaccharides
S. Kobayashi · M. Ohmae 159

In Vitro Enzyme-Induced Vinyl Polymerization
A. Singh · D. L. Kaplan 211

Author Index Volumes 101–194 225

Subject Index 251

Synthesis of Phenol Polymers Using Peroxidases

Matthias Reihmann · Helmut Ritter (✉)

Institute of Organic and Macromolecular Chemistry II,
Heinrich-Heine University of Düsseldorf, Gebäude 26.33, Ebene 00,
Universitätsstraße 1, 40225 Düsseldorf, Germany
H.Ritter@uni-duesseldorf.de

1	Introduction	2
1.1	The Potential of Enzymes as Catalysts in Organic Reactions	2
1.2	The Potential of Enzymes in Polymer Chemistry	3
2	Polymerization of Phenols Using Peroxidases	4
2.1	Definitions	4
2.2	The Principle of the Peroxide-Catalyzed Polymerization	5
2.3	Mechanism of the Peroxidase Catalysis	5
2.3.1	Peroxidases in Nature	5
2.3.2	The Catalytic Cycle of Horseradish Peroxidase	7
2.3.3	Side Reactions in the Catalytic Cycle	10
2.4	Mechanism of Phenol Polymer Formation	13
2.5	Influence of the Solvent Composition	14
3	Polymerization of Phenols Using Horseradish Peroxidase	18
3.1	Unsubstituted Phenol	18
3.2	p-Substituted Phenols	20
3.3	m-Substituted Phenols	32
3.4	o-Substituted Phenols, Bisphenols, and other Phenols	36
4	Polymerization of Phenols Using Other Peroxidases	39
5	Polymerization of Phenols Using Model Complexes	41
6	Outlook	44
References		45

Abstract This chapter deals with the peroxidase-catalyzed polymerization of phenol and its derivatives. Such a polymerization needs an oxidation reagent, normally hydrogen peroxide. Since this type of polymerization is believed to be a potential environmentally friendly alternative process for the production of phenol–formaldehyde resins, it has been investigated thoroughly by many research groups. Typical research challenges in this field are investigations regarding the polymerization mechanism, the mechanism of enzyme catalysis, the structure of the resulting phenol polymers, and, needless to say, the synthesis of new materials. A comprehensive overview is given that covers the progress made during recent years, starting from the synthesis of simple resins obtained from unsubstituted phenol and leading up to the production of polyaromatic materials having multiple reactive groups in the side chain. One key for successful work in this field is to understand the characteristics of the enzyme. However, it is also important to be familiar with the influences of different solvents and various concentrations of the reactants (monomer, enzyme, hydrogen peroxide) on the yield, molecular weight, and structure of the resulting polymers. In other words, peroxidase-catalyzed polymerization is an interdisciplinary area covering different fields, mainly biochemistry, organic chemistry, and polymer chemistry for the synthesis of phenol polymers, as well as physical chemistry to understand their properties. This chapter tries to emphasize this aspect.

Keywords Enzyme catalysis · Horseradish peroxidase · Green chemistry · Phenol polymer · Polyphenol

Abbreviations
AOT	Bis(2-ethylhexyl)sodium sulfosuccinate
BOD	Bilirubin oxidase
BPA	Poly(isopropylidenediphenol) resin
CP/MASS	Cross polarization/magic angle sample spinning
DIMEB	Heptakis(2,6-di-O-methyl)-β-cyclodextrin
FT-IR	Fourier-transform infrared
HEPES	N-(2-Hydroxyethyl)piperazine-N'-(2-ethanesulfonic acid)
HOMO	Highest occupied molecular orbital
HPLC	High-performance liquid chromatography
HRP	Horseradish peroxidase
LUMO	Lowest unoccupied molecular orbital
MALDI-TOF	Matrix-assisted laser desorption/ionization time-of-flight
MMA	Methyl methacrylate
MPL	Laccase from *Myceliophthore*
PCL	Laccase from *Pycnoporus coccineus*
PEG	Polyethylene glycol
PPO	Poly(phenylene oxide)
RAMEB	Randomly 2,6-dimethylated β-cyclodextrin
SBP	Soybean peroxidase
TEED	N,N,N',N'-tetraethylethylenediamine
TVL	Laccase from *Trametes versicolor*

1
Introduction

1.1
The Potential of Enzymes as Catalysts in Organic Reactions

Without doubt, enzyme-catalyzed syntheses are becoming more and more important in contemporary chemical synthesis [1–11]. Several years ago, the use of enzymes for synthetic purposes was still restricted to several specialized groups of biochemistry. The number of commercially available enzymes is increasing every year and new methods in biochemistry, including genetic engineering, are expected to provide extremely pure enzymes at acceptable prices [12–14]. Thus, the use of enzyme catalysis is of growing interest for synthetic chemists and eventually for large-scale industrial production [15–23]. The main reasons for the increasing acceptance of isolated or even whole-cell enzymes in organic and polymer chemistry are the environmentally friendly and resource-saving conditions enabled by the enzyme catalysis: mild reaction conditions (in general room temperature, atmospheric pressure, neutral pH [6, 17, 19, 20, 24–30]) and high selectivity (often high enantio-, regio-, and chemoselectivity [30–34]). Furthermore, enzymes are nontoxic, they are accessible from natural renewable resources, and they offer catalyst recyclability [12, 16, 20, 22, 32]. In addition, the use of enzymes offers the possibility of synthesizing new materials which are not, or at least not effectively, available by conventional methods.

1.2
The Potential of Enzymes in Polymer Chemistry

According to the advantages mentioned above, enzyme-catalyzed polymer syntheses are also of increasing interest. Typical examples of enzymatically produced macromolecules are the synthesis of polysaccharides and polyesters using hydrolases, the production of biopolymers like cellulose, xylan, chitin, or lignin by hydrolases or peroxidases, and the synthesis of polyaromatic

Table 1 Typical enzymes used in polymer chemistry

Enzyme class	Enzyme type	Typical substrates to build polymers
Oxidoreductases	Peroxidases	Phenols, anilines, vinyl compounds
	Laccases	Phenol, 1-naphthol, acrylamide
	Tyrosinase	Chitosan, poly(4-hydroxystyrene), lignin
	Glycose oxidase	Vinyl compounds
	Xanthine oxidase	Acrylamide
Transferases	Phosphorylase	D-Glucosyl phosphate
Hydrolases	Glycosidase	Saccharide fluorides
	Chitinase	Chitobiose oxazoline
	Hyaluronidase	Hyaluronic acid, hyaluronic acid saccharides
	Lipase	Lactones, esters, carbonates, anhydrides, halogenated alcohols, hydroxy acids
	Protease	Amino acid esters
	PHB depolymerase	ε-Caprolactone
	Epoxide hydrolase	Glycidol
Ligases	Cellulase	Uridine diphosphate glucose
	Chitinsynthase	Uridine N-acetylglycosamine
	Glycosyl transferase	Nucleoside mono- and diphosphates
Lyases		
Isomerases		

resins from phenols and anilines via peroxidase catalysis [7, 8, 11, 14, 16–20, 22, 35]. Some polymerization reactions catalyzed by these enzymes are listed in Table 1.

We will focus in the following paragraphs on the oxidation of phenols catalyzed by peroxidases [18–20, 35, 36] and the potential of this reaction in polymer chemistry. One advantage of this type of polymerization is the possibility of producing phenolic resins without using toxic formaldehyde, which is a necessary reagent for the production of conventional phenol-formaldehyde resins like Resol or Novolac [19, 35–38]. Another application, which was studied intensively, is the treatment of wastewaters. Peroxidases are able to catalyze the polymerization of toxic aromatic contaminants of wastewater. The resulting polymers can be separated from the aqueous solution by simple filtration [39, 40]. Furthermore, the peroxidase-catalyzed polymerization of substituted phenols and anilines can lead to functional polymers which are otherwise difficult to synthesize, such as conducting polymers [41–43], redox polymers [44–46], or polymers with polymerizable [47, 48] or photocross-linkable [49, 50] groups in the side chain.

2
Polymerization of Phenols Using Peroxidases

2.1
Definitions

Two important terms have to be defined clearly, since both are frequently found in the literature: "enzymatic polymerization" and "polyphenol". Enzymatic polymerization means the chemical synthesis of polymers in vitro via nonbiosynthetic pathways, catalyzed by an isolated enzyme [16, 35]. In contrast, common macromolecules in nature are synthesized via enzyme catalysis in vivo. All synthetic procedures using living cells or organisms (e.g., *Escherichia coli*) and which require fermentation are not enzymatic polymerizations in this sense.

The expression polyphenol is assigned in organic chemistry to an aromatic compound with two or more hydroxyl functions, such as a flavonoid compound. In contrast, in publications dealing with the enzyme-catalyzed polymerization of phenol derivatives, this term is often used to name the polymeric products. Although this expression is not incorrect for these materials, its use can lead to confusion, especially if polymers made from flavonoid compounds are discussed. In order to avoid misunderstanding, in this article the expression "phenol polymer" is used instead of polyphenol to name all polymers produced from phenols by peroxidase catalysis.

2.2
The Principle of the Peroxide-Catalyzed Polymerization

The principle of the polymerization of phenols catalyzed by a peroxidase is explained in the following text for the enzyme horseradish peroxidase (HRP). The mechanism of the HRP catalysis is fairly well understood and has been the subject of many investigations [24, 51–54]. HRP catalyzes the one-electron oxidation of phenols by a peroxide to form the corresponding phenoxy radicals. Usually, hydrogen peroxide is used as oxidizing reagent. During this process, two water molecules are formed [55] (Eq. 1):

$$2PhOH + H_2O_2 \rightarrow 2PhO^{\cdot} + 2H_2O \qquad (1)$$

The resulting phenoxy radicals react to form polymers via subsequent recombination and radical transfer steps, as will be discussed later. The overall polymerization can therefore be written as follows (Scheme 1):

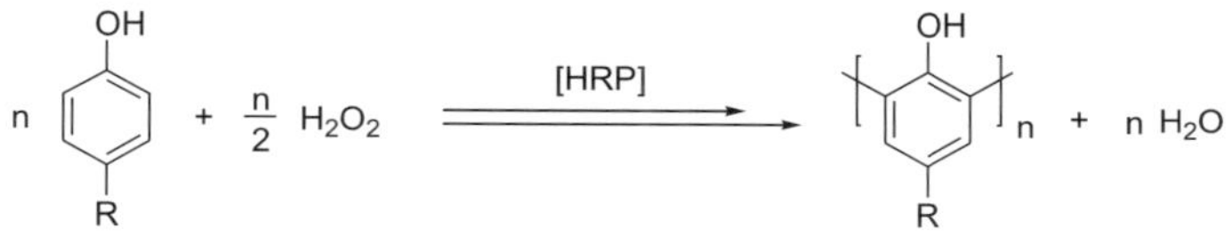

Scheme 1 Polymerization of phenols catalyzed by horseradish peroxidase (HRP)

2.3
Mechanism of the Peroxidase Catalysis

2.3.1
Peroxidases in Nature

Peroxidases are found in nature in plants, microorganisms, and higher organisms. Common peroxidases and their various biological functions are shown in Table 2 [24]. Peroxidases are able to catalyze the oxidation of aromatic compounds, the oxidation of heteroatoms, epoxidation, and the enantioselective reduction of racemic hydroperoxides [12, 16–20, 22, 24, 35]. Typical reactions catalyzed by peroxidases are listed in Table 3 [24].

For the synthesis of phenol polymers by peroxidase catalysis, the oxidation of aromatic electron donors in the presence of peroxides is of importance. Typical peroxides used in combination with peroxidases are hydrogen peroxide, alkyl peroxides, and benzyl peroxide. Typical substrates are electron-rich aromatic compounds like phenols and anilines. This function of peroxidases was first discovered in 1855 with the oxidation of guaiacol by peroxidases in the presence of hydrogen peroxide [56]. More systematic investigations and

Table 2 Common peroxidases and their biological functions

Peroxidase	Catalyzed reaction	Biological function
Horseradish peroxidase	$2\,ArH + H_2O_2 \rightarrow Ar\bullet + 2\,H_2O$	Biosynthesis of plant hormones
Catalase	$2\,H_2O_2 \rightarrow H_2O + O_2$	Detoxification of hydrogen peroxide
Cytochrome C Peroxidase	$2\,Cc(II) + H_2O_2 \rightarrow Cc(III)\bullet + 2\,H_2O$	Cytochrome C metabolism
Lignin peroxidase	$2ArH + H_2O_2 \rightarrow Ar\bullet + 2H_2O$	Degradation of lignin
Chloroperoxidase	$2ArH + 2Cl + H_2O_2 \rightarrow ArCl + 2H_2O$	Biosynthesis of caldariomycin
Myeloperoxidase	$H_2O_2 + Cl + H_3O\bullet \rightarrow ClO + 2H_2O$	Antimicrobial
Lactoperoxidase	$2ArH + H_2O_2 \rightarrow Ar\bullet + 2H_2O$	Antimicrobial

Table 3 Peroxidase-catalyzed reactions

Reaction	Reaction scheme	Typical substrates
Electron transfer	$2\,ArH \xrightarrow[-ROH]{ROOH} Ar\text{-}Ar + H_2O$	Phenols, anilines
Disproportionation	$H_2O_2 \rightarrow 2\,H_2O + O_2$	H_2O_2
Sulfoxidation	$R_1\text{-}S\text{-}R_2 \xrightarrow[-ROH]{ROOH} R_1\text{-}S(=O)\text{-}R_2$	Thioanisole
Epoxidation	$R_1\text{-}CH=CH\text{-}R_2 \xrightarrow[-H_2O]{H_2O_2}$ epoxide	Alkenes
Demethylation	$R_1R_2N\text{-}CH_3 \xrightarrow[-ROH]{ROOH} R_1R_2N\text{-}H + HCHO$	N-N-Dimethylaniline
Dehydrogenation	$2\,(HO)_2C(COOH)_2 \xrightarrow[-2H_2O]{O_2} 2\,(O=)_2C(COOH)_2$	Dihydroxyfumaric acid
Hydroxylation	Ar-OH + HOOC-CH(OH)-COOH $\xrightarrow[-H_2O]{O_2}$ hydroxylated product	L-Tyrosine, adrenaline
α-Oxidation	$R_1CH(R_2)CHO \xrightarrow[-HCOOH]{O_2} R_1C(=O)R_2$	Aldehydes

probably the first definition of the reactivity of a nonhydrolytic acting enzyme by its purpurogallin number were presented in 1917 [57].

Most peroxidases are heme enzymes and contain the ferric protoporphyrin IX (protoheme) group with an iron atom in their active center (Fig. 1). Only a few other peroxidases known so far have other metal centers and/or different prosthetic groups. Glutathione peroxidase, for example, has a selen-

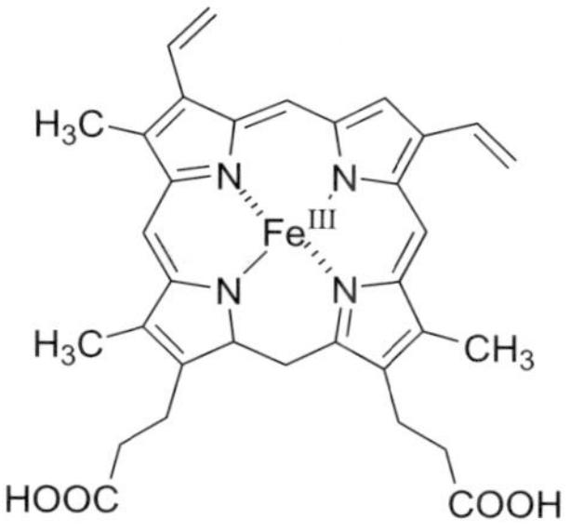

Fig. 1 The ferric protoporphyrin IX group

ium, bromoperoxidase a vanadium, and manganese peroxidase and flavoperoxidase have manganese atoms in their active centers [24, 53, 54]. Also, HRP contains the protoheme group with an Fe(III) atom in the active center. The group is held in position by electrostatic interaction of one propionic acid group of the heme and a lysine residue (Lys174) of the apoprotein. The molecular mass of HRP is about 40 000 Da [53].

2.3.2
The Catalytic Cycle of Horseradish Peroxidase

The first step in the oxidation of phenols catalyzed by HRP is the formation of a precursor complex of hydrogen peroxide (or an organic hydroperoxide) with the enzyme (Scheme 2, **4**). Next, the elimination of water leads to the so-called compound I (Scheme 2, **5**), with an oxyferryl group in the active center surrounded by porphyrin in the form of a Π radical cation. The Fe(III) of the active center of HRP is oxidized to Fe(IV) while the porphyrin structure loses one electron [24, 55, 58–60]. Compound I can be seen as well as Fe(V) in the active center surrounded by a neutral porphyrin [18]. Compound I oxidizes the first equivalent of phenol by oxidoreduction [61]. The phenol coordinates in the active center and is oxidized via one-electron transfer to the corresponding phenoxy radical cation, while the porphyrin radical cation is reduced to the neutral porphyrin. After deprotonation of the phenoxy radical cation, the resulting phenoxy radical leaves the active center. Hence, this mechanism is called in biochemistry an irreversible ping-pong mechanism [39, 62]. The proton of the phenoxy radical is abstracted by a base, probably distal histidine (His – 42) to form the so-called compound II (Scheme 2, **6**) [24, 55, 63]. This species oxidizes the second equivalent phenol. The phenol is oxidized to the phenoxy radical by one-electron transfer to the Fe(IV), so that the native enzyme is regenerated (Scheme 2, **3**). The proton of the phenol forms water as a leaving group with the oxygen of the oxyferryl group and the proton which was bound to the His – 42. Thus, in one catalytic cycle, two equivalents of a phenol are oxidized by one equivalent of

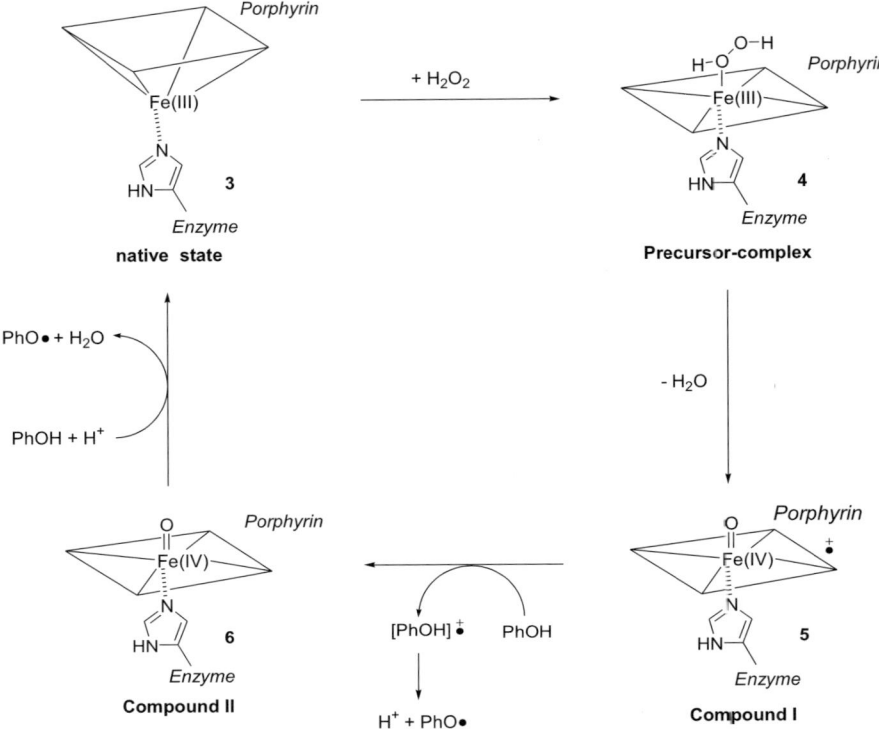

Scheme 2 The catalytic cycle of horseradish peroxidase

hydrogen peroxide producing two phenoxy radicals and one water molecule (Scheme 2) [55, 60].

2.3.2.1
Mechanism of Compound I Formation

The detailed mechanism of the formation of compound I by HRP and hydrogen peroxide was the subject of many investigations for more than 30 years. In an initial step, the enzyme turns the hydrogen peroxide into a better nucleophile by abstraction of one proton to the His – 42 and the peroxide coordinates side-on to the fivefold-coordinated Fe(III) atom (Scheme 3, **8**) [59, 60]. Next, a heterolytic cleavage of the O – O bond is introduced by a push–pull mechanism of the two histidine residues in the active center, the distal His – 42, and the proximal His – 170. First, the protonated His – 170 facilitates the formation of the iron peroxide bond through the effect of its positive charge (Scheme 3, **9**) [24, 53, 55, 63]. The proton transfer to an adjacent basic amino acid residue (Scheme 3, **10**) leads to a charge inversion and hence to the cleavage of the O – O bond of the oxywater complex formed by the proton donation of Arg – 38

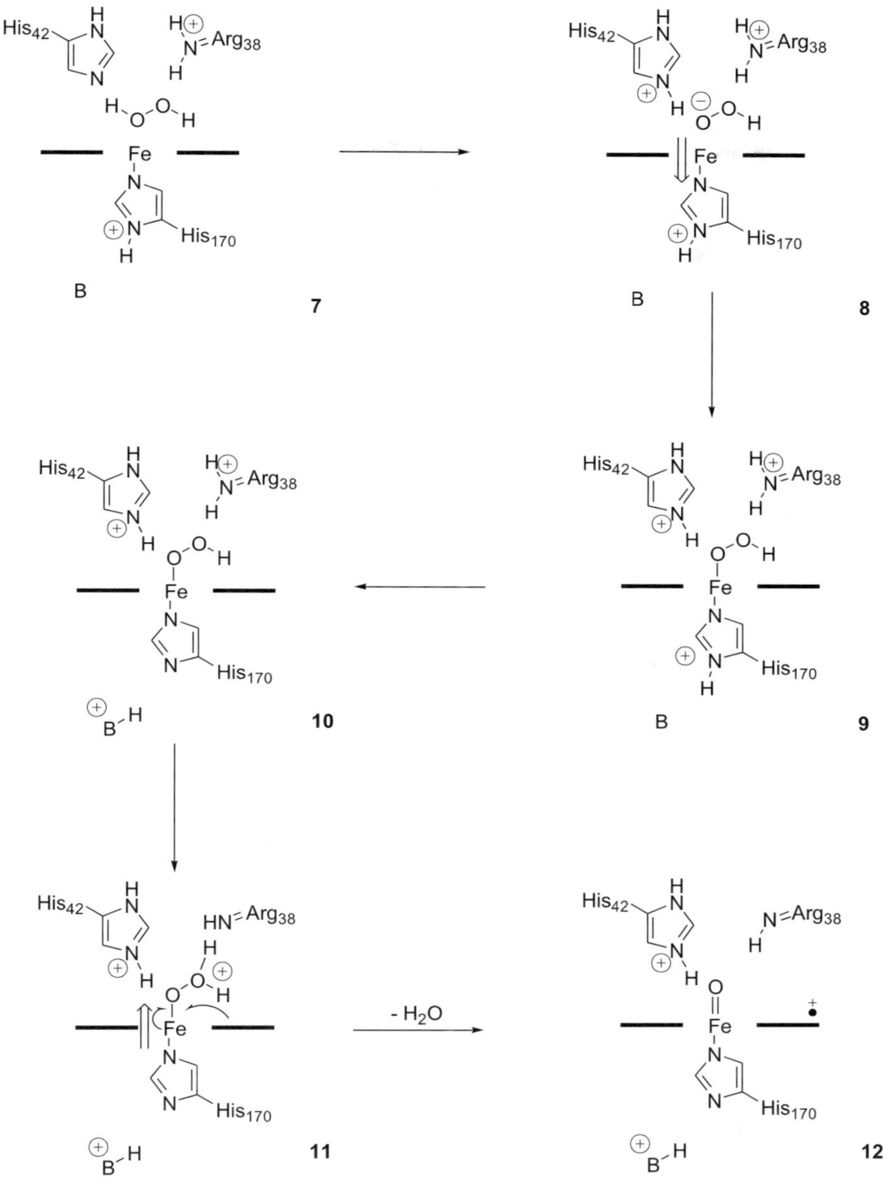

Synthesis of Phenol Polymers Using Peroxidases

Scheme 3 Mechanism of compound I formation

(Scheme 3, **11**). The cleavage of the O – O bond liberates water as leaving group and compound I is formed (Scheme 3, **12**) [24, 55, 64, 65].

In some cases, the formation of compound I was accompanied by the oxidation of an amino acid residue instead of the porphyrin structure [53].

Furthermore, recent molecular dynamics simulations have shown that this mechanism probably takes place only under neutral or basic conditions [63]. In this case His – 42 can function as a proton acceptor and Arg – 38 as a proton donor to build the oxywater complex (Scheme 3). In acidic conditions, however, the distal His – 42 is present in the protonated form and therefore the peroxide ligand is bound more flexibly. A dynamic exchange of the oxygen atoms bound to the iron atom can be formulated [63]. Thus, in acidic conditions the Arg – 38 reacts as the proton donor to the oxygen, which is bound to the iron atom. Exchange of the oxygen atoms then leads to the oxywater complex. The proton of the other oxygen is accepted by a water molecule [63].

2.3.2.2
Mechanism of Compound II Formation and Phenol Oxidation

Following the previously described formation of compound I, the first equivalent of phenol coordinates in the active center (Scheme 3, **13**) [24, 55, 66] and is oxidized accompanied by simultaneous reduction of the porphyrin radical cation (Scheme 4, **14**) [24, 54, 64, 65]. The proton is accepted by His – 42 and the resulting phenoxy radical leaves the active center. The resulting neutral compound II coordinates a second equivalent of phenol in the active center (Scheme 4, **14**) [24, 54, 64, 65]. Redox reaction leads to a second phenoxy radical and the Fe(IV) is reduced back to the native Fe(III) state (Scheme 4, **15**) [24, 54, 64, 65]. With the His – 42 donating the proton back (Scheme 4, **16**), water is formed (Scheme 4, **17**), which leaves the active center [24, 55, 67].

The investigation of the reaction of compound II with *o*-diphenols also revealed a two-step mechanism, in which the first step corresponds to the formation of an enzyme–substrate complex, and the second to electron transfer from the substrate to the iron atom. The size and the hydrophobicity of the substrate control their access to the hydrophobic binding site of HRP, but the electron density in the hydroxyl group of C-4 was found to be the most important feature for the electron-transfer step [68].

2.3.3
Side Reactions in the Catalytic Cycle

Successful work with peroxidases requires knowledge about the possible side reactions in the catalytic cycle. The most important are shown in Scheme 5 [39]. The pathway from **18** to **21** covers the normal catalytic cycle. If the local phenol concentration is too low and/or the local concentration of hydrogen peroxide is too high, compound I is converted into an intermediate (Scheme 5, **23**) [39, 69]. This intermediate can follow three different paths of decomposition. First it can react back to the na-

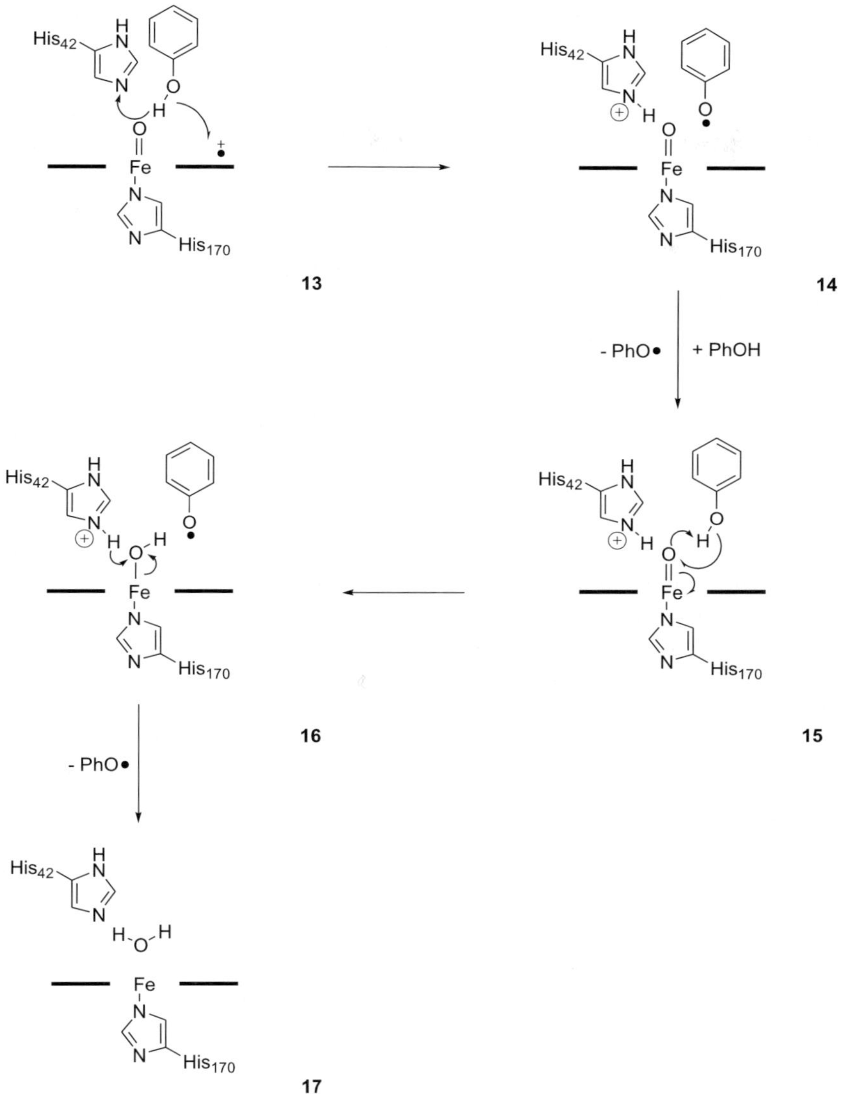

Scheme 4 Mechanism of compound II formation and phenol oxidation

tive enzyme by decomposing the hydrogen peroxide to water and oxygen (Scheme 5, **25**) [39, 69]. This pathway corresponds to a weak catalase effect. It decelerates the reaction and the process consumes hydrogen peroxide, which is hence lost for the further polymerization process. This reaction can be avoided by dosing the hydrogen peroxide to the reaction over a large time interval, since phenols are much better reductants than hydro-

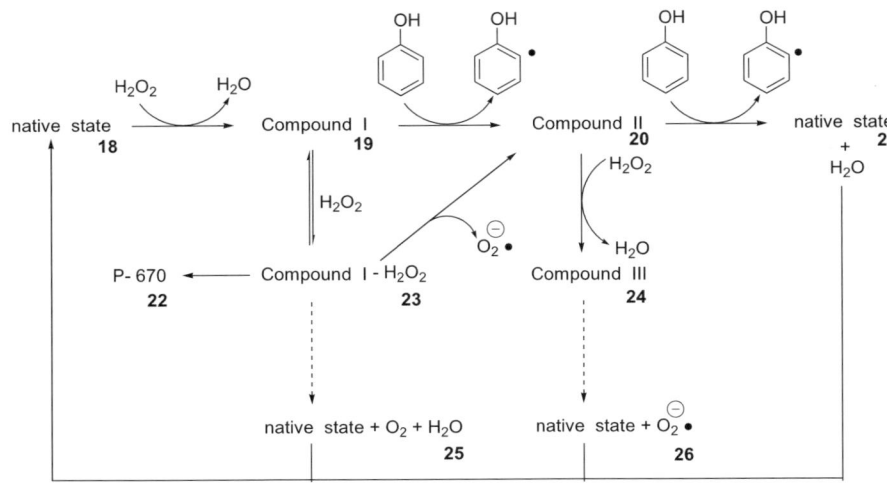

Scheme 5 Side reactions in the peroxidase catalytic cycle

gen peroxide. Alternative pathways for the intermediate **23** are the formation of compound II by generation of an oxygen radical anion or the formation of the verdohemoprotein P-670 (sometimes referred as compound IV) (Scheme 5, **22**) [39, 69]. The latter leads to an irreversible inhibition of the enzyme.

Imbalance between the concentrations of phenol and hydrogen peroxide can cause a second side reaction if the rapid conversion of compound II is hindered. In this case, compound II reacts with excess hydrogen peroxide to form compound III, which is not catalytically active (Scheme 5, **24**) [39, 69]. This reaction consumes hydrogen peroxide, but it is not an irreversible inhibition of the enzyme, because this species regenerates the native enzyme in a very slow reaction by forming a hydrogen peroxide radical anion (Scheme 5, **26**) [39, 69]. However, this process is relatively slow [39].

Another risk of permanent inhibition of the enzyme is the attack of radicals on amino acid residues near the active center. This danger of deactivation of the enzyme increases with increasing enzyme-to-substrate ratio and increasing phenol concentration. Another problem, which is not connected to the enzyme cycle, appears due to the nature of the phenol polymerization process. The phenol polymers usually precipitate out of the solution when they reach higher molecular weights. During the precipitation, the polymer chains may adsorb or trap enzyme molecules [39, 40, 70]. Thus, many studies have been made to protect the enzyme from deactivation, and several additives have been found to reduce the loss of enzyme activity via mechanisms that have not been fully investigated. One of the most frequently used protection additives is polyethylene glycol (PEG) [70].

2.4
Mechanism of Phenol Polymer Formation

As previously described, the function of the enzyme is to produce the phenoxy radicals under mild reaction conditions. The resulting phenoxy radicals are able to form polymers via recombination processes if a suitable reaction medium (solvent mixture composition), pH value (buffer systems), and kind of substrate are chosen. Some phenols, for example, are known to react preferentially via oxidation to form *ortho*-diketones [68] or Pummerer ketones [52], and are therefore not suitable for polymerization. Other phenols are known to prefer dimerization reactions, for instance caffeic acid [71].

The oxidative polymerization of phenols is basically a polycondensation reaction. A model for the polymerization was developed, which divides the

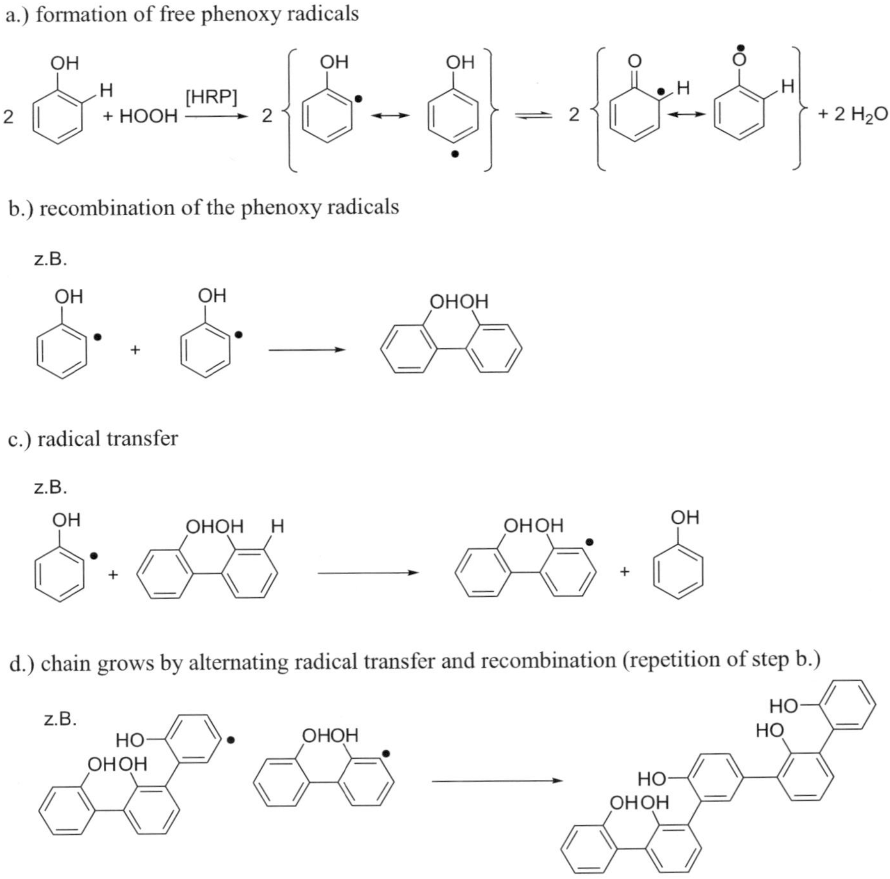

a.) formation of free phenoxy radicals

b.) recombination of the phenoxy radicals

c.) radical transfer

d.) chain grows by alternating radical transfer and recombination (repetition of step b.)

Scheme 6 Mechanism of phenol polymer formation

polymerization into four steps (Scheme 6) [36, 37, 72, 73]. The first step is the formation of phenoxy radicals by the HRP-catalyzed oxidation of phenols (Scheme 6a). It is the only step that is controlled by the enzyme kinetics. The phenoxy radicals form dimers by recombination (Scheme 6b). Apparently, at the beginning of the reaction, practically all phenols are converted to dimers [74–76]. When the concentration of free phenoxy radicals is decreasing, an electron-transfer reaction is more likely than further recombination [74]. This leads to the formation of oligomer radicals (Scheme 6c), which then form oligomers of even higher molecular weight by recombination (Scheme 6d). The radical transfer reaction of a phenoxy radical and an oligomer regenerates a phenol monomer, which can be oxidized again by HRP to initiate new radical transfer reactions. When the phenoxy radical is not reacting fast enough in a recombination or a radical transfer step, oxidation may take place leading to ketone structures [74, 76, 77].

2.5
Influence of the Solvent Composition

Most peroxidase-catalyzed polymerizations of phenols studied so far have been carried out in a mixture of organic solvent and aqueous buffer solution. Under these conditions, the phenol polymers precipitate out of the solution. The effect of the solvent composition on the yield and the molecular weight of the resulting polymers was the subject of many studies [37, 78, 79]. A polymerization in pure buffer solution is usually not possible, because most phenol derivates are insoluble in aqueous buffer solution [37, 78]. Thus, during the polymerization process, the dimers and trimers formed at the early stage of the reaction precipitate immediately out of the reaction, preventing the formation of high molecular weight materials. In pure organic solvent, however, the enzyme activity is significantly reduced and the yield of polymer is negligible [78, 80]. Various mixtures of organic solvents with aqueous buffer solutions have been tested for the polymerization of phenols; for example, 1,4-dioxane, acetone, and N,N-dimethylformamide for the polymerization of p-phenylphenol in the presence of phosphate buffer [37], methanol with phosphate buffer for the polymerization of phenol [81, 82], and ethanol with N-(2-hydroxyethyl)piperazine-N'-(2-ethanesulfonic acid) (HEPES) for the polymerization of m-cresol [79]. An early comprehensive study of the polymerization behavior of various aromatic compounds can be found in the literature, where the effect of the ratio of 1,4-dioxane to buffer (phosphate and HEPES) on the yield of the resulting polymers was shown (Table 4) [78].

These studies have demonstrated that a solvent consisting of about 80 vol % 1,4-dioxane and 20 vol % phosphate buffer at neutral pH is a good compromise between yield and molecular weight of the resulting phenol polymers [37, 78]. In a recent work, the enzyme activity was evaluated in var-

Table 4 Effect of solvent/buffer ratios on yield of poly(p-phenylphenol)

Solvent/buffer	Yield [%]	Reaction time [h]
75% Dioxane/25% PO_4	73	2
80% Dioxane/20% PO_4	49	2
90% Dioxane/10% PO_4	24	18
95% Dioxane/5% PO_4	2.2	18
75% Dioxane/25% HEPES	71	2
80% Dioxane/20% HEPES	93	2
90% Dioxane/10% HEPES	29	2
95% Dioxane/5% HEPES	negligible	2

ious solvent compositions. According to this study, besides 1,4-dioxane the use of methanol (60 or 80 vol %) as solvent for polymerization is also advantageous [80]. It is possible to control the molecular weight by varying the solvent composition: with more organic solvent in the mixture, the molecular weight of the resulting polymers increases, but the yield normally decreases simultaneously.

The reason that a mixture of 80 vol % 1,4-dioxane and 20 vol % phosphate buffer is such a good compromise for the polymerization of phenols is probably up to the formation of aggregates consisting of phenols and 1,4-dioxane. With this solvent composition, the mole fraction of 1,4-dioxane is equal to that of water [83]. Experimental data have been published which clearly indicate the formation of phenol aggregates in a 1,4-dioxane/phosphate buffer mixture [83, 84]. The stabilization of these aggregates is best with a composition of 80 vol % 1,4-dioxane and 20 vol % phosphate buffer [83]. It seems logical that the formation of aggregates could support the recombination of radicals and the electron-transfer step.

An alternative way of performing the peroxidase-catalyzed polymerization of phenols is the use of a reverse micellar system. A frequently used system is the ternary mixture water/isooctane/bis(2-ethylhexyl)sodium sulfosuccinate (AOT) or a biphasic system using water/isooctane [85–89]. In general, the shape and the molecular weight of polymers gained by polymerization in reversed micelles can be regulated by adjusting the surfactant concentration. Such behavior was also found for the HRP-catalyzed polymerization of p-ethylphenol. Increasing the AOT content by about a factor of 10 resulted, according to this study, in the formation of phenol polymers with 20-fold higher molecular weight [85].

The HRP-catalyzed polymerization of phenols can also be carried out as a dispersion polymerization in 1,4-dioxane/buffer mixtures with poly(ethylene glycol), poly(vinyl alcohol), and poly(methyl vinyl ether) as stabilizers. The dispersion polymerization of phenol and its derivatives leads to the formation of relatively monodisperse polymer particles with a typical diameter

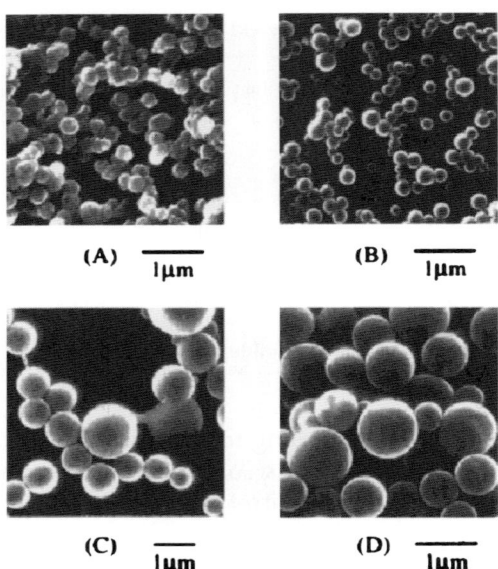

Fig. 2 SEM photographs of polymer particles from **A** phenol, **B** m-cresol, **C** p-cresol, **D** p-phenylphenol

around 250 nm. SEM photographs of such particles from phenol and some derivatives are shown in Fig. 2 [20].

A new method for polymerizing hydrophobic phenol derivatives in 100% aqueous buffer without organic solvent is the use of cyclodextrins as carrier molecules [48, 90–92]. Cyclodextrins are ring-shaped molecules, which have a hydrophilic surface and a hydrophobic cavity [93–95]. This cavity can be adjusted to the reaction conditions: while α-cyclodextrin, consisting of six α-glucose molecules, fits with unsubstituted phenol [92, 96], most p-substituted and m-substituted phenols can be included in β-cyclodextrin, which consists of seven α-glucose molecules [48, 90]. The even bigger γ-cyclodextrin, consisting of eight α-glucose molecules, works well with biphenyl systems. The selective or random methylation of the OH groups in the 2 and 6 positions of the cyclodextrin (DIMEB, RAMEB) improves the solubility of especially β-cyclodextrins in water even further [93–95]. Thus, a hydrophobic phenol forms water-soluble host–guest complexes in water. These complexes can be polymerized via peroxidase catalysis (Fig. 3).

The main advantage of this system is that an organic solvent is not needed to polymerize the phenols and therefore the whole process is of more environmental benefit. Another important point is that the phenol polymers can be synthesized in 100% buffer solution, enabling the enzyme to work at maximum activity in its natural medium. Therefore, the polymerization yield and conversion can be increased [48]. Finally, the use of cyclodextrin often in-

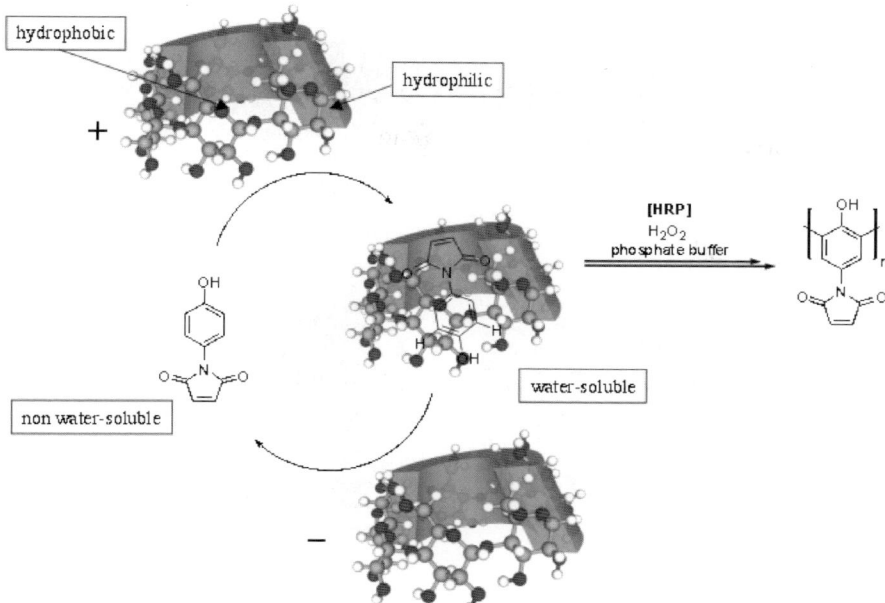

Fig. 3 HRP-catalyzed oxidative polymerization of a water-soluble phenol–cyclodextrin complex

creases the utility of enzymes in organic synthesis, because the cyclodextrins release the monomers in a controlled manner and inhibition of the enzyme due to high concentrations of substrate can be avoided [97].

An interesting way to realize the polymerization in ordered two-dimensional films is to react phenols substituted with a longer tail in the *para* position on a Langmuir trough. For instance, thin films from *p*-tetradecyloxyphenol and phenol have been produced using the Langmuir–Blodgett technique. The monomeric monolayer at the air/water surface was polymerizable by HRP in the subphase [98, 99].

Very recently, it was demonstrated that the polymerization of phenols can be performed within polyelectrolyte microcapsules [100]. This procedure is briefly presented in Fig. 4 [100]. First, polyelectrolyte microcapsules are prepared with layer-by-layer assemblies on small cores. The cores are removed by a suitable solvent and the pH is adjusted in a way that increases the permeability of the shell, allowing penetration of the enzyme inside the capsules. The capsules are closed by increasing the pH, leaving the enzyme trapped inside the capsule. Under these conditions, the shell is penetrable for the phenol molecules, but not for the enzyme or the resulting polymers, due their high molecular weight. Thus, biocatalyst-rich microvolumes can be realized in a continuous aqueous bulk phase. The capsule diameter may be adjusted from 100 nm to tens of microns depending on the tem-

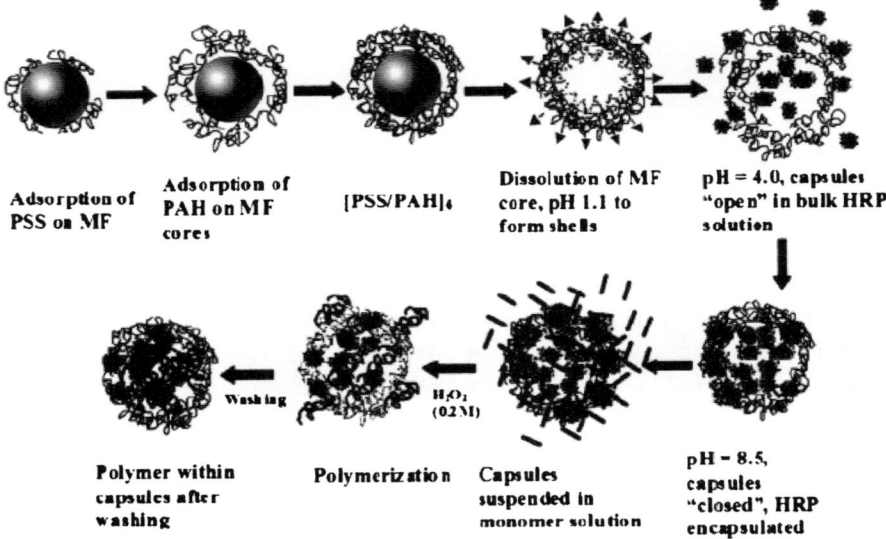

Fig. 4 Enzyme-catalyzed polymerization of phenols within polyelectrolyte microcapsules. MF = melamine formaldehyde; PSS = poly(sodium 4-styrenesulfonate); PAH = poly(allylamine) hydrochloride

plate core. So far, 4-(2-aminoethyl)phenol (tyramine) has been polymerized by this system to yield fluorescent polymeric products. In another work, the HRP-catalyzed polymerization of 4-hydroxyphenylacetic acid was successfully applied for the shell modification of such microcapsules. Due to the fluorescent nature of the phenolic polymer coating of the capsules, the wall structure and the attached layers could be studied by confocal image analysis [101].

3
Polymerization of Phenols Using Horseradish Peroxidase

3.1
Unsubstituted Phenol

Phenol (**27**) is the simplest and most important phenolic compound for industrial applications. The oxidative polymerization of phenol with a conventional catalyst normally affords insoluble products with uncontrolled structure [102]. In contrast, the polymerization of phenol via peroxidase catalysis enables the formation of soluble powdery materials, which consist of phenylene (**28**) and oxyphenylene (**29**) units (Scheme 7) [37, 38, 103]. The ratio of phenylene to oxyphenylene units was reported to be partially controllable

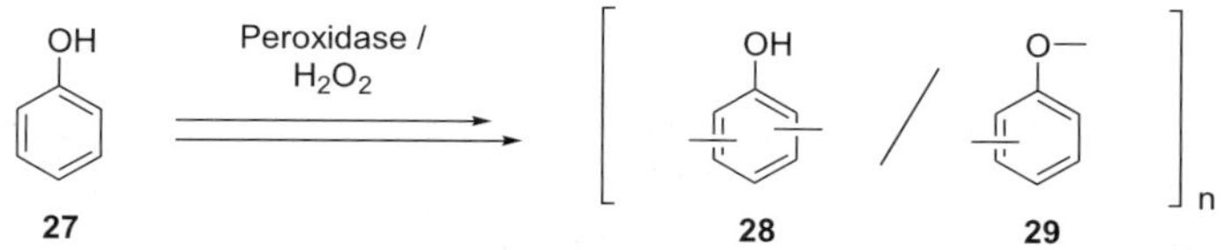

Scheme 7 Polymerization of phenol by peroxidase catalysis

(with a phenylene rate ranging from 32 to 66%) by changing the solvent composition and the pH value [82]. These results suggest that the single electron of the phenoxy radical is stabilized in altered positions of the phenol when different solvent compositions are used. The produced phenol polymers showed high thermal stability (5% weight loss at 350 °C followed by the onset of rapid degradation at 378 °C under argon).

An interesting system for polymerizing phenol under similar conditions, but via controlled enzymatic in situ production of hydrogen peroxide from glucose, can be realized using the bienzymatic system horseradish peroxidase and glucose oxidase [104]. The polymerization of phenol was also recently performed in mixtures of 1,4-dioxane, methanol, or DMF with distilled deionized water, without using buffer salts, which usually tend to contaminate the precipitating polymers [80]. The polymerization of phenol was also realized in aqueous buffer solution in the absence of any organic solvent. A catalytic amount of α-cyclodextrin was used as carrier to bring the phenols into solution, thus leading to soluble phenol polymers in high yields [92]. A new and interesting approach to influencing the coupling regioselectivity of the polymerization of phenol by use of PEG as template was recently presented [105]. The addition of PEGs with molecular weight between 400 and 20 000 g/mol enabled the polymerization of phenol in aqueous buffer solution. PEGs with larger or smaller molecular weight were found to work inadequately. The precipitated product was found to be a complex of the phenol polymer and PEG, with the phenol polymer having a content of phenylene units around 90%. The polymerization of phenol in methanol/buffer mixture without PEG afforded only a content of phenylene units between 40 and 70% [82]. UV-VIS analysis supported the scenario shown in Fig. 5, where the phenols are arranged around the PEG due to hydrogen bonding, which should enhance the regioselectivity during the recombination of the phenoxy radicals.

The technique of using templates during the polymerization of phenols, thiols, and anilines was also used successfully to produce imprinted polymers. Besides phenol, various other monomers (p-substituted phenols, aniline, 4-aminophenyl ethanol, and thiol) were polymerized in buffer solution in the presence of Cu(II), Ni(II), or Fe(III). The corresponding polymers displayed selectivity in binding to the metal, which was used in the imprinting step against the other three metals used for the screening [106]. It has

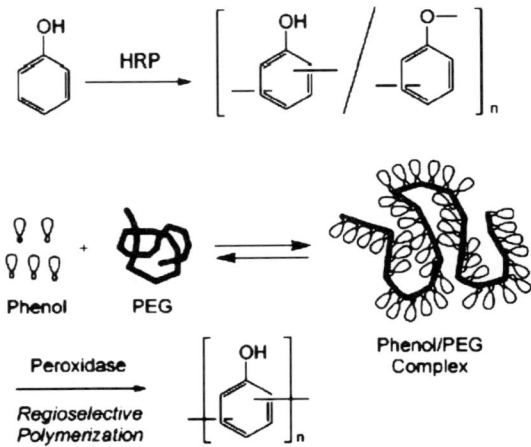

Fig. 5 Polymerization of phenol in water using PEG as template

to be mentioned in this connection that even without using such template techniques, phenol polymers demonstrated an enormous potential for the realization of metal ion sensors [107].

3.2
p-Substituted Phenols

Most research in the field of peroxidase-catalyzed polymerization of phenols has been carried out with *p*-substituted phenols. In this case, the reactive *para* position of the phenol is blocked, so that the recombination of the phenol should take place mainly at the *ortho* positions of the phenols [88]. Thus, the structure of the resulting phenol polymers should not be as complicated as that for unsubstituted phenol. Scheme 8 shows the recombination process of a *p*-substituted radical (30). The radical form 31, having the radical stabilized at one of the two *ortho* positions, has the corresponding tautomeric form 36 and is the most stable (Scheme 8). Thus, the recombination should occur preferentially at the *ortho* positions. The mesomeric structure 32, having the radical in the *meta* positions, is less stabilized. The tautomeric structure 34, having the radical in the *para* position, should be more stable than 32 but less stable than structure 31. The mesomeric form is 33, having the radical stabilized at the oxygen atom (Scheme 8).

While the linkage structure of the phenol polymers made from *p*-substituted phenols was not investigated in detail in most cases, recent studies have shown clear evidence that the polymerization of *p*-substituted phenols leads preferentially to *ortho–ortho* coupling during the initial stage of the polymerization [46, 76]. In contrast, it was also reported that the polymerization of *p*-substituted phenols may lead to polymers that consist (according to

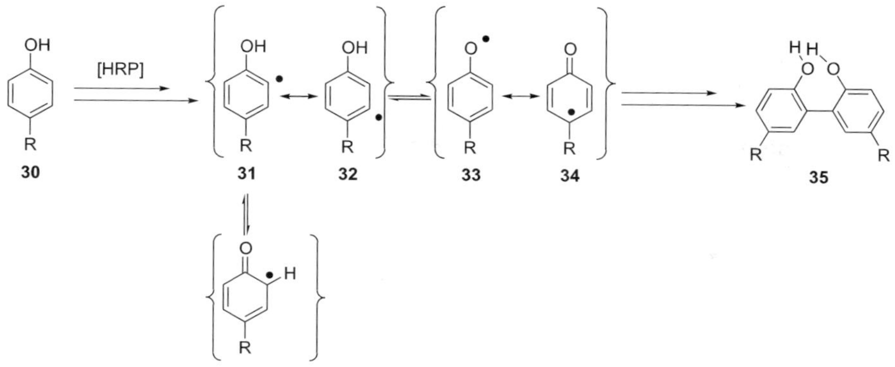

Scheme 8 Recombination process of a *p*-substituted radical

NMR integrations and IR analysis) of phenylene and oxyphenylene units [47]. This NMR analysis can be performed by comparing the area of aromatic protons to a standardized signal of a side-chain group [47]. It is important to also take into account the presence of aromatic protons at the chain ends, otherwise the calculation can lead to false conclusions [46]. According to recent studies, polymers of 4-chlorophenol consist mainly of *ortho–ortho*-linked phenols, forming a helical structure. However, these polymers tend to oxidatively cross-link, forming ether linkages via a not fully investigated mechanism. Thus, in the IR spectrum of phenol polymers, IR absorptions corresponding to aromatic ether structures may arise when the polymers are stored at room temperature [108]. On the other hand, mechanistic studies of the polymerization of 4-hydroxybenzenesulfonic acid have shown clear evidence that C – O – C coupling can indeed be the dominant linkage structure of the resulting polymers—at least during the polymerization of phenols with electron-withdrawing groups. This result was found by monitoring the polymerization process by ^{1}H and ^{13}C NMR spectroscopy during incremental addition of H_2O_2 [75]. Even for the polymerization of *p-t*-butylphenol, the analysis of separated dimers and trimers from the resulting products revealed that two sorts of dimers were formed during the polymerization process, which were assigned to phenylene and oxyphenylene linkages in these dimers [109].

The influence of the *para* substituent on the polymerization behavior was systematically investigated. Phenols with electron-withdrawing groups were found to be poor substrates for HRP-catalyzed polymerization, while electron-rich phenols having electron-donating residues were readily polymerized under the same conditions. The *p*-substituted phenols showed in the same study higher reaction rates than *m*-substituted phenols [110]. When polymerizing *p-n*-alkylphenols, the yield increased on increasing the chain length from a methyl to a pentyl substituent. The maximum molecular weight was already reached with the propyl substituent. A further increase up to heptyl did not lead to a further increase of the yield, while the molecular weight decreased only slightly [89]. In contrast, using a reverse micellar system (isooctane, AOT, water) for the polymerization of *p-n*-alkylphenols, the maximum yield was already reached with *p*-ethylphenol and decreased with increasing chain length [89]. On the other hand, the molecular weight of the resulting phenol polymers was found to increase on increasing the length of the alkyl group. The same behavior was found when the polymerization was carried out in a biphasic system of isooctane and water [111].

The simplest *p*-alkylphenol is *p*-cresol (**35**), which has a methyl substituent. One of the first detailed studies of the HRP-catalyzed oxidation of *p*-cresol was reported in 1976 [51]. Recently, a detailed in situ NMR analysis revealed details of the coupling mechanism of the *p*-cresol polymerization. ^{1}H NMR and ^{1}H-^{1}H gCOSY 2D NMR analysis suggested that *ortho–ortho* coupling (**43**) is the dominant coupling mechanism at the initial stage of the polymerization. The consumption of dimer accelerated only after the complete conversion of the monomer in the reaction mixture. After a reaction time of about 75 min, the formation of Pummerer's ketone (**44**) was observed. These ketonic species are formed from *ortho–para*-coupled dimers by intramolecular Michael addition. They are probably not able to participate in the further polymerization process and remain as side products (Scheme 9) [76]. Experiments with 4-propylphenol have revealed that the formation of Pummerer's ketone may be suppressed at lower temperatures [112].

One of the most investigated *p*-substituted phenols is *p*-phenylphenol (**45**). Working with this compound offers the following advantages: it is electron rich with a high-level HOMO and its molecular weight is relatively high. Thus, it is a very suitable substrate for HRP, and high molecular weight polymers are easy to synthesize. The molecular weight of poly(*p*-phenylphenol) synthesized with the HRP/H_2O_2 system in 1,4-dioxane/buffer mixtures can reach 26 000 g/mol [113]. However, it was reported that the resulting polymers are not completely soluble in organic solvents like DMF, due to either the presence of fractions of high molecular weight or cross-linking processes. Differential scanning calorimetry (DSC) thermograms of poly(*p*-phenylphenol) often indicated thermal cross-linking or branching reactions [78]. Since *p*-phenylphenol has a conjugated ring system, the phenoxy radicals, generated at the beginning of the polymerization, are very stable. This is a further reason why high molecular weight polymers can be realized in good yields with this monomer. However, the polymerization of *p*-phenylphenol lacks precise structure control. The reason is that the single electron of the *p*-phenylphenol radical can be stabilized preferentially at the *ortho* positions, but also at the *para* position of the phenyl substituent (Scheme 10, **46**). Therefore, different kinds of linkage structures are possible, leading to a complex polymer structure (Scheme 10, **47–49**) [78].

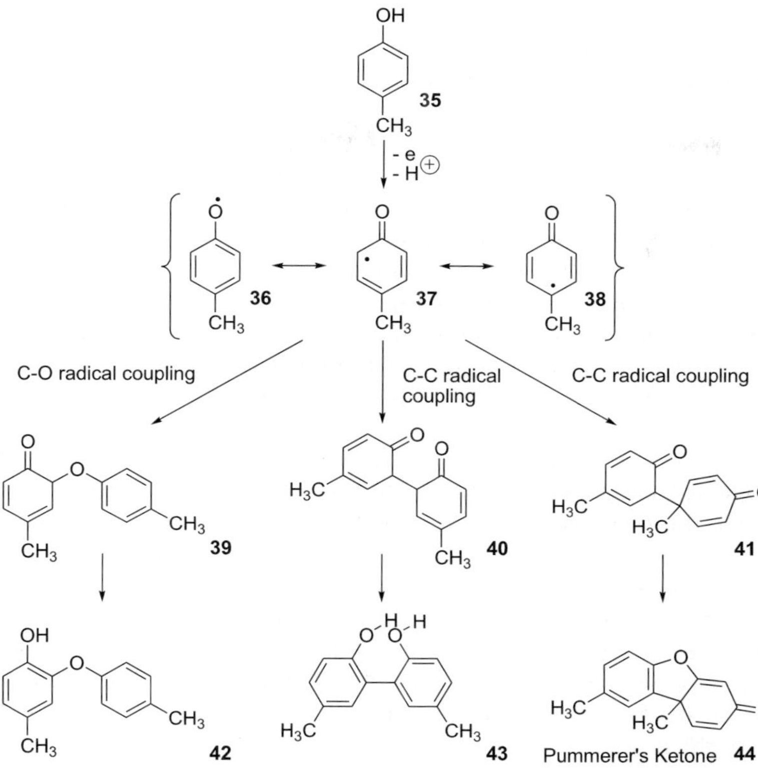

Synthesis of Phenol Polymers Using Peroxidases

Scheme 9 Coupling mechanism of *p*-cresol polymerization

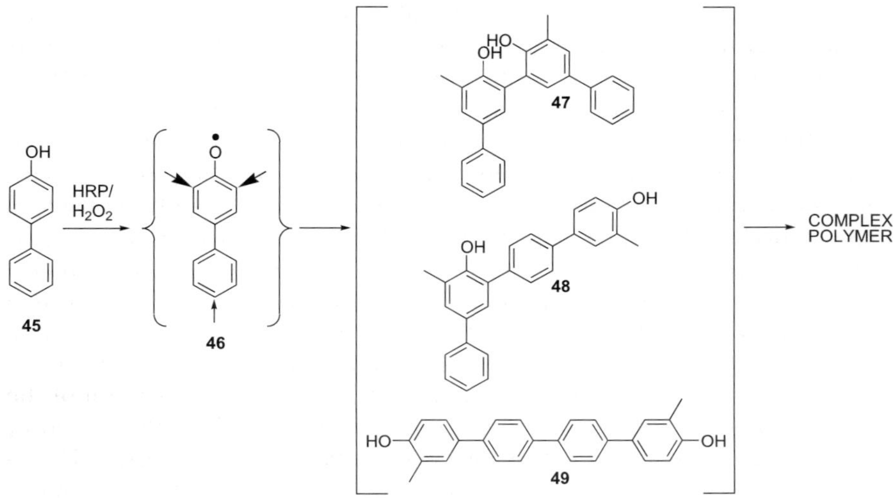

Scheme 10 Polymerization of *p*-phenylphenol

The structure of a poly(p-phenylphenol) made by polymerization in 1,4-dioxane/buffer mixture was investigated by ^{13}C solid-state CP/MASS NMR. The results showed that an o-substituted product is the major constituent of the polymer. No signals corresponding to phenoxy ether linkages were observed in the FT-IR spectrum [78]. The evolution of the molecular weight during the polymerization of p-phenylphenol was recently monitored as a function of the amount of H_2O_2 added by MALDI-TOF mass spectroscopy [114]. Even the normalized distributions of each repeat unit as a function of H_2O_2 added could be displayed by this technique. According to this study, at the beginning mainly dimers and trimers are formed. At an H_2O_2 to monomer ratio of 0.3, the amount of trimers was maximal and decreased as the trimers reacted to higher molecular weights. The overall molecular weight increased quickly during the initial stage and continued up to 60% of the total amount of H_2O_2 until a plateau was reached. This plateau was due to the fact that the phenol polymers precipitate out of the solution when their molecular weight is high enough. The polymerization was carried out in acetone/buffer mixture. A ratio of 50 vol% acetone and 50 vol% 0.01 M sodium phosphate buffer was found to give the best compromise between molecular weight and polydispersity of the resulting polymers [114].

Many efforts have been made to study the polymerization of halogenated phenols. The focus of attention is not only the production of phenol polymers in this case. It is believed that the HRP-catalyzed polymerization of toxic aromatic compounds, for example 4-chlorophenol, can be developed into a sophisticated method to clean aromatic pollutants from wastewater, as the resulting polymers can be removed from the wastewater easily by filtration. Several models for the kinetics of the HRP-catalyzed polymerization and precipitation of 4-chlorophenol have been developed [39]. However, while peroxidase catalysis was found to effectively reduce the concentration of phenols in the aqueous mixture, an accumulation of toxic, soluble products occurred during the polymerization process [40]. Several fluorinated phenols have been investigated as well. HRP is able to catalyze the polymerization of 4-fluorophenol in methanol/buffer mixtures. However, due to the electron-withdrawing fluoro substituent, this compound is a poor substrate to HRP. The resulting polymers had a molecular weight of not more than 700 g/mol and the yield could not be increased over 37%. Based on IR analysis the materials were said to consist of phenylene and oxyphenylene linkages. Contact angle measurements have been tried with these materials, but a sufficient film formation was not achieved [115].

Many publications have shown the outstanding potential of the HRP-catalyzed phenol polymerization for the synthesis of functional phenol polymers. An exceptional feature of the peroxidase catalysis is the possibility of polymerizing chemoselectively phenols having double or triple bonds in the side chain. One of the first examples of this feature was the polymerization of 2-(4-hydroxyphenyl)ethyl methacrylate (Scheme 11, **50**). The HRP-

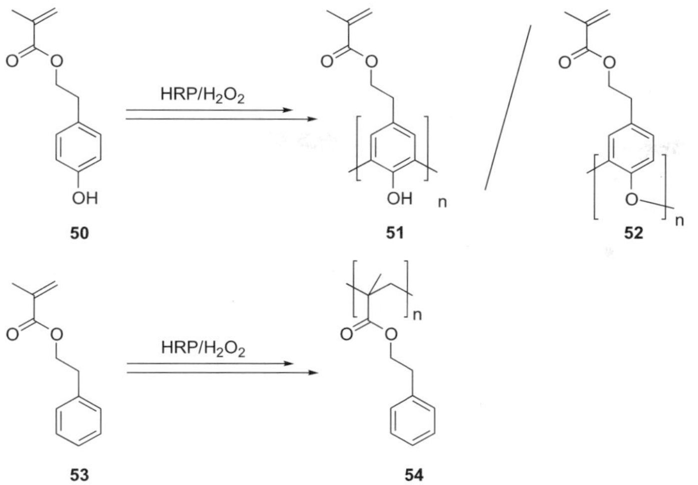

Scheme 11 Polymerization of 2-(4-hydroxyphenyl)ethyl methacrylate

catalyzed oxidative polymerization of this monomer using hydrogen peroxide in a solution of acetone/acetate buffer (pH 7, 50 : 50 vol %) resulted in the formation of a phenol polymer having the methacrylic groups unaffected in the side chain [47]. The polymer was said to consist of phenylene and oxyphenylene linkages, based on simplified ^{1}H NMR integrations related to the phenyl moiety without taking the end groups into account. Interestingly, the HRP/H$_2$O$_2$ system is able to catalyze the polymerization of the vinyl bond under the same conditions, if no phenol structure is present. The practically similar 2-phenylethyl methacrylate (53), just missing the OH group, was radically polymerized with the HRP/H$_2$O$_2$ system to give the polymethacrylate. With conventional AIBN initiation, even the (4-hydroxyphenyl)ethyl methacrylate (50) was polymerized to the polymethacrylate via radical polymerization [47]. Thus, the HRP-catalyzed polymerization of a phenol occurs chemoselectively even in the presence of a polymerizable vinyl bond, providing an easy way to synthesize directly a functional phenol polymer (macromonomer) having polymerizable vinyl bonds in the side chain.

This observation can be explained by the appearance of different kinds of electrophilic or more nucleophilic radicals. The electron-rich phenol is attacked by highly electrophilic radicals, while the methacrylic group is selectively attacked by less electrophilic radicals (e.g., R – O$^{\bullet}$, C$^{\bullet}$), which are generated for instance during an AIBN initiation. The recombination of two phenoxy radicals is for that reason more likely than the attack of a phenoxy radical on the electron-poor methacrylic double bond [48].

The two other vinyl monomers presented in Scheme 12, 4′-hydroxy-N-methacryloyl anilide (55) and N-methacryloyl-11-aminoundecanoyl-4-hydroxy anilide (56), were polymerized chemoselectively with the HRP/H$_2$O$_2$

system, and polymers were also prepared from 4-hydroxyphenyl-N-maleimide (57) [48]. All three monomers were polymerized in the form of their cyclodextrin complexes in aqueous buffer solution only. The resulting phenol polymers having the unaffected double bonds in the side chain acted as macromonomers, which could be copolymerized subsequently with MMA or styrene (Scheme 12) [48]. Even polymers containing the reactive vinylcyclopropane group have been successfully polymerized in aqueous medium using the cyclodextrin technique. The vinylcyclopropane groups were not attacked during the polymerization and hence the phenol polymers were subsequently cross-linked by radical initiation with AIBN [116]. The idea of chemoselec-

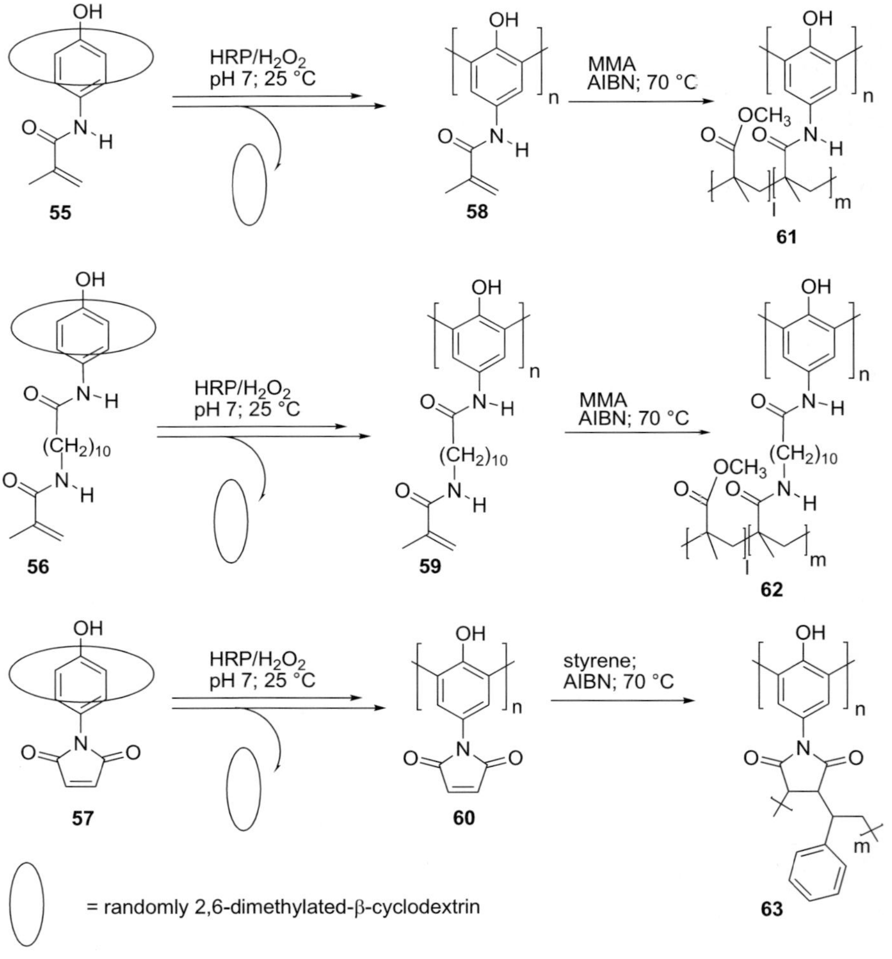

Scheme 12 Polymerization of vinyl monomers and macromonomers

tive polymerization has also been extended to triple bonds. An example is the polymerization of *m*-ethynylphenol, which will be discussed in the next section.

All phenol polymers having free vinyl groups in the side chain (**50**, **56**, and **57**) could be furthermore subjected to thermal curing due to cross-linking through the methacrylic groups [47, 48]. A multi-methacrylate oligomer was also prepared by polymer-analogous functionalization of a poly(isopropylidenediphenol) (BPA) resin. Such materials could be of potential interest for the formulation of dental composites as direct esthetic restorative materials [15].

Thermal cross-linking of phenol polymers was also achieved by copolymerization of two different functional phenols. A copolymer of 4-hydroxyphenyl-*N*-maleimide (**57**, Scheme 12) and a furanosyl derivative (*N*-(4-hydroxyphenyl)-2-furamide (**64**, Scheme 13) was subjected to irreversible cross-linking by thermally induced [4 + 2] cycloaddition (Diels–Alder cycloaddition). The copolymerization behavior of these two phenols was studied by following the monomer concentrations by HPLC, the increasing molecular weight by GPC, and the polymer composition by MALDI-TOF mass spectroscopy. It was found that the more electron rich and sterically less demand-

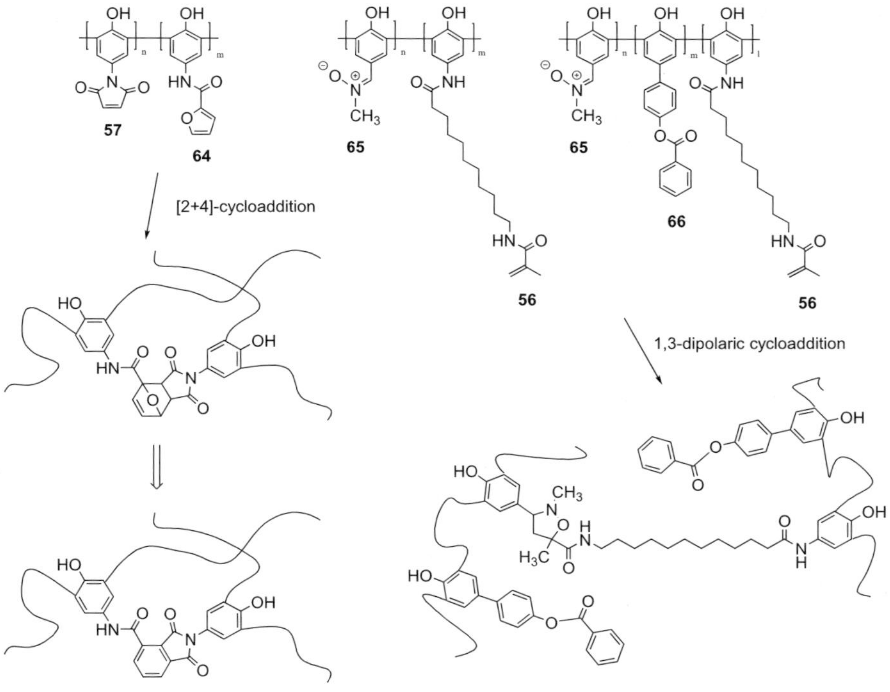

Scheme 13 Thermal cross-linking of phenol polymers

ing furanosyl derivative (**64**) was polymerized preferentially at the beginning of the reaction, while the maleimide compound (**57**) was built into the copolymers more at the end of the reaction [74]. The analysis also revealed details of the polymerization mechanism. The furanosyl derivative reacted first practically completely to the dimer. Next, the radical transfer led to homopolymers of **64** and copolymers with a dominating amount of **64** incorporated. Since the radical transfer step regenerates the monomeric phenol molecule, the concentration of **64** was found to be roughly constant for a long time after the formation of the dimers. Later, the polymerization of the maleimide compound **57** exhibited an increasing reaction rate.

Another way to realize thermally cross-linkable phenol copolymers was reached via the 1,3-dipolaric cycloaddition reaction between phenols having a vinyl group (**55**, **56**) and phenols bearing a nitrone group (**65**), which act as a 1,3-dipole [74]. These materials were readily cross-linked at room temperature.

Altogether, there have been only a few studies published dealing with the copolymerization behavior of distinct phenols, and usually the characterization of the copolymers was not fully examined. An early study of copolymerizations between different phenols and anilines can be found, wherein the copolymer compositions were characterized by elemental analysis [78]. In addition, monomeric phenols have been copolymerized with phenol polymers. This procedure offers, for example, an interesting way to turn lignin, a polymeric by-product from the pulp and paper industry, into a technical material. Lignin was reacted with phenol in an HRP-catalyzed copolymerization to produce lignin phenolic resins [117].

Despite the polyaromatic structure of phenol polymers, which goes along with their strong UV absorption, many photo reactions have been performed with phenol polymers. Using the HRP/H_2O_2 system it was possible to realize directly photoreactive phenol homopolymers containing nitrone groups (**65**) in the side chain [118]. Irradiation of films of this material induced the formation of three-membered ring structures, leading to oxaziridines in the side chain. In other words, the refractive index of these polymer films could be changed by irradiation. Such polymers are not available by polymerization of vinyl compounds having nitrone groups in the side chain, because cyclopolymerization is the preferred reaction under these conditions [119]. The HRP-catalyzed polymerization of 4-hydroxybenzaldehyde methylnitrone is therefore the only polymerization method involving radical species, which enables direct homopolymerization. New phenol polymers might also have potential for the production of new photoresists. Photocross-linkable phenol polymer films have been realized by polymerization of phenols having cinnamoyl groups attached in the *para* position via ester or amide groups [49]. The amides were found to undergo photochemically induced cross-linking via [2 + 2] cycloaddition faster than the corresponding esters. This might be due to preordering of the molecules in the polymer film by hydrogen bonding. Another innovative application is the synthesis of phenol polymers with azobenzene groups, which have been used for optical surface relief gratings (SRGs). Polymers from 4-phenylazophenol were prepared with the HRP/H_2O_2 system. A molecular weight (M_w) of about 3000 g/mol was achieved, and the yield was about 80% [50]. Despite the low degree of polymerization of about 15, it was reported that these materials were sufficient to produce thin films of good optical quality. Irradiation of these films revealed interesting photoanisotropic properties enabling the fabrication of high efficiency SRGs [120].

Enzymatically produced phenol copolymers have been successfully applied in the synthesis of polymer–CdS nanocomposites. Such nanocomposites are a convenient way to overcome the tendency of nanometer-sized particles to agglomerate due to their large surface-to-volume ratio. Copolymers consisting of 4-ethylphenol and 4-hydroxythiophenol, two mutually incompatible monomers in terms of solubility, were prepared by HRP-catalyzed oxidative copolymerization in a reverse micellar environment [43]. The molecular weight of the copolymers increased with increasing 4-hydroxythiophenol content (Table 5). In a subsequent step, CdS was attached via the thiol bonds to the copolymer. The resulting microspherical copolymers containing the CdS nanoparticles displayed fluorescence characteristics without low-energy emissions, which are usually associated with surface recombinations. The materials were stable in their solution behavior and luminescent properties.

The HRP-catalyzed polymerization of phenols was found to be a convenient way to produce redox polymers and conducting (electronically conducting and ionically conducting) polymers. Besides the interest in electronic conductive polyanilines [121], many efforts have been made to produce ionically conductive phenol polymers for battery applications. A classic effort is the synthesis of poly(hydroquinone) for use as a redox polymer. Typically, poly(quinone)s are prepared via chemical or electrochemical methodologies [122, 123]. Both processes produce a large amount of by-products and lead to complex polymer structures. The first alternative pathway to produce poly(hydroquinone) by peroxidase catalysis was based on a multienzymic

Table 5 Copolymer molecular weight and CdS content as a function of monomer composition

Monomer composition (% 4-hydroxythiophenol)	M_n	M_w	Polydispersity (M_n/M_w)	CdS content in polymer (wt%)
0	1650	2300	1.4	0
10	1750	2550	1.5	3.2
30	2500	3490	1.4	5.4
50	4390	10775	2.2	7.2

strategy (Scheme 14) [44]. The enzyme β-glucuronidase was used to catalyze the transfer of β-D-glucose to hydroquinone under aqueous conditions. The resulting arbutin (**69**) was polymerized using horseradish peroxidase (HRP) or soybean peroxidase (SBP) in various buffer mixtures and yielded water-soluble phenol polymers with molecular weights ranging from 1600 to 3200 g/mol. The quantitative deglycosylation of the poly(arbutin) was subsequently performed in aqueous HCl. The structure of the poly(arbutin) was considered to consist of *ortho–ortho* couplings, and therefore the structure of the corresponding poly(hydroquinone) should be more regular. This may explain why the enzymatically prepared poly(hydroquinone) was soluble in THF, DMSO, DMF, acetone, and methanol, in contrast to the electrochemically prepared poly(hydroquinone), which was found to be insoluble in these solvents.

A chemoenzymatic way to produce poly(hydroquinone) was achieved by enzymatic oxidative polymerization of 4-hydroxyphenyl benzoate, followed by alkaline hydrolysis of the resulting polymer [45]. HRP and SBP were used as enzymes. The molecular weight of the resulting poly(4-hydroxyphenyl benzoate) varied between 1100 and 2400 g/mol. The structure was said to consist of phenylene and oxyphenylene moieties, which was found by IR analysis and titration of the residual amount of phenolic groups in the polymer. Other phenol polymers have shown their potential for electronic applications as well. Besides hydroquinone, catechol has also been used as substrate for peroxidase-catalyzed polymerization. The molecular weights of the reac-

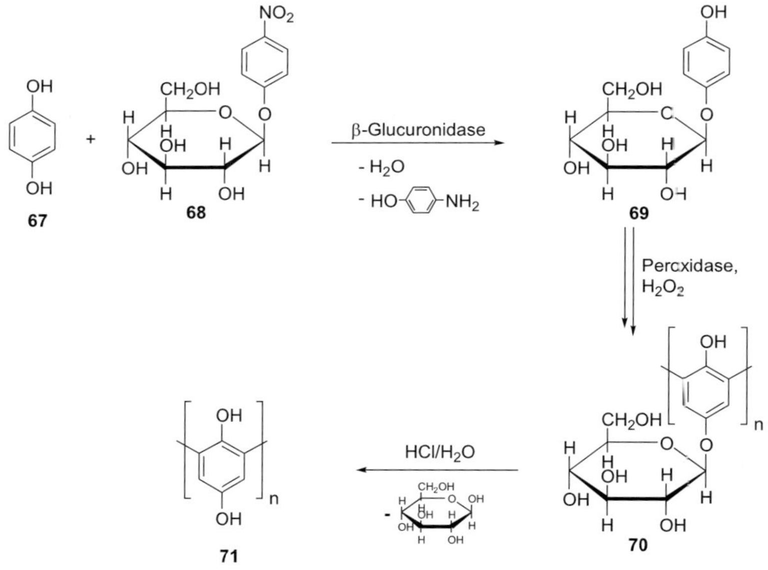

Scheme 14 Polymerization of hydroquinone

tion products were not determined. Based on IR and elemental analysis, the structure of the poly(catechol) was found to be a mixture of phenylene and oxyphenylene units. The polymer possessed low conductivity but had a stable reversible redox behavior during cyclic voltammetry measurements. The iodine-labeled poly(catechol) was reported to have a low electrical conductivity from 10^{-6} to 10^{-9} S/cm [124]. In contrast, an iodine-doped thin film of a copolymer consisting of phenol and tetradecyloxyphenol displayed a conductivity of 10^{-2} S/cm [99]. Polymers of hydroquinone mono-oligo(ethylene glycol) ether have been prepared by HRP catalysis. They were lithiated and mixed with PEG. Thin films of this material had a high ionic conductivity of 4×10^{-5} S/cm [125]. Furthermore, the surface resistivity of a simple p-phenylphenol polymer doped with nitrosylhexafluorophosphate, for example, can reach 10^5 Ω [113].

Recently, the synthesis of phenol polymers starting from natural phenol monomers have attracted more and more interest. The use of natural phenols offers an easy way to polymerize chiral materials. A good example is the polymerization of tyrosine and its derivatives. Tyrosine ethyl ester and methyl esters in the form of their hydrochlorides (72) were polymerized using the HRP/H_2O_2 system in high concentrations of buffer in aqueous solution (Scheme 15) [126]. The precipitation of tyrosine polymers made from both entantiomeric forms and a racemic mixture was observed. The molecular weight varied between 1500 and 4000 g/mol, depending on the monomer and buffer concentrations. The stereoconfiguration of the amino acid did not influence the molecular weight or yield remarkably according to this study. The polymer structure was estimated to be a mixture of phenylene and oxyphenylene units. The poly(tyrosine ester)s were furthermore converted into a new class of poly(tyrosine) by alkaline hydrolysis of the ester groups. The resulting poly(amino acid) was soluble only in water. Additionally, the enzyme HRP also initiated the oxidative homopolymerization of N-acetyltyrosine and the

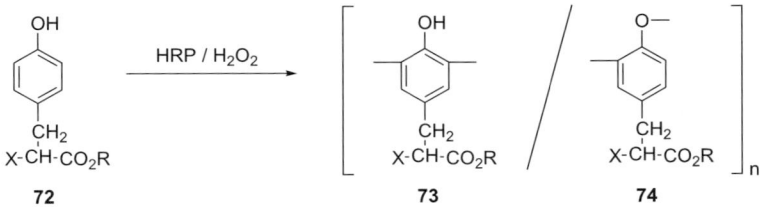

72-1 : L-Form, R=Et, X=$NH_3^+Cl^-$
72-2a: L-Form, R=Me, X=$NH_3^+Cl^-$
72-2b: D-Form, R=Me, X=$NH_3^+Cl^-$
72-2c: D,L-Form, R=Et, X=$NH_3^+Cl^-$
72-4 : L-Form, R=H, X=NHC(=O)CH_3

Scheme 15 Polymerization of tyrosine derivatives

copolymerization of N-acetyltyrosine and arbutin (**69**). In this case it was necessary to collect the homopolymer by dialysis and the copolymer by reprecipitation due to the solubility of the polymers in the buffer polymerization medium [126].

Tyrosine derivatives have also been polymerized readily under micellar conditions [127]. These studies were performed with both enantiomeric forms of the amphiphilic decyl esters of tyrosine (DEDT, DELT), which self-assemble above their critical micelle concentration in aqueous buffer solution into rod- or platelike-shaped fibers. The critical micelle concentrations were found to vary in dependence on pH due to the changing protonation level of the α-amino groups. The self-assembled structures were polymerized by addition of HRP and H_2O_2. The polymerized fibers were more robust than the unpolymerized fibers. While the diameter of the fiber structures remained nearly unchanged, their length increased in the order of hundreds of microns and the surface appearance changed from a rough fibrillar structure to a smooth surface without evident substructure [127]. However, the polymerization process led to a different organization of the molecules. Interestingly, it was reported that the stereoconfiguration of the tyrosine decyl ester had a significant influence on the reaction rate. The monomer in the D configuration displayed, according to the second-order kinetic fit, nearly twice the reaction rate of the L-tyrosine derivative [128]. Copolymers of DELT and L-tyrosineamide produced different self-assembled structures. The unpolymerized comonomer mixtures formed mostly smooth, featureless, and amorphous aggregates on a gold-coated mica surface. The copolymerization induced the formation of regular hemispherically shaped interacting aggregates, perhaps due to more structure organization.

3.3
m-Substituted Phenols

Compared to *p*-substituted phenols, only a few studies have been carried out with *m*-substituted phenols. When an *m*-substituted phenol is used as monomer for peroxidase-catalyzed polymerization, the enzyme-generated radicals are stabilized preferentially in the *ortho* and *para* positions. This implies that the structures of phenol polymers resulting from *m*-substituted phenols are less controllable. For example, the thermal properties of poly(*m*-cresol) were compared to those of poly(*p*-ethylphenol). The latter, made from a *p*-substituted phenol, displayed a large exotherm at 116 °C in the DSC thermogram, an indication of thermal cross-linking. In contrast, the poly(*m*-cresol) was stable with about 10% weight loss at 300 °C in a nitrogen atmosphere [79]. This difference can be explained by the fact that the polymerization of *m*-cresol, having three positions for bond formation on the phenylene ring, results in a cross-linked polymer. The monomer units in the poly(*p*-ethylphenol) are, on the other hand, mainly linked at the *ortho* positions, as the *para* position is blocked by the ethyl group. As a result, this polymer is not expected to be cross-linked [79].

The HRP-catalyzed polymerization of *m*-alkylphenols, especially *m*-cresol, has been performed in aqueous organic solvents and under reversed micellar conditions [78, 79, 129–131]. The polymerization of *m*-cresol using standard conditions (HRP/H_2O_2 in 1,4-dioxane/buffer 80 : 20 vol %) led to polymers that were almost insoluble in common organic solvents [129, 130]. However, several equivolume mixtures of phosphate buffer and organic solvents have been applied successfully to produce THF-soluble polymers from *m*-cresol with molecular weights between 1000 and 3000 g/mol [131]. While *m*-ethylphenol was also readily polymerized under the same conditions, HRP did not catalyze the polymerization of *m*-isopropylphenol and *m*-*t*-butylphenol. These two monomers can, however, be polymerized with SBP as will be discussed later. During the polymerization of *m*-cresol, the monomer conversion and the development of the molecular weight were estimated as a function of addition of H_2O_2. It was found that the monomer conversion increased gradually as a function of added H_2O_2, whereas the molecular weight was almost constant during the polymerization. This was explained by the hypothesis that the resulting soluble dimers and oligomers reacted much faster than the monomer, and that the precipitated polymer did not react any more during the polymerization. This idea was supported by the fact that the calculated HOMO levels of all possible occurring dimers (Scheme 16, **75a–e**) were larger than that of *m*-cresol itself, thus resulting in a higher reactivity of the dimers in the oxidative polymerization reaction [131].

Intensive studies with UV-VIS and MALDI-TOF analysis opened new insights into the effect of the solvent composition on the polymerization of

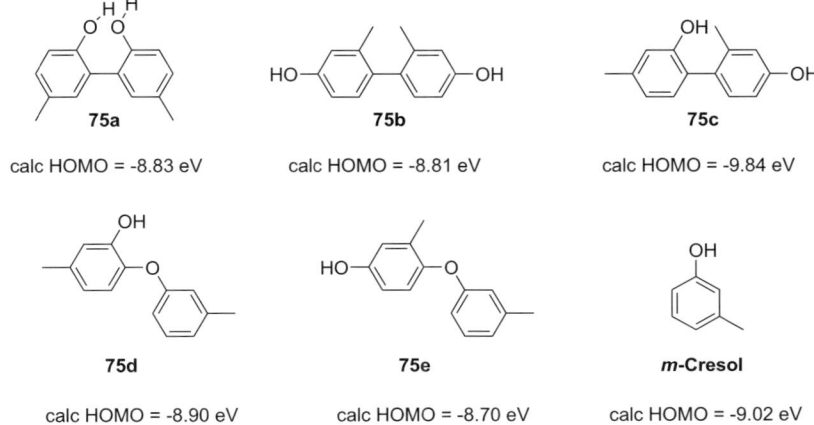

Scheme 16 Calculated HOMO levels of *m*-cresol dimers

m-cresol [84]. By these techniques, strong hints were found that the molecular weight of the resulting polymers and the yield are related to the self-association of *m*-cresol in mixtures of organic solvents and water. While, for example, clusters of *m*-cresol were not formed in 1,4-dioxane, addition of water remarkably promoted the self-association of *m*-cresol. On the other hand, clusters of *m*-cresol formed favorably in methanol, but were readily hydrated with an increasing water content in the mixture. These effects were in

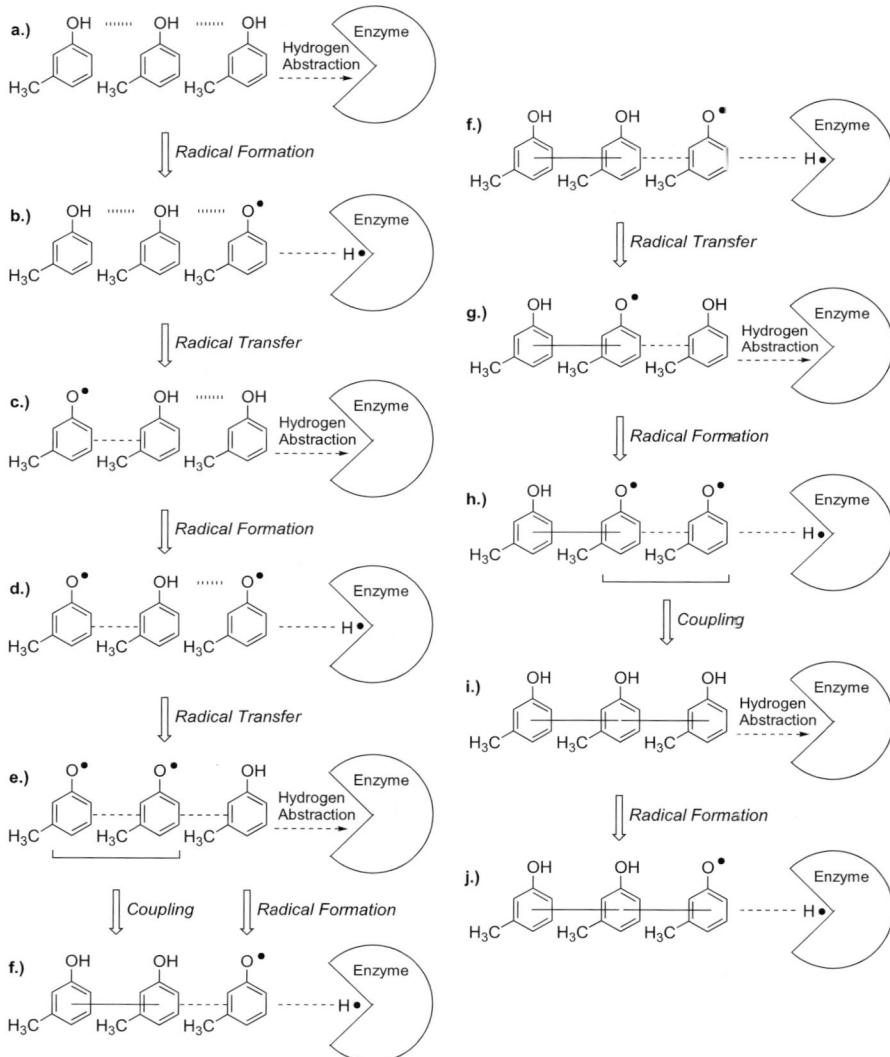

Fig. 6 Clustering of *m*-cresol as a possible key factor during HRP-catalyzed polymerization

agreement with the polymerization behavior. Thus, the polymer is obtained in high yields when *m*-cresol clusters are formed in solution, while hydration of these clusters will restrain the polymerization. An explanation for this behavior is shown in Fig. 6 [84]. The formation of clusters of *m*-cresol near an enzyme molecule enables the continuous generation of phenoxy radicals through the radical transfer process in the cluster, and also the recombination process of the radicals is promoted by the self-aggregation.

The polymerization of *m*-cresol was also carried out successfully in ethanol/buffer mixtures [79]. The molecular weight was controllable by changing the solvent composition. Enzyme kinetics revealed the effect of the solvent on the enzyme activity and substrate portioning between bulk solvent and the enzyme. While the polymer solubility increased with increasing ethanol content in the solvent mixture, the enzyme activity went through a maximum (around 20 v/v % ethanol content) with increasing amount of ethanol. This was similar to the results found previously for aqueous 1,4-dioxane/buffer mixtures. The enzyme showed no activity at 100% ethanol content, but it was found to be not irreversibly deactivated. Hence, gradual addition of buffer led to nearly complete restoration of the enzyme activity.

Furthermore, *m*-cresol and *m*-halogenated phenols have been polymerized in aqueous buffer solution only in the presence of 2,6-di-*O*-methyl-β-cyclodextrin [90]. In the absence of cyclodextrin, only insoluble materials in very low yields were obtained under aqueous conditions from these phenols [81]. The structure of the host–guest complexes between cyclodextrin and the phenols was characterized by 2D NMR [48, 90] and the association constants of the cyclodextrin complexes were determined by the Benesi–Hildebrand method. Furthermore, 3-fluorophenol, 3-chlorophenol, and 3-bromophenol were successfully polymerized in various aqueous organic solvents such as methanol, acetone, or isopropanol [115].

Functionalized polymers can be prepared from *m*-substituted phenols: the HRP-catalyzed oxidative polymerization of *m*-ethynylphenol (**76**) using hydrogen peroxide in a solution of methanol/phosphate buffer (pH 7, 50 : 50 vol %) under air resulted in a phenol polymer having the ethynyl groups unaffected in the side chain [132]. In contrast, a reaction with a conventional oxidation catalyst (Cu(I)Cl/TEED) led to coupling of the acetylene moiety to produce bis(3-hydroxyphenyl)butadiyne (**78**, Scheme 17) [132].

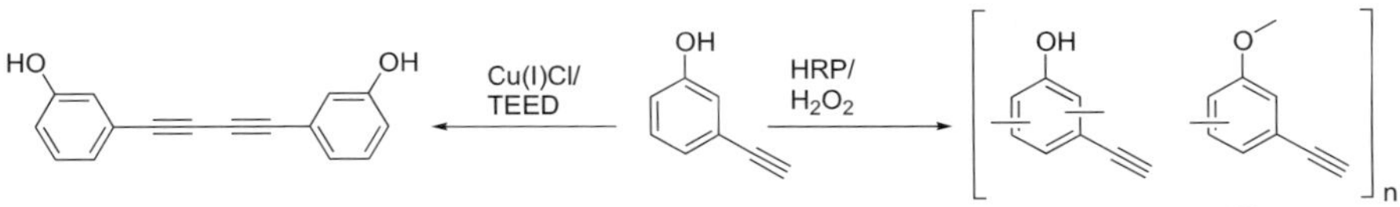

Scheme 17 Polymerization of *m*-ethynylphenol

3.4
o-Substituted Phenols, Bisphenols, and other Phenols

At first sight, o-substituted phenols do not seem to be suitable substrates for producing polymeric materials by peroxidase catalysis. However, o-methoxyphenols (guaiacols), for instance, have been converted to oligomeric materials (dimers to pentamers) by peroxidase catalysis (Scheme 18) [133]. SBP, an enzyme which will be discussed in the next section, was used for this reaction instead of HRP.

A different type of polymer can be realized by polymerization of 2,6-dimethylphenol. The polymerization leads to poly(phenylene oxide) (PPO), a material of technical importance. This type of synthesis was performed earlier using a copper/amine catalyst system [102, 134]. PPO consisting exclusively of oxy-1,4-phenylene units was realized using HRP/H_2O_2 as the catalyst system [135]. It is also possible to produce PPO via an alternative pathway from syringic acid (**79**). This type of HRP-catalyzed oxidative polymerization involves elimination of carbon dioxide and hydrogen to give PPO with molecular weights up to 13 000 g/mol (Scheme 19) [136, 137]. In addition, the isomer 4-hydroxy-3,5-dimethylbenzoic acid can be used to produce PPO. It was found by mass spectroscopy that the polymers possessed exclusively the structure presented in Scheme 18 with the carboxylic acid and the hydroxyl group as end groups. The polymerization of unsubstituted 4-hydroxybenzoic acid was not observed under similar reaction conditions. An example of the polymer-analogous modification of such enzymatically prepared PPO plastics is their demethylation with boron tribromide to yield poly(oxy-2,6-dihydroxy-1,4-phenylene) [138].

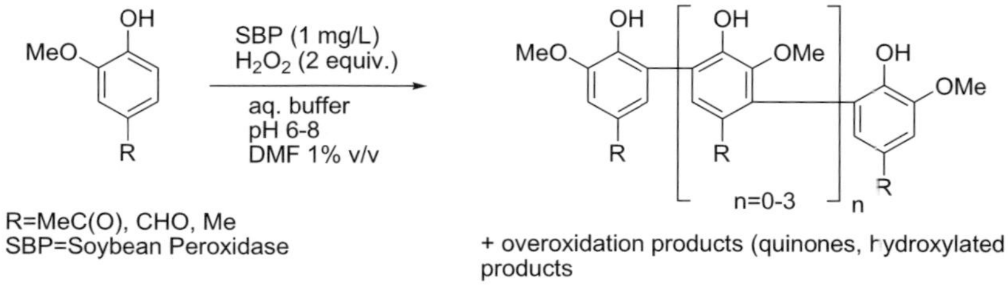

Scheme 18 Conversion of o-methoxyphenols to oligomers

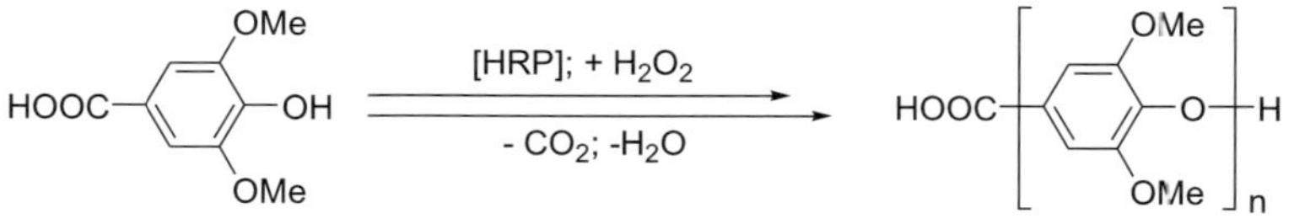

Scheme 19 Polymerization of syringic acid to form PPO

Synthesis of Phenol Polymers Using Peroxidases

Several bisphenol derivatives were reported to polymerize through peroxidase catalysis to give soluble polymers, even though these monomers are bifunctional. For example, the HRP-catalyzed oxidative polymerization

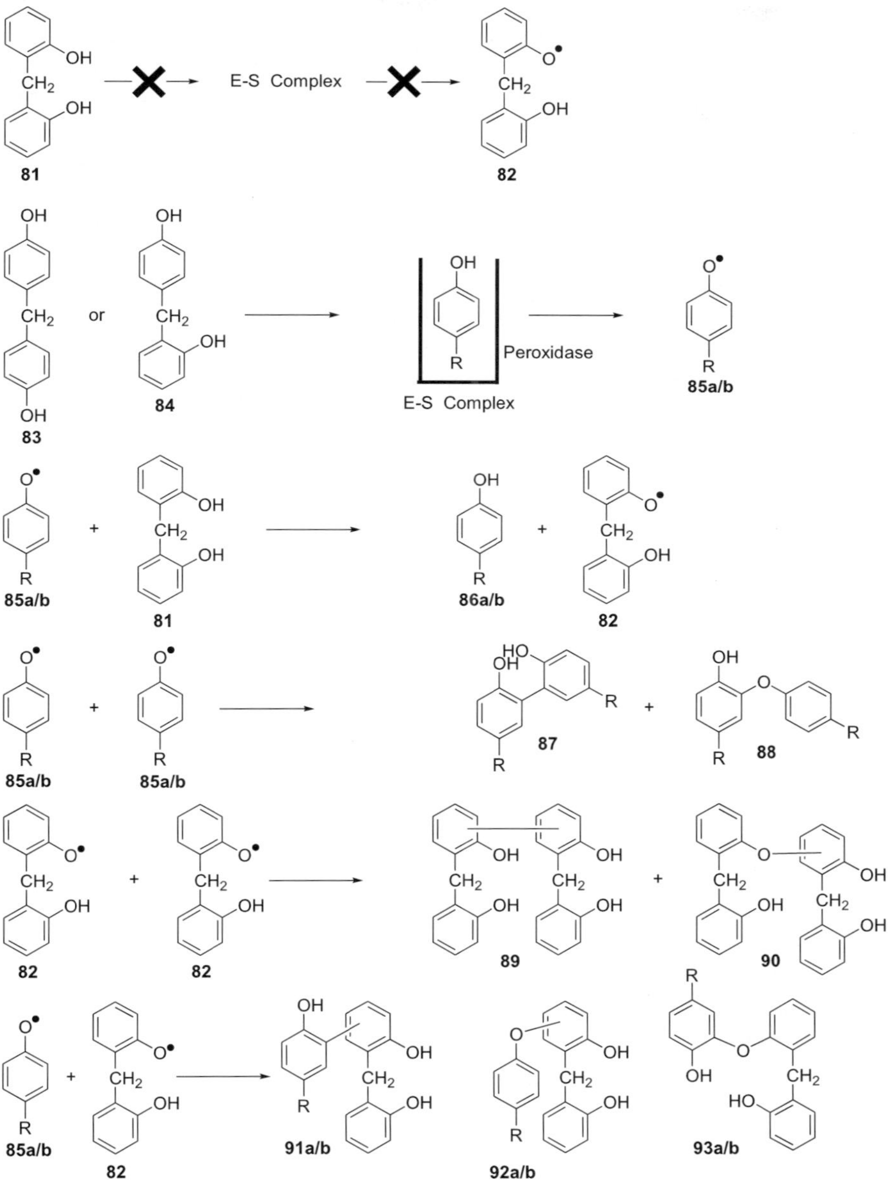

Scheme 20 Radical transfer reactions between enzymatically polymerizable and nonpolymerizable monomers

of 4,4′-oxybisphenol yielded α-hydroxy-ω-hydroxyoligo(oxy-1,4-phenylene). During the reaction, the redistribution and/or rearrangement of a quinone–ketal intermediate took place, involving the elimination of hydroquinone to give oligo(oxy-1,4-phenylenes) [139]. Thermally curable polymers have been synthesized from bisphenol A and 4,4′-biphenol [140, 141]. The treatment of bisphenol F, an industrial product consisting of 2,2′-, 2,4′-, and 4,4′-dihydroxydiphenylmethanes, with the HRP/H_2O_2 system led to the quantitative formation of a highly reactive prepolymer for curing via copolymerization of all three isomers. While the pure isomers 2,4′- and 4,4′-dihydroxyphenylmethane were converted into the homopolymer in high yield with a peroxidase/H_2O_2 system, no homopolymerization of 2,2′-dihydroxyphenylmethane occurred. However, the industrial mixture bisphenol F was polymerized nearly quantitatively. Thus, these findings suggest that radical transfer reactions between both enzymatically polymerizable monomers and nonpolymerizable monomers took place (Scheme 20) [142].

Fluorescent polymers have been realized by the oxidative polymerization of 2-naphthol (**94**) [143]. The polymerization was carried out in a reverse micellar system using AOT/isooctane to give the polymer in single and interconnected microspheres. The frequency of the AOT carbonyl IR vibrations was shifted as a function of the monomer concentration, thus proving the penetration of the monomers into the interfacial region of the reversed mi-

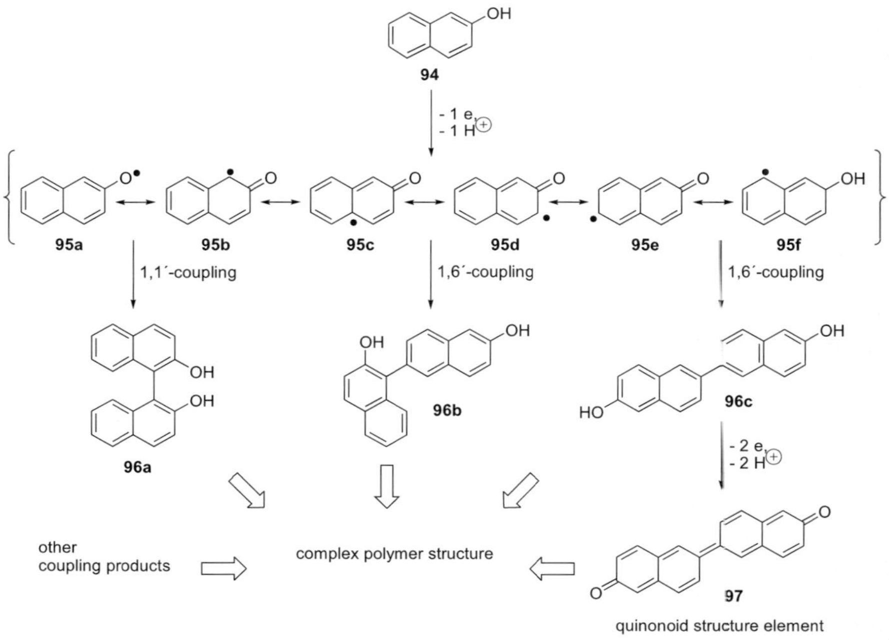

Scheme 21 Polymerization of 2-naphthol

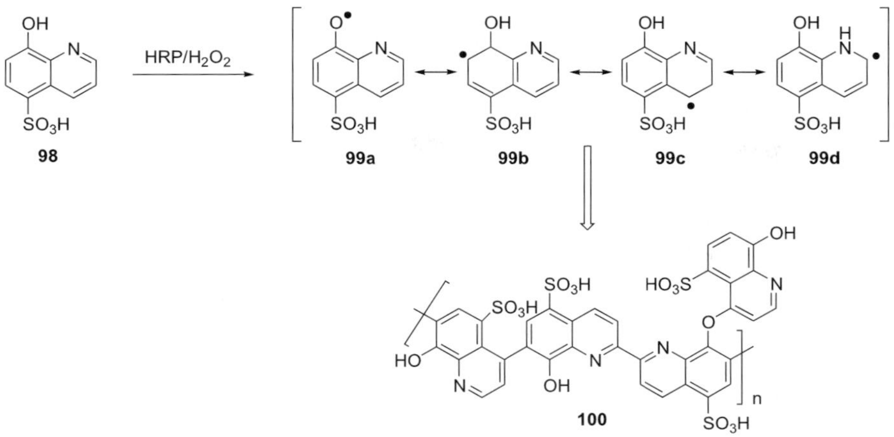

Scheme 22 Polymerization of 8-hydroxyquinoline-5-sulfonate

celles. Besides the fluorescence characteristic of the 2-naphthol chromophore, the polymer showed an additional well-resolved fluorescence attributed to extended quinonoid structures (**97**) at the end of or within the aromatic polymer chain (Scheme 21).

Higher-substituted two-ring structures have also been investigated. The polymerization of 8-hydroxyquinoline-5-sulfonate (**98**) was studied by in situ NMR spectroscopy [144]. The change of the resonances of the five observable protons allowed the use of NMR to follow directly which sites of the monomer participated in the oxidative coupling during the polymerization. Positions 2, 4, and 7 were the preferred sites for the polymerization. The lower preference for position 4 can be ascribed to the steric hindrance posed by the sulfonate group. Furthermore, ^{13}C NMR data suggested the existence of C – O – C linkages as well as C – C linkages in the polymer (Scheme 22) [144].

4
Polymerization of Phenols Using Other Peroxidases

So far, most peroxidase-catalyzed oxidative polymerizations have been carried out using the enzyme horseradish peroxidase (HRP). Another useful peroxidase that catalyzes the oxidative polymerization of phenols is soybean peroxidase (SBP). While the use of either HRP or SBP may often lead to similar products and results [77], the enzyme activity, yield, and molecular weight of the resulting polymers can also sometimes depend strongly on the type of enzyme used for the polymerization process. For example, SBP was found to be superior to HRP for the efficient polymerization of bisphenol A [140], but the polymerization of phenol with SBP afforded a lower polymer yield than that using HRP under the same reaction conditions [145]. Various *m*-substituted phenols have been polymerized with HRP and SBP under the same reaction conditions, for example *m*-alkylphenols, *m*-halogenated phenols, and *m*-phenylphenol [131]. The HRP-catalyzed polymerization of *m*-cresol afforded the polymer almost quantitatively, while the yield decreased enormously using *m*-ethylphenol. The polymerization of *m*-isopropyl- and *m-t*-butylphenol was not initiated by HRP. In the case of the analogous SBP-catalyzed polymerizations, the yield increased with increasing substituent volume. These data suggest that for *m*-substituted phenols, HRP has a higher activity toward monomers with smaller substituents, while SBP is preferable for larger-volume substituents.

A new way to realize polymers from *m*-substituted phenols having a bulky substituent via HRP catalysis is polymerization in the presence of a redox mediator [146]. Cardanol is a phenol derivative from a renewable resource, which has a very bulky substituent in the *meta* position: a C15 unsaturated hydrocarbon chain with one to three double bonds. Thus, cardanol was polymerized by SBP, while HRP did not initiate this polymerization [147]. However, the HRP-catalyzed polymerization took place in the presence of *N*-ethyl phenothiazine and phenothiazine-10-propionic acid. The structure and properties of the resulting polymers have been reported to be similar to those made by SBP catalysis. Thus, the two presented phenothiazine derivatives most likely act as a mediator for electron transfer between HRP and cardanol, as demonstrated in Fig. 7 [146].

Furthermore, SBP was often successfully applied in the synthesis of functional polymers. A good example is the synthesis of a phenol polymer containing thymidine, which may be of interest in terms of biorecognition. In a first step, a lipase was used to realize the regioselective acylation of thymidine at the 5'-hydroxyl group. The following polymerization of the phenolic nucleoside was catalyzed by SBP to give the thymidine-containing polymer **104** with a yield of 70% and a M_n of 21 700 g/mol (Scheme 23) [148].

Also laccases, oxidoreductases with a Cu atom in the active center, have been successfully applied to the synthesis of phenol polymers. For example, phenol and several derivatives were polymerized by laccase from *Pycnoporus coccineus* (PCL) in aqueous organic solvents to yield the corresponding polymers [149]. The same enzyme was used to produce PPO from syringic acid (**79**) and from 2,6-dimethylphenol. Laccase from *Myceliophthore* (MPL) was found to catalyze the polymerization of syringic acid to PPO [135–137]. Lac-

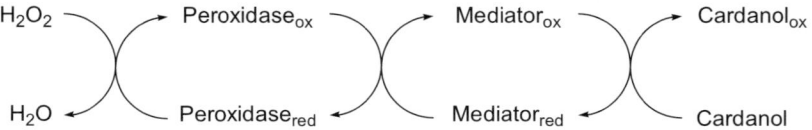

Fig. 7 The proposed role of mediators in the peroxidase-catalyzed oxidation of cardanol

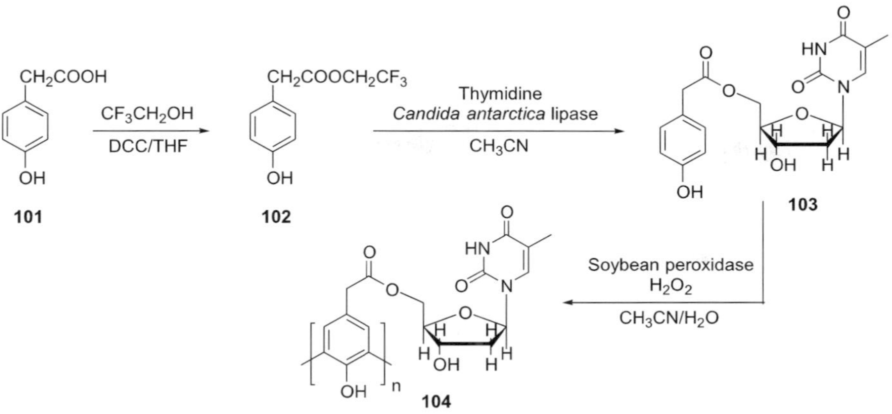

Scheme 23 Synthesis of a phenol polymer containing thymidine

case from *Trametes versicolor* (TVL) catalyzed the oxidative polymerization of 1-naphthol [150], and 1,5-dihydroxynaphthalene was polymerized by the Cu-containing bilirubin oxidase (BOD). Finally, tyrosinase was used for several cross-linking reactions, for example for the cross-linking of soluble lignin fragments to give insoluble materials [151].

5
Polymerization of Phenols Using Model Complexes

The peroxidase-catalyzed polymerization of phenol and aniline derivatives has shown a lot of potential for the synthesis of new functional polymers under mild conditions, and for the environmentally friendly synthesis of phenol polymers without using toxic formaldehyde, as was shown in the previous paragraphs. The main drawback of this type of synthesis are therefore the cost and the handling of the enzyme, although peroxidases are widely distributed in nature and HRP is extensively used commercially.

Many efforts have been made to replace the enzyme by synthetic metal complexes. A simple way to do so is the use of hematin, which is the oxidized form of the heme group. The free heme group is unstable and rapidly oxidized to hematin, and is therefore not of interest for model complexes. Ethylphenol was successfully polymerized by hematin in a DMF/buffer mixture [152]. The pH was varied between 4.0 and 11.0 and the yield increased with increasing pH value. The results in terms of molecular weight and yield were at pH 11 comparable to those of the polymers obtained by HRP catalysis at pH 7. Based on UV-VIS and Mössbauer spectroscopy, a mechanism was proposed for the hematin-catalyzed polymerization which is comparable to that proposed for HRP (Scheme 24) [152].

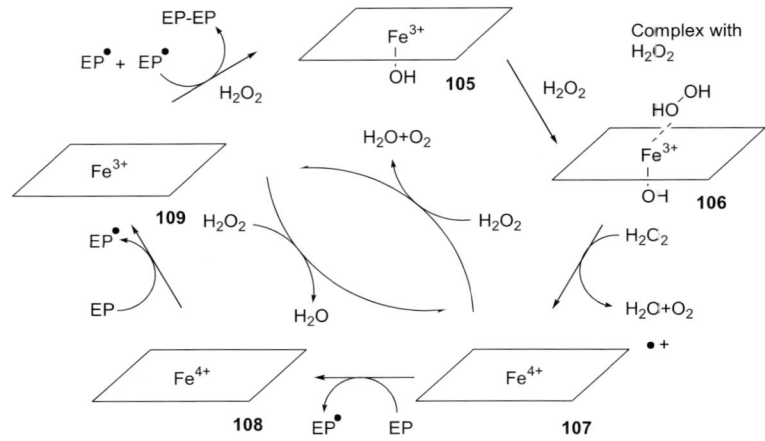

Scheme 24 Hematin-catalyzed polymerization of ethylphenol (EP)

Another compound which has been found to somewhat imitate the active site of peroxidases is the commercially available Fe(II)-salen catalyst. This catalyst was used successfully to produce phenol polymers, which could be of interest for industrial production [153, 154]. For example, cardanol can be polymerized by the Fe(II)-salen catalyst [155]. Due to the unsaturated bonds in the side chain of the cardanol components, the resulting polymers could be thermally cured, or cured by use of cobalt naphthenate to give brilliant films with a high-gloss surface. This reaction proves that reactive prepolymers can be synthesized from renewable resources (cardanol is the main component obtained by thermal treatment of cashew nutshell liquid). This process could be a true alternative to conventional phenol–formaldehyde resins (Scheme 25) [155]. Other non-heme iron complexes have been found to

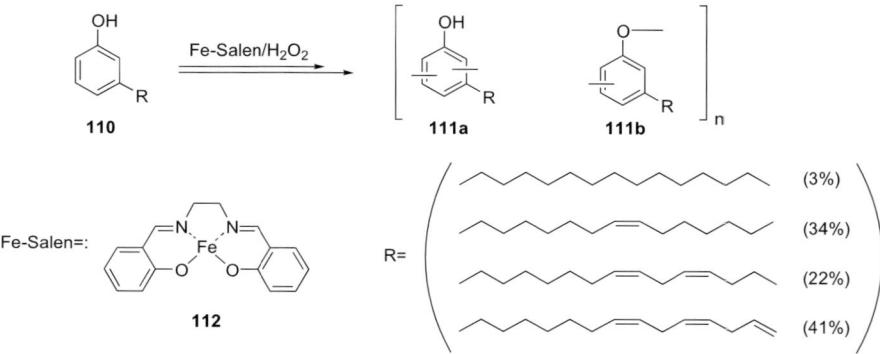

Scheme 25 Polymerization of cardanol using Fe(II)-salen catalyst

act in H_2O_2-dependent oxidations as well, acting as a typical Fenton reagent or through metal-based mechanisms [156].

Another polymer from partially renewable resources was synthesized by first reacting 6-hydroxy-1-naphthoic acid via acylation with alcohols from linseed oil or fish oil. Polymerization using the Fe-salen catalyst gave the corresponding polynaphthols in high yields. The resulting polymers were again cured by cobalt naphthenate or thermal treatment to give brilliant films with high hardness and glossy surface [157]. The use of the Fe(II)-salen catalyst can be combined with peroxidase catalysis. Polymers from *m*-cresol or bisphenol A, which have been prepared by peroxidase catalysis, can be cross-linked by the Fe-salen catalyst. Under suitable reaction conditions (appropriate concentrations of substrate and catalyst), it was reported that the phenol polymers can even be coupled without cross-linking to produce ultrahigh molecular weight phenol polymers [158]. This approach has been extended to other phenol polymers as well. Polymers from phenol, *o*- and *m*-substituted phenols, bisphenol, and triphenol were subjected to oxidative coupling by the Fe-salen/H_2O_2 system. On the other hand, this reaction scarcely proceeded with polymers made from *p*-substituted phenols [159]. Additionally, poly(amino acids) were reacted with *p*-(2-aminoethyl)phenol (tyramine). The resulting functionalized poly(amino acids) were converted to soluble high molecular weight poly(amino acids) and further to cross-linked gels by Fe-salen- or HRP-induced oxidative coupling [160].

Regioselective oxidative polymerization of 4-phenoxyphenol (**113**) was realized by a tyrogenase model complex to yield poly(1,4-phenylene oxide) (Scheme 26) [161, 162]. The polymerization was performed in toluene under oxygen. The molecular weight (M_n) ranged from 700 to 3800 g/mol, which was significantly less than that with a conventional CuCl/TEED catalyst.

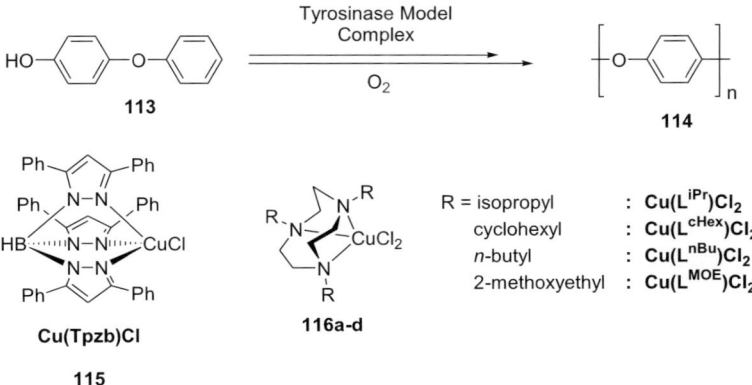

Scheme 26 Oxidative polymerization of 4-phenoxyphenol

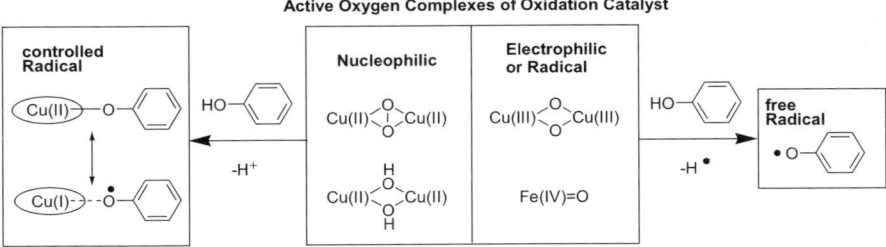

Fig. 8 Working hypothesis for the control of the phenoxy radical coupling by the oxidation catalyst

The results were compared to those for an oxidative polymerization using HRP/H_2O_2 in a 1,4-dioxane/buffer mixture, but the resulting materials were insoluble in DMF.

Although a conventional CuCl/TEED catalyst is known to catalyze the oxidative polymerization of 4-phenoxyphenol under dioxygen, this catalyst could not control the regioselectivity during the recombination step of the phenoxyradicals. The improvement of the presented tyrogenase complexes (**115, 116a–d**) lies in their ability to form controlled radicals. Since the tyrogenase complexes are more nucleophilic than the CuCl/TEED catalyst, they abstract only protons and not hydrogen atoms from the 4-phenoxyphenol molecule, to form stable intermediates. As a result, the regioselectivity is regulated in the subsequent coupling step (Fig. 8) [162].

6
Outlook

During the last ten years, many research results have shown that oxidative polymerization catalyzed by peroxidases is a convenient, resource-saving, and environmentally friendly method for synthesizing phenol polymers. In contrast to the conventional synthesis of phenol–formaldehyde resins, the peroxidase-catalyzed polymerization of phenol proceeds under mild reaction conditions (room temperature, neutral pH). The polymerization of toxic phenols has promising potential for the cleaning of wastewaters. Moreover, the polymerization of phenols from renewable resources is expected to attract much attention in times of worldwide demand for the replacement of petroleum-derived raw materials. Besides the environment-protecting aspects of this innovative type of polymerization, the enzyme-catalyzed polymerization represents a convenient method to realize new types of functional polyaromatic polymers. Phenol polymers made by peroxidase catalysis should have much potential for electronic and optical applications. The synthesis of functional phenol polymers is facilitated by the fact that polymerizable groups and reactive groups are usually not affected by the enzyme-catalyzed polymerization of phenols.

It is expected that new types of functional phenol polymers with interesting properties in terms of thermal stability or optical and electronic properties will be presented in the next few years. New examples of phenol polymers from natural resources will hopefully broaden the interest for this type of polymerization from a commercial point of view. In this connection, the finding of suitable highly active model complexes could be of increasing interest in reducing the cost of the oxidative polymerization of phenol derivatives.

References

1. Ritter H (1997) In: Arshady R (ed) Desk reference of functional polymers, syntheses, and applications. American Chemical Society, Washington, DC, p 103
2. Whitesides GM, Wong CH (1994) Angew Chem 106:851
3. Jones JB (1986) Tetrahedron 42:3351
4. Faber K (1992) Biotransformations in organic chemistry. Springer, Berlin Heidelberg New York
5. Sananiello E, Ferraboschi P, Griseni P, Manzocchi A (1992) Chem Rev 92:1071
6. Wong CH, Whitesides GM (1994) Enzyme catalysis in organic synthesis. Pergamon, Trowbridge
7. Theil F (1995) Chem Rev 95:2203
8. Drauz K, Waldmann H (1995) Enzyme catalysis in organic synthesis. Wiley, Weinheim
9. Schmidt RD, Verger R (1998) Angew Chem 110:1694
10. Bornscheuer UT, Kazlauskas RJ (1998) In: Rehm JH, Reed G, Puhlet V, Stadler PJW, Kelly DR (eds) Biotechnology series, vol 8a. Wiley, Weinheim, p 37
11. Gross RA, Kumar A, Kalra B (2001) Chem Rev 101:2097
12. Carrea G, Riva S (2000) Angew Chem 112:2312
13. Klibanov AM (1990) Acc Chem Res 23:114
14. Takayama S, McGarvey GJ, Wong CH (1997) Chem Soc Rev 26:407
15. Tiba A, Culbertson BM (1999) J Macromol Sci Pure Appl Chem 36:489
16. Kobayashi S, Shoda S, Uyama H (1995) Adv Polym Sci 121:1
17. Kobayashi S, Shoda S, Uyama H (1996) In: Salomone JC (ed) The polymeric materials encyclopedia. CRC, Boca Raton, p 2102
18. Kobayashi S, Shoda S, Uyama H (1997) In: Kobayashi S (ed) Catalysis in precision polymerization. Wiley, Chichester, Chap 8
19. Kobayashi S, Uyama H, Ohmae M (2001) Bull Chem Soc Jpn 74:613
20. Kobayashi S (1999) J Polym Sci A Polym Chem 37:3041
21. Gross RA, Kaplan DL (1998) In: Swift G (ed) ACS Symp Ser. American Chemical Society, Washington, DC, p 684
22. Kobayashi S, Uyama H (1998) In: Schlüter AD (ed) Material science and technology—synthesis of polymers. Wiley, Weinheim, Chap 16
23. Joo H, Yoo YH, Dordick JS (1998) Korean J Chem Eng 15:362
24. Adam W, Lazarus M, Saha-Möller CR, Weichhold O, Hoch U, Häring D, Schreier P (1999) In: Scherper T (ed) Advances in biochemical engineering/biotechnology. Springer, Berlin Heidelberg New York

25. Kazlauskas RJ, Weissfloch ANE (1995) J Org Chem 60:6959
26. Halling PJ (1994) Enzyme Microb Technol 16:178
27. Zaks A, Klibanov AM (1985) Proc Natl Acad Sci USA 82:3192
28. Blackwood AD, Curran LJ, Moore BD, Halling PJ (1994) Biochim Biophys Acta 1206:161
29. Xu K, Klibanov AM (1996) J Am Chem Soc 118:9815
30. Reetz MT, Zonta A, Schimossek K, Liebeton K, Jaeger KE (1997) Angew Chem 109:2961
31. Bornscheuer UT, Kazlauskas RJ (1999) Hydrolases in organic synthesis—regio- and stereoselective biotransformations. Wiley, Weinheim
32. Wescott CR, Klibanov AM (1991) J Am Chem Soc 113:3166
33. Terradas F, Teston-Hanry M, Fitzpatrick PA, Klibanov AM (1993) J Am Chem Soc 115:390
34. Faber K, Ottolina G, Riva S (1995) Trends Biotechnol 13:63
35. Kobayashi S, Uyama H, Ohmae M (2001) Chem Rev 101:3793
36. Nayak PL (1998) Des Monomers Polym 1:259
37. Dordick J, Marletta MA, Klibanov AM (1986) Proc Natl Acad Sci USA 1986:83
38. Uyama H, Kurioka H, Sugihara J, Kobayashi S (1996) Bull Chem Soc Jpn 69:189
39. Nicell JA (1994) J Chem Technol Biotechnol 60:203
40. Ghioureliotis M, Nicell JA (1999) Enzyme Microb Technol 25:185
41. Alva KS, Kumar J, Marx KA, Tripathi S (1997) Macromolecules 30:4024
42. Tripathy SK (1999) Chem Eng News 77:68
43. Premachandran R, Banerjee S, John VT, McPherson GL, Akkara JA, Kaplan DL (1997) Chem Mater 9:1342
44. Wang P, Martin BD, Parida S, Rethwisch DG, Dordick JS (1995) J Am Chem Soc 117:12885
45. Tonami H, Uyama H, Kobayashi S, Rettig K, Ritter H (1999) Macromol Chem Phys 200:1998
46. Reihmann M, Ritter H (2002) J Macromol Sci Pure Appl Chem 39:1369
47. Uyama H, Lohavisavapanich C, Ikeda R, Kobayashi S (1998) Macromolecules 31:554
48. Reihmann MH, Ritter H (2000) Macromol Chem Phys 201:798
49. Reihmann MH, Ritter H (2000) Macromol Chem Phys 201:1593
50. Viswanathan NK, Kim DY, Bian S, Williams J, Liu W, Li L, Samuelson L, Kumar J, Tripathy SK (1999) J Mater Chem 9:1941
51. Hewson WD, Dunford B (1976) J Biol Chem 251:6036
52. Hewson WD, Dunford B (1976) J Biol Chem 251:6043
53. Poulos TL (1993) Curr Opin Biotechnol 4:484
54. Deurzen MPJv, Rantwijk Fv, Sheldon RA (1997) Tetrahedron 53:13183
55. Dunford HB (1991) In: Everse J, Everse KE, Grisham MB (eds) Peroxidases in chemistry and biology, vol 1. CRC, Boca Raton
56. Schönbein CF (1856) Verh Nat Ges Basel 1:467
57. Willstätter R, Stoll A (1919) Ann Chem 21:416
58. Uyama H, Walda S, Kobayashi S (1999) Chem Lett :893
59. Baek HK, Waert HEv (1992) J Am Chem Soc 114:718
60. Lalot T, Brigodiot M, Maréchal E (1999) Polym Int 48:288
61. Hewson WD, Hager LP (1979) J Biol Chem 254:3182
62. Cornish-Bowden A, Wharton CW (1998) Enzyme kinetics. IRL, Oxford
63. Filizola M, Loew GH (2000) J Am Chem Soc 122:18
64. Dawson JH (1988) Science 240:433

65. Anni H, Yonetani T (1992) In: Sigel H, Sigel A (eds) Metal ions in biological systems: degradation of environmental pollutants by microorganisms and their metalloenzymes. Marcel Dekker, New York, p 219
66. Rodríguez-Lopez JN, Gilabert MA, Tudela J, Thorneley RNF, García-Cánovas F (2000) Biochemistry 39:13201
67. Ortiz de Montellano PR (1992) Annu Rev Pharmacol Toxicol 32:89
68. García-Moreno M, Moreno-Conesa M, Rodríguez-López JN, García-Cánovas F, Varón R (1999) Biol Chem 380:689
69. Baynton KJ, Bewtra JK, Biswas N, Taylor KE (1994) Biochim Biophys Acta 1206:272
70. Nakamoto S, Machida N (1992) Water Res 26:49
71. Matsumoto K, Takahashi H, Miyake Y, Fukuyama Y (1999) Tetrahedron Lett 40:3185
72. Job D, Dunford HB (1976) Eur J Biochem 66:607
73. Dec J, Bollag JM (1991) Arch Environ Contam Toxicol 29:1561
74. Reihmann MH, Ritter H (2001) Macromol Biosci 1:85
75. Liu W, Cholli AL, Kumar J, Tripathy S, Samuelson L (2001) Macromolecules 34:3522
76. Sahoo SK, Liu W, Samuelson LA, Kumar J, Cholli AL (2002) Macromolecules 35:9990
77. Mejias L, Reihmann MH, Sepulveda-Boza S, Ritter H (2002) Macromol Biosci 2:24
78. Akkara JA, Senecal KJ, Kaplan DL (1991) J Polym Sci A Polym Chem 29:1561
79. Ayyagari M, Akkara JA, Kaplan DL (1998) In: Gross RA, Kaplan DL, Swift G (eds) ACS Symp Ser: enzymes in polymer synthesis, vol 684. American Chemical Society, Washington, DC, Chap 6
80. Akita M, Tsutsumi D, Kobayashi M, Kise H (2001) Biosci Biotechnol Biochem 65:1581
81. Oguchi T, Tawaki S, Uyama H, Kobayashi S (2000) Bull Chem Soc Jpn 73:1389
82. Mita N, Oguchi T, Tawaki S, Uyama H, Kobayashi S (2000) Polymer Prepr 41:223
83. Liu W, Ma L, Wang ID, Jiang SM, Cheng YH, Li TJ (1995) J Polym Sci A Polym Chem 33:2339
84. Oguchi T, Wakisaka A, Tawaki S, Tonami H, Uyama H, Kobayashi S (2002) J Phys Chem B 106:1421
85. Rao AM, John VT, Gonzalez RD, Akkara JA, Kaplan DL (1993) Biotechnol Bioeng 41:531
86. Ayyagari MS, Marx KA, Tripathy SK, Akkara JA, Kaplan DL (1995) Macromolecules 28:5192
87. Rojo M, Gómez M, Estrada P (2001) J Chem Technol Biotechnol 76:69
88. Banerjee S, Ramanair P, Wu K, John VT, McPherson G, Akkara JA, Kaplan D (1998) In: Gross RA, Kaplan DL, Swift G (eds) ACS Symp Ser: enzymes in polymer synthesis, vol 684. American Chemical Society, Washington, DC, Chap 7
89. Uyama H, Kurioka H, Sugihara J, Komatsu I, Kobayashi S (1997) J Polym Sci A Polym Chem 35:1453
90. Tonami H, Uyama H, Kobayashi S, Reihmann M, Ritter H (2002) e-Polymers no 003
91. Mejias L, Schollmeyer D, Sepulveda-Boza S, Ritter H (2002) Macromol Biosci 3:395
92. Mita N, Tawaki S, Uyama H, Kobayashi S (2002) Macromol Biosci 2:127
93. Saenger W (1980) Angew Chem 92:343
94. Szejtli J (1997) J Mater Chem 7:575
95. Wenz G (1994) Angew Chem 106:851
96. Huang MJ, Watts JD, Bodor N (1997) Int J Quantum Chem 65:1135
97. Harper JB, Easton CJ, Lincoln SF (2000) Curr Org Chem 4:429
98. Akkara JA, Ayyagari M, Bruno FF, Samuelson L, John VT, Karayigitoglu C, Tripathy SK, Marx KA, Rao DVGLN, Kaplan DL (1994) Biomimetics 2:331
99. Bruno FF, Akkara JA, Kaplan DL, Tripathy SK (1995) Ind Eng Chem Res 34:4009

100. Ghan R, Shutava T, Patel A, John VT, Lvov Y (2004) Macromolecules 37:4519
101. Shutava T, Zheng Z, John VT, Lvov Y (2004) Biomacromolecules 5:914
102. Hay AS (1998) J Polym Sci A Polym Chem 36:505
103. Uyama H, Kurioka H, Kaneko I, Kobayashi S (1999) Macromol Rapid Commun 20:401
104. Uyama H, Kurioka H, Kobayashi S (1997) Polym J 27:190
105. Kim Y-J, Uyama H, Kobayashi S (2003) Macromolecules 36:5058
106. Cui A, Singh A, Kaplan DL (2002) Biomacromolecules 3:1353
107. Wu X, Kim J, Dordick JS (2000) Biotechnol Prog 16:513
108. Reihmann M (2002) PhD-thesis, University of Mainz
109. Mita N, Maruichi N, Tonami H, Nagahata R, Tawaki S, Uyama H, Kobayashi S (2003) Bull Chem Soc Jpn 76:375
110. Xu YP, Huang GL, Yu YT (1995) Biotechnol Bioeng 47:117
111. Ayyagari M, Akkara JA, Kaplan DL (1996) Acta Polym 47:193
112. Wu X, Liu W, Nagarajan R, Kumar J, Samuelson LA, Cholli AL (2004) Macromolecules 37:2322
113. Dordick J, Marletta MA, Klibanov AM (1987) Biotechnol Bioeng 30:31
114. Xu P, Kumar J, Samuelson L, Cholli AL (2002) Biomacromolecules 3:889
115. Ikeda R, Maruichi N, Tonami H, Tanaka H, Uyama H, Kobayashi S (2000) J Macromol Sci Pure Appl Chem 37:983
116. Pang Y, Ritter H, Tabatabai M (2003) Macromolecules 36:7090
117. Liu J, Yang F, Xian M, Qiu L, Ye L (2001) Macromol Chem Phys 202:840
118. Heinenberg M, Reihmann MH, Ritter H (2000) Des Monomers Polym 3:501
119. Heinenberg M, Ritter H (1999) Macromol Chem Phys 200:1792
120. Bian S, Liu W, Williams J, Samuelson L, Kumar J, Tripathy SK (2000) Chem Mater 12:1585
121. Liu W, Kumar J, Tripathi S, Senecal KJ, Samuelson L (1999) J Am Chem Soc 121:71
122. Etori H, Kanbara T, Yamamoto T (1994) Chem Lett 461
123. Kanbara T, Miyazaki Y, Yamamoto T (1995) J Polym Sci A Polym Chem 33:999
124. Dubey S, Singh D, Misra RA (1998) Enzyme Microb Technol 23:432
125. Mandal BK, Walsh CJ, Sooksimuang T, Behroozi SJ (2000) Chem Mater 12:6
126. Fukuoka T, Tachibana Y, Tonami H, Uyama H, Kobayashi S (2002) Biomacromolecules 3:768
127. Marx KA, Alva KS, Sarma R (2000) Mater Sci Eng C11:155
128. Sarma R, Alva KS, Marx KA, Tripathy SK, Akkara JA, Kaplan DL (1996) Mater Sci Eng C4:189
129. Kurioka H, Uyama H, Kobayashi S (1994) Macromol Rapid Commun 15:507
130. Uyama H, Kurioka H, Komatsu I, Sugihara J, Kobayashi S (1995) Bull Chem Soc Jpn 68:3209
131. Tonami H, Uyama H, Kobayashi S, Kubota M (1999) Macromol Chem Phys 200:2365
132. Tonami H, Uyama H, Kobayashi S (2000) Biomacromolecules 1:149
133. Antoniotti S, Santhanam L, Ahuja D, Hogg MG, Dordick JS (2004) Org Lett 6:1975
134. Hay AS, Blanchard HS, Endres GF, Eustance JW (1959) J Am Chem Soc 81:6335
135. Ikeda R, Sugihara J, Uyama H, Kobayashi S (1996) Macromolecules 29:8072
136. Ikeda R, Sugihara J, Uyama H, Kobayashi S (1998) Polym Int 47:295
137. Uyama H, Ikeda R, Yaguchi S, Kobayashi S (2000) ACS Symp Ser 764:113
138. Ikeda R, Uyama H, Kobayashi S (1997) Polym Bull 38:273
139. Fukuoka T, Tonami H, Maruichi N, Uyama H, Kobayashi S, Higashimura H (2000) Macromolecules 33:9152

140. Kobayashi S, Uyama H, Ushiwata T, Sugihara J, Kurioka H (1998) Macromol Chem Phys 199:777
141. Kobayashi S, Kurioka H, Uyama H (1996) Macromol Rapid Commun 17:503
142. Uyama H, Maruichi N, Tonami H, Kobayashi S (2002) Biomacromolecules 3:187
143. Premachandran RS, Banerjee S, Wu X-K, John VT, McPherson GL, Akkara J, Ayyagari M, Kaplan D (1996) Macromolecules 29:6452
144. Alva KS, Samuelson L, Kumar J, Tripathy S, Cholli AL (1998) J Appl Polym Sci 70:1257
145. Uyama H, Kurioka H, Komatsu I, Sugihara J, Kobayashi S (1995) Macromol Rep A32:649
146. Won K, Kim YH, An ES, Lee YS, Song BK (2004) Biomacromolecules 5:1
147. Ikeda R, Tanaka H, Uyama H, Kobayashi S (2000) Polym J 32:589
148. Wang P, Dordick JS (1998) Macromolecules 31:941
149. Mita N, Tawaki S-I, Uyama H, Kobayashi S (2003) Macromol Biosci 3:253
150. Aktas N, Kibarer G, Tanyolac A (2000) J Chem Technol Biotechnol 75:840
151. Guerra A, Ferraz A, Cotrim AR, da Silva FT (2000) Enzymol Microb Technol 26:315
152. Akkara JA, Wang J, Yang D-P, Gonsalves KE (2000) Macromolecules 33:2377
153. Tonami H, Uyama H, Kobayashi S, Higashimura H, Oguchi T (1999) J Macromol Sci Pure Appl Chem 39:719
154. Tonami H, Uyama H, Oguchi T, Higashimura H, Kobayashi S (1999) Polym Bull 42:125
155. Ikeda R, Tanaka H, Uyama H, Kobayashi S (2000) Macromol Rapid Commun 21:496
156. Mekmouche Y, Ménage S, Toia-Duboc C, Fontecave M, Galey J-B, Lebrun C, Pécaut J (2001) Angew Chem 113:975
157. Tsujimoto T, Uyama H, Kobayashi S (2004) Macromolecules 37:1777
158. Fukuoka T, Uyama H, Kobayashi S (2003) Macromolecules 36:8213
159. Fukuoka T, Uyama H, Kobayashi S (2004) Macromolecules 37:5911
160. Fukuoka T, Uyama H, Kobayashi S (2004) Biomacromolecules 5:977
161. Higashimura H, Fujisawa K, Moro-oka Y, Kubota M, Shiga A, Terahara A, Uyama H, Kobayashi S (1998) J Am Chem Soc 120:8529
162. Higashimura H, Kubota M, Shiga A, Fujisawa K, Moro-oka Y, Uyama H, Kobayashi S (2000) Macromolecules 33:1986

Enzymatic Synthesis and Properties of Polymers from Polyphenols

Hiroshi Uyama[1] · Shiro Kobayashi[2,3] (✉)

[1]Department of Materials Chemistry, Graduate School of Engineering, Osaka University, 565-0871 Suita, Japan

[2]Department of Materials Chemistry, Graduate School of Engineering, Kyoto University, 615-8510 Kyoto, Japan

[3]*Present address:*
R & D Center for Bio-based Materials, Kyoto Institute of Technology, Matsugasaki, Sakyo-ku, 606-8585 Kyoto, Japan
kobayash@kit.ac.jp

1	Introduction	52
2	**Enzymatic Oxidative Coupling of Catechol Derivatives**	52
2.1	Polymerization of Catechols	52
2.2	Curing of Urushiol and Its Analogues	53
3	**Enzymatic Oxidative Polymerization of Flavonoids**	55
3.1	Catechins	56
3.2	Other Flavonoids	58
3.3	Properties of Polymeric Flavonoids	59
4	**Enzymatic Synthesis of Biopolymer-Polyphenol Conjugates**	62
4.1	Chitosan Conjugates	62
4.2	Poly(amino acid) Conjugates	62
4.3	Conjugates of Synthetic Polymers	64
5	**Concluding Remarks**	65
	References	65

Abstract Enzymatic synthesis and properties of polymers from polyphenols are reviewed. Catechol derivatives were subjected to oxidative polymerization and curing. High-performance coatings were prepared from urushiol and its analogues using laccase as a catalyst. Flavonoids were oxidatively polymerized by polyphenol oxidase, laccase, and peroxidase to produce the flavonoid polymers. The conjugation of polyphenols on amine-containing polymers took place via the enzyme catalysis. The flavonoid-containing polymers showed good antioxidant properties and enzyme inhibitory effects.

Keywords Enzymatic oxidative polymerization · Flavonoid · Urushi · Curing · Conjugate

Abbreviations
AAPH 2,2′-azobis(2-amidinopropane)dihydrochloride
AP acetone powder
ChC *Clostridium histolyticum* collagenase

CNSL	cashew nut shell liquid
DP	degree of polymerization
EGCG	epigallocatechin gallate
HRP	horseradish peroxidase
LDL	low-density lipoprotein
ML	*Myceliophthora* laccase
MMP	matrix metalloproteinases
PCL	*Pycnoporus coccineus* laccase
PL	poly(ε-lysine)
POSS	polyhedral oligomeric silsesquioxane
PPO	polyphenol oxidases
ROS	reactive oxygen species
SP	starch-urea phosphate
XO	xanthine oxidase

1
Introduction

For the last decades, enzymatic synthesis of phenolic polymers has been extensively investigated [1–10]. In living cells, various oxidoreductases play an important role in maintaining the metabolism of living systems. So far, several oxidoreductases—peroxidase, laccase, bilirubin oxidase etc.—have been reported to catalyze an oxidative polymerization of phenol derivatives, and among them, peroxidase is most often used. The enzymatically synthesized phenolic polymers are expected to become an alternative to conventional phenolic resins, which have limitations of their preparation and use due to concerns over the toxicity of formaldehyde.

In the previous chapter "Synthesis of Phenol Polymers Using Peroxidases", the enzymatic oxidative polymerization of monophenolic derivatives is described. This chapter deals with the enzymatic synthesis and properties of polymers from polyphenols, compounds having more than two hydroxyl groups on the aromatic ring(s). In particular, cured phenolic polymers (artificial urushi) and flavonoid polymers are examined from the standpoint of the enzymatic synthesis of functional materials.

2
Enzymatic Oxidative Coupling of Catechol Derivatives

2.1
Polymerization of Catechols

The oxidative coupling of catechol proceeded in the presence of a peroxidase catalyst [11], during which an unstable *o*-quinone intermediate may be formed, followed by complicated reaction pathways to give poly(catechol).

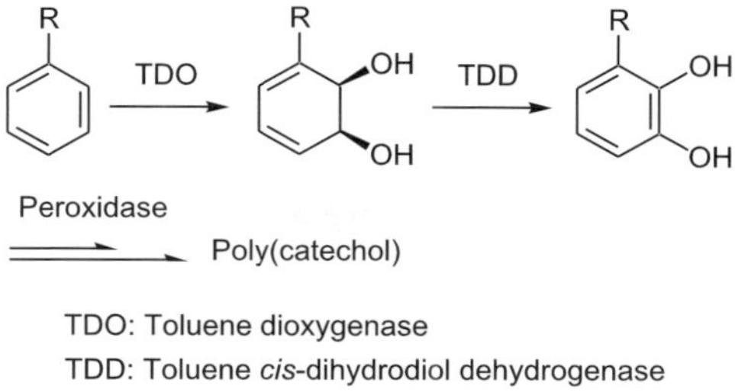

Fig. 1 Synthesis of poly(catechol) by multienzymatic processes

The iodine-labeled polymer showed low electrical conductivity in the range of 10^{-6} to 10^{-9} S cm^{-1}.

Laccase was also a good catalyst for the oxidative polymerization of catechol [12]. The polymerization using the laccase enzyme produced by a culture of *Trametes versicolor* was conducted batch-wise in an aqueous acetone to give a polymer with molecular weight of less than 1000.

Synthesis of poly(catechol) was demonstrated by multienzymatic processes (Fig. 1) [13]. Aromatic compounds were converted to catechol derivatives by the catalytic action of toluene dioxygenase and toluene cis-dihydrodiol dehydrogenase, followed by the peroxidase-catalyzed polymerization to give the polymer with molecular weight of several thousands.

2.2
Curing of Urushiol and Its Analogues

Urushi is a typical Japanese traditional coating that displays excellent toughness and brilliance over a long period. In the early days of the 20th century, pioneering works by Majima revealed that the important components of urushi are "urushiols," whose structure is a catechol derivative directly linked to unsaturated hydrocarbon chains consisting of a mixture of monoenes, dienes, and trienes at the 3- or 4-position of catechol [14, 15]. Film-forming of urushiols proceeds in air at room temperature without organic solvents; hence, urushi seems very desirable as a coating material from an environmental standpoint. In vitro enzymatic hardening reaction of catechol derivatives bearing an unsaturated alkenyl group at the 4-position of the catechol ring was carried out using *Pycnoporus coccineus* laccase (PCL) as a catalyst to yield a crosslinked film with excellent dynamic viscoelasticity [16].

Fast-drying hybrid urushi has been developed [17]. Kurome urushi is reacted with silane-coupling agents possessing an amino, epoxy, or isocynate group, resulting in a shorter curing time for urushi.

So far, few modeling studies of urushi have been attempted, mainly due to difficulties in the chemical synthesis of the urushiol. New urushiol analogues have been developed by a convenient synthetic process for the preparation of

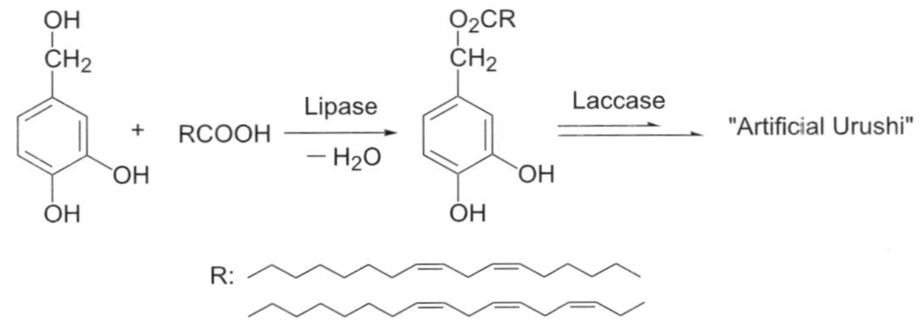

Fig. 2 Enzymatic synthesis of artificial urushi

"artificial urushi" (Fig. 2) [18–21]. The urushiol analogues were synthesized using lipase as a catalyst in a single step. These compounds were cured using PCL as a catalyst in the presence of acetone powder (AP, an acetone-insoluble part of the urushi sap containing mainly polysaccharides and glycoproteins) under mild reaction conditions without the use of organic solvents, yielding a brilliant film ("artificial urushi") with a high-gloss surface. Starch-urea phosphate (SP), a synthetic material, was also available as a substitute for AP for in vitro enzymatic curing of the urushiol analogues, although the film hardness was smaller than that obtained in the presence of AP. The use of SP as the third component provided the artificial urushi from exclusively synthetic compounds.

The curing monitoring by FT-IR showed that the curing of the urushiol analogue proceeds via oxidative coupling of the phenol moiety and the subsequent autoxidation of the unsaturated group in the side chain. The storage modulus (E') and dissipation factor ($\tan \delta$) of the cured film showed that the film had homogeneous structure and good viscoelasticity, and that its characteristics were similar to those of natural urushi (Fig. 3).

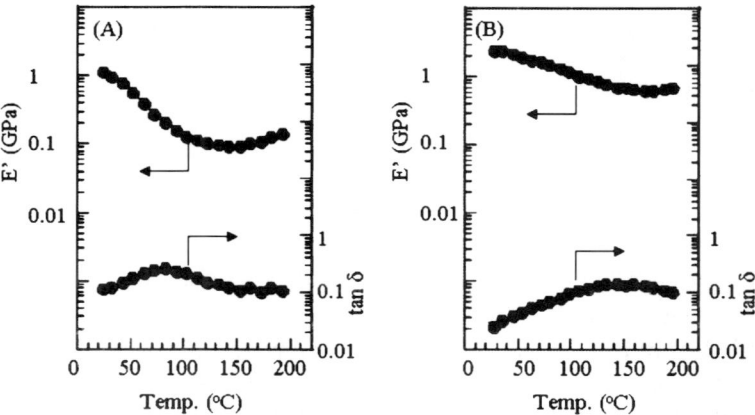

Fig. 3 Dynamic viscoelasticity of (A) artificial urushi from an urushiol analogue having linolenic acid moiety in the presence of AP and (B) natural urushi

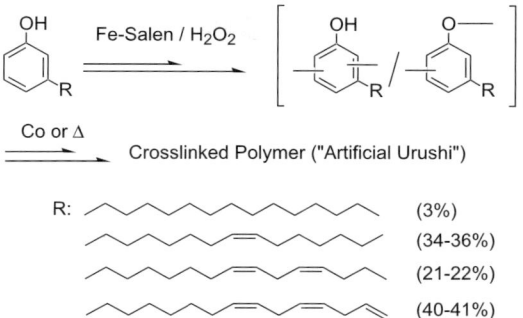

Fig. 4 Synthesis and curing of polyCNSL

Furthermore, new crosslinkable polyphenols based on model "urushi" were designed and synthesized by the oxidative polymerization of other urushiol analogues using Fe-salen, a model catalyst of peroxidase [22–24]. Such polyphenols were readily cured to give crosslinked polymeric films with high-gloss surface.

Cardanol, a main component obtained by the thermal treatment of cashew nut shell liquid (CNSL), is a phenol derivative having a meta substituent of a C15 unsaturated hydrocarbon chain with one to three double bonds as the major. Another new crosslinkable polyphenol was synthesized by the Fe-salen-catalyzed oxidative polymerization of CNSL (Fig. 4) [25, 26]. The polymerization of CNSL in 1,4-dioxane or without solvent resulted in a soluble polymer with a molecular weight of several thousand in good yields. The curing of polyCNSL proceeded by oxidation catalyzed by cobalt naphthenate in air or in a thermal treatment (150 °C for 2 h) to yield a yellowish transparent film, which is also an "artificial urushi" in a broad sense. The resulting crosslinked film exhibits good elastic properties comparable with cardanol-formaldehyde coating materials. FT-IR monitoring of the curing of polyCNSL showed that the crosslinking mechanism is similar to that of the oil autoxidation.

3
Enzymatic Oxidative Polymerization of Flavonoids

Bioactive polyphenols are present in a variety of plants and are used as important components of human and animal diets [27]. Flavonoids are a broad class of low-molecular-weight secondary plant polyphenolics, which are benzo-γ-pyrone derivatives consisting of phenolic and pyrane rings. Flavonoids are usually subdivided according to their substituents into flavanols, flavones, flavanones, chalcones, and anthocyanidines.

Recently, there has been increasing interest in flavonoids due to their biological and pharmacological activity including antioxidant, anti-carcinogenic, probiotic, anti-microbial, and anti-inflammatory properties [28–31]. These properties are potentially beneficial in preventing diseases and protecting the stability of the genome. Many of these activities have been related to their antioxidant actions.

3.1
Catechins

Green tea is derived from *Camellia sinensis*, an evergreen shrub of the *Theaceae* family. Most of the polyphenols in green tea are flavanols, commonly known as catechins; the major catechins in green tea are (+)-catechin, (–)-epicatechin, (–)-epigallocatechin, (–)-epicatechin gallate, and (–)-epigallocatechin gallate (EGCG) (Fig. 5). Numerous biological activities have been reported for green tea and its contents—among them, the preventive effects against cancer are most notable [32–34].

Catechin was aerobically subjected to an oxidative coupling by plant polyphenol oxidase (PPO) to give polymers [35], whose analytical properties were similar to those obtained by autoxidation with hydrogen peroxide [36] and to tannins extracted from *Acacia catechu*. The main reaction sequence involves formation of o-quinone, followed by the coupling of this unstable intermediate.

The oxidative coupling of catechin by PPO extracted from grapes was reported [37]. In case of coupling at a pH below 4, the result contained mostly colorless products, whereas yellow compounds were formed at the higher pH. From the product mixtures, eight dimer fractions corresponding to the major products formed at pH 3 and pH 6 were isolated and analyzed by

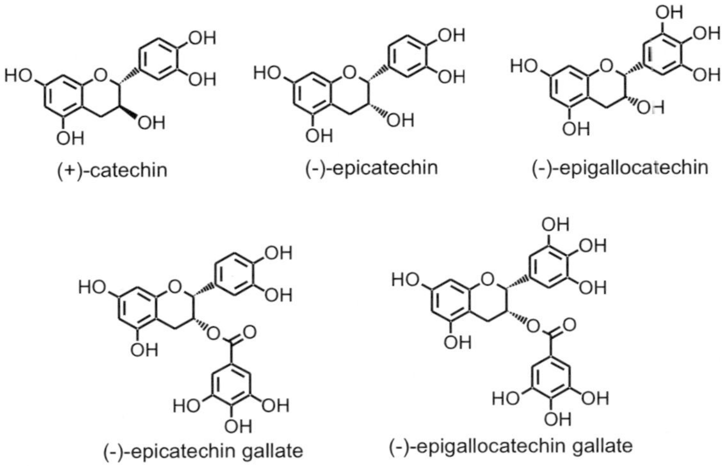

Fig. 5 Structure of the main components of green tea catechins

NMR, UV-vis, and mass spectroscopies [38]. A detailed NMR analysis provided a structural hypothesis for five resulting products. The colorless dimers are dehydrodicatechins of type B with C – O and C – C interflavin linkages. Two yellow dimers correspond to dehydrodicatechin A and to a structure of quinone-methide-type.

Mushroom PPO was used for the coupling of catechin with an excess of phloroglucinol, in which the *o*-quinone intermediate of catechin was trapped by phloroglucinol to form the biphenyl group at the 6' position of catechin [39]. The PPO-catalyzed reaction of catechin and mesquitol resulted in the construction of the atropisomers (*R*)- and (*S*)-mesquitol-[5 → 8]-catechins as the only identified products.

The oxidative coupling of catechin can be carried out in a peroxidase/hydrogen peroxide system. The coupling by horseradish peroxidase (HRP) produced oligomers with a degree of polymerization (DP) of less than 5 [40]. Dehydrodicatechin A was isolated and identified from the reaction mixture. A kinetic study of the oxidation of catechin by peroxidase from strawberries (*Fragaria* × *ananassa*) showed highly efficient catalytic activity of the enzyme at a low concentration of hydrogen peroxide [41].

The reaction products obtained by the HRP-catalyzed oxidative coupling of catechin were analyzed by reversed-phase and size-exclusion chromatographies [42]. The products were dehydrodicatechin A and oligomers with DP equal or greater than 5. The joint use of both chromatographies permits the qualitative and quantitative identification of the product mixtures.

Comparative study on the products obtained by the peroxidase-catalyzed and PPO-catalyzed oxidative coupling of catechin was reported [43]. Both enzymes produced the similar products. In the PPO-catalyzed oxidation, catechin was almost consumed, and the oligomers were mainly formed. On the other hand, a significant amount of catechin remained, and the amount of oligomer fractions was small when peroxidase was used as a catalyst. The low catalytic activity of peroxidase may be due to the inhibition of the enzyme by the resulting products.

Two unusual dimers formed by the oxidation of catechin using an HRP catalyst were isolated and identified. One is the dicarboxylic acid compound with a C – C linkage between the C-6' of the B-ring and the C-8'' of the D-ring. The diacid was formed by ortho cleavage of the E-ring. Another is dimer of C – C linkage between the C-2 of the C-ring and the C-8'' of the D-ring [44]. In the laccase-catalyzed oxidative coupling, new catechin-hydroquinone adducts were formed.

In the enzymatic polymerization of phenol derivatives, a mixture of hydrophilic organic solvent and buffer is often used as a medium for the efficient production of polymers [45–47]. The HRP-catalyzed polymerization of catechin was carried out in an equivolume mixture of 1,4-dioxane and buffer (pH 7) to give a polymer with a molecular weight of 3.0×10^3 in 30% yield [48]. Using methanol as cosolvent improved the polymer yield and molecular weight [49].

In the polymerization of catechin using laccase derived from *Myceliophthora* (ML) as catalyst, the reaction conditions were examined in detail [50]. A mixture of acetone and acetate buffer (pH 5) was suitable for the efficient synthesis of soluble poly(catechin) with higher molecular weight. The mixed ratio of acetone greatly affected the yield, molecular weight, and solubility of the polymer. The polymer synthesized in 20% acetone showed low solubility in *N,N*-dimethylformamide (DMF), whereas the polymer obtained in acetone content of less than 5% was completely soluble in DMF. In the UV-vis spectrum of poly(catechin) in methanol, a broad peak centered at 370 nm was observed. In alkaline solution, this peak was red-shifted, and the peak intensity became larger than that in methanol. In the ESR spectrum of the enzymatically synthesized poly(catechin), a singlet peak at $g = 1.982$ was detected, whereas the catechin monomer possessed no ESR peak.

3.2
Other Flavonoids

Many flavonoids are subjected to enzymatic oxidation. Quercetin, one of the most famous flavonoids, has received much attention due to its interesting biological effects. The oxidation of quercetin by PPO was examined [51]. In reaction monitoring by UV-vis spectroscopy, the peak at 291 nm increased, and the peak at 372 nm decreased. The oxidation was inhibited by the specific inhibitor for PPO (4-hexylresorcinol).

Rutin (Fig. 6) is one of the most commonly found flavonol glycosides identified as vitamin P with quercetin and hesperidin. The oxidation of rutin by an HRP/hydrogen peroxide system was reported [52]. The oxidation at

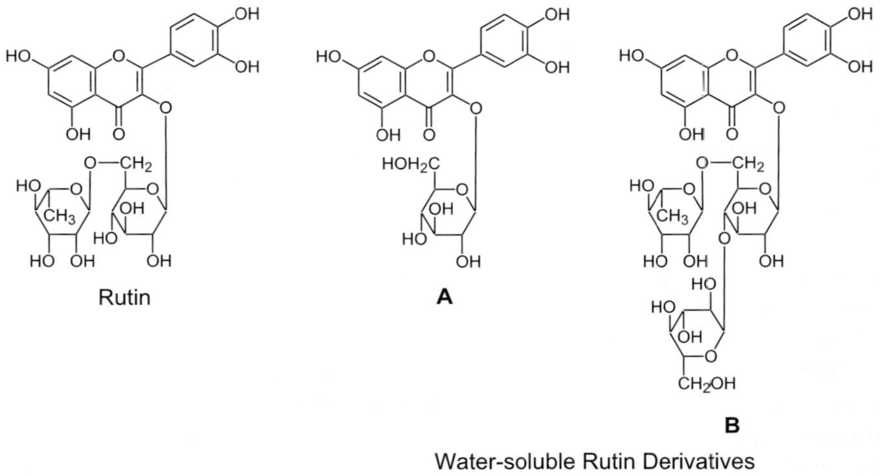

Fig. 6 Structure of rutin and its water-soluble derivatives

pH 7.8 produced an ascorbate-reducible compound, which is supposed to be an o-quinone derivative.

An oxidative polymerization of rutin using ML as a catalyst was examined in a mixture of methanol and buffer to produce a flavonoid polymer [53]. Under selected conditions, the polymer with molecular weight of several thousand was obtained in good yields. The resulting polymer was readily soluble in water, DMF, and dimethyl sulfoxide (DMSO), although rutin monomer showed very low water solubility. UV measurement showed that the polymer had broad transition peaks around 255 and 350 nm in water, which were red-shifted in an alkaline solution. ESR measurement showed the presence of a radical in the polymer.

A water-soluble rutin derivative consisting of **A** and **B** (20 : 80 mol %, Fig. 6) is a commercially available product. This derivative was also polymerized by the laccase catalyst; the laccase-catalyzed polymerization in an equivolume mixture of methanol and acetate buffer produced a polymer with molecular weight of 1×10^4 in a high yield.

Peroxidase-catalyzed polymerization of various flavonoids was investigated in 1,4-dioxane/pH 8 buffer (50 : 50 vol %) [48]. Flavonols (quercetin and rutin) as well as isoflavones (diadzein and 5,6,4'-trihydroxyisoflavone) were subjected to enzymatic oxidative coupling to produce flavonoid polymers. The polymerization of diadzein resulted in a polymer soluble in DMF and DMSO in a high yield. Soybean peroxidase also catalyzed the polymerization of diadzein.

3.3
Properties of Polymeric Flavonoids

Oxidative stress triggered by reactive oxygen species (ROS) in the human body is a contributing factor to the pathogenesis of neurodegenerative disorders such as cerebral ischemia/reperfusion injury and trauma, as well as chronic conditions such as Parkinson's disease and Alzheimer's disease. ROS attacks biological molecules such as lipids, proteins, enzymes, DNA, and RNA, leading to cell or tissue injury associated with degenerative diseases. Superoxide anions are one of the most typical ROS formed during normal aerobic metabolism and by activated phagocytes. The reduction of molecular oxygen to superoxide anion by xanthine oxidase (XO), generating hydroxyl radicals and uric acid, is an important physiological pathway. However, an excess of superoxide anions are capable of damaging biomacromolecules such as those mentioned above, which are attacked by ROS.

Superoxide anion scavenging activity of enzymatically synthesized poly(catechin) was evaluated. Poly(catechin), synthesized using an HRP catalyst, greatly scavenged superoxide anion in a concentration-dependent manner, and at 200 µM almost completely scavenged a catechin unit concentration (Fig. 7) [49]. The laccase-catalyzed synthesized poly(catechin) also

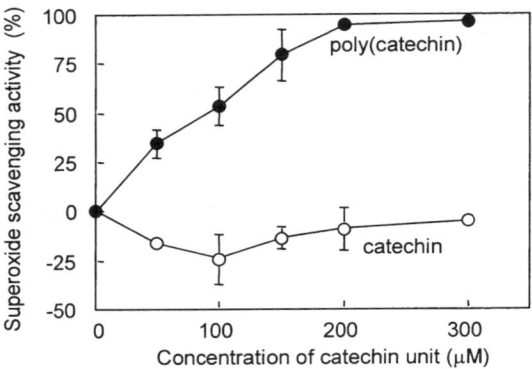

Fig. 7 Superoxide anion scavenging activity of catechin and poly(catechin)

showed excellent antioxidant properties [50]. Conversely, catechin showed pro-oxidant property in concentrations lower than 300 µM. These results demonstrate that the poly(catechin) possesses much higher potential for superoxide anion scavenging compared with intact catechin.

EGCG is a major ingredient of green tea possessing powerful antioxidant properties and cancer-chemopreventive properties due to the actions of radical scavenging, enzyme inhibition, and metal chelation. The polymer obtained by laccase-catalyzed oxidative coupling of EGCG showed much higher superoxide anion scavenging activity than the EGCG monomer and enzymatically synthesized poly(catechin) [54]. The antioxidant activity of rutin was greatly amplified by the laccase-catalyzed oxidative coupling; poly(rutin) also showed a much higher scavenging capacity toward superoxide anion than rutin [53].

Oxidation of low-density lipoprotein (LDL) leads to its enhanced uptake by macrophages, which is believed subsequently to result in foam cell formation, one of the first stages of atherogenesis. Therefore, antioxidants that protect LDL against oxidation are potentially anti-atherogenic compounds. Although the mechanism for in vivo oxidation of LDL has not been established, free radical autoxidation may be a consideration. The LDL protection was evaluated by the addition of 2,2′-azobis(2-amidinopropane)dihydrochloride (AAPH) in the presence of LDL and polymerized flavonoids. Poly(catechin), obtained using an HRP catalyst, showed much greater inhibition activity against LDL oxidation in a concentration-dependent manner, compared to the catechin monomer [49]. These data imply that a structure of the enzymatically synthesized poly(catechin) is much more capable of inhibiting oxidation of LDL than that of the monomer. In the case of poly(rutin) synthesized using a laccase catalyst, the inhibition effect against the AAPH-induced oxidation lasted longer compared to the inhibition effect of the monomer [53].

Protection effects of rutin and poly(rutin) against endothelial cell damage caused by AAPH were reported [53]. AAPH induces peroxidation only in lipid membrane of cells during the first step. The addition of AAPH caused significant cell death due to oxidative injury. However, poly(rutin) enhanced cell viability with higher protection effects against the oxidative damage than was offered by the rutin monomer at low concentration. In particular, in high concentration, the polymer exhibited a further increase in protection relating to a concentration increase. In contrast, the monomer induced fatal cytotoxicity by itself at the same concentration. These results imply that poly(rutin) is a more potent chain-breaking antioxidant when scavenging free radicals in an aqueous system than the monomer.

XO is not only an important biological source of reactive oxygen species but also the enzyme responsible for the formation of uric acid associated with gout leading to painful inflammation in the joints. The XO-inhibition effect due to enzymatically synthesized poly(catechin) increased with an increasing concentration of catechin units, while the monomeric catechin showed an almost negligible inhibition effect [49, 50]. This markedly amplified XO-inhibition activity of poly(catechin) was considered to be due to effective multi-valent interaction between XO and the condensed catechin units in the poly(catechin).

PolyEGCG also showed excellent inhibition of XO. The XO-inhibition effect of EGCG monomer was quite low, with an inhibition of less than about 5% over a range of tested concentrations (Fig. 8) [54]. In contrast, polyEGCG showed greatly amplified XO inhibition effects in a concentration-dependent manner. Moreover, the inhibition of polyEGCG was higher than that of allopurinol, a frequently used commercial inhibitor for gout treatment. Thus, polyEGCG is expected as one of leading candidates of therapeutic molecules against various diseases induced by free radicals and/or enzymes including gout. Kinetic analysis showed that polyEGCG is an uncompetitive inhibitor of XO.

Mutans streptococci are the major pathogenic organisms of dental caries in humans. The pathogenicity is closely related to production of extracellular water-insoluble glucans from sucrose by glucosyltransferase and acid release from various fermentable sugars. Poly(catechin) obtained by HRP catalyst

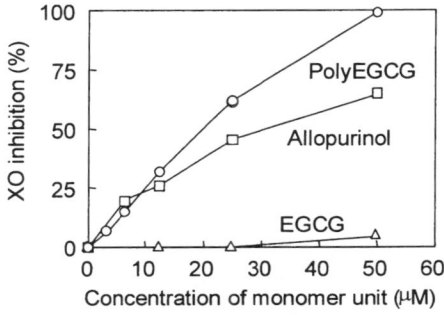

Fig. 8 XO inhibition of EGCG, polyEGCG, and allopurinol

in a phosphate buffer (pH 6) markedly inhibited glucosyltransferase from *Streptococcus sorbrinus* 6715, whereas the inhibitory effect of catechin for this enzyme was very low [40].

4
Enzymatic Synthesis of Biopolymer-Polyphenol Conjugates

4.1
Chitosan Conjugates

PPO was used as catalyst for modification of natural polymers. The enzymatic treatment of a chitosan film in the presence of PPO and phenol derivatives produced a new material of chitosan derivatives [55]. During the reaction, unstable *o*-quinones were formed, which then reacted with chitosan to yield modified chitosans. In the enzymatic treatment of *p*-cresol with a low concentration of chitosan (< 1%), the reaction solution was converted into a gel [56].

A similar treatment in the presence of chlorogenic acid produced modified chitosan soluble under both acidic and basic conditions [57]. The PPO-catalyzed reaction of 3,4-dihydroxyphenylethylamine (dopamine) provided water-resistant adhesive properties to chitosan [58]. A chitosan derivative modified with hydroxy or dihydroxybenzaldehyde was crosslinked by PPO to give stable and self-sustaining gels [59].

This enzymatic modification of chitosan was applied for the synthesis of a chitosan–catechin conjugate. The formation of a Michael-type adduct and/or Schiff base was proposed during the PPO-catalyzed conjugation of catechin with chitosan [60]. Rheological measurement demonstrates that the resulting conjugate behaves as an associative thickener.

PPO catalyzed the conjugation of chitosan and proteins. A peptide-modified chitosan derivative was synthesized by the PPO-catalyzed reaction of chitosan with acid-hydrolyzed casein [61]. Even in the low concentration of this peptide-modified chitosan, high viscosities and shear-thinning properties were observed. The 1% solution of the peptide-modified chitosan behaved as weak gel. PPO was used to conjugate chitosan with green fluorescent protein (GFP) [62]. The resulting GFP–chitosan conjugate showed pH-responsive properties. Furthermore, this conjugate was selectively deposited onto a micropatterned surface in response to an applied voltage.

4.2
Poly(amino acid) Conjugates

Poly(ε-lysine) (PL) is a biopolymer produced from culture filtrates of *Streptomyces albulus* and shows good antimicrobial activity against Gram-positive and -negative bacteria; thus, it is widely used as an additive in food industry.

A new inhibitor against disease-related enzymes, collagenase, hyaluronidase, and xanthine oxidase was developed by the conjugation of catechin on poly(ε-lysine) using ML as a catalyst (Fig. 9) [63].

Matrix metalloproteinases (MMPs)—typically collagenase and gelatinase—degrade and remodel structural proteins in the extracellular matrix. Since MMPs play an essential role in the homeostasis of the extracellular matrix, an imbalance in their expression or activity may have important consequences in various pathologies. Thus, MMPs have recently become interesting targets for drug design in the search of novel anticancer, antiarthritis, and other pharmacological agents useful in the management of inflammatory processes. The PL-catechin conjugate showed greatly amplified concentration-dependent inhibition activity against bacterial collagenase from *Clostridium histolyticum* (ChC) on the basis of the catechin unit (Fig. 10), which is considered to be due to effective multi-valent interaction between ChC and the catechin unit in the conjugate. The kinetic study suggests that this conjugate is a mixed-type inhibitor for ChC.

Hyaluronidase is an enzyme that catalyzes hydrolysis of hyaluronic acid and is often involved in a number of physiological and pathological processes. Potent hyaluronidase inhibitors have antiallergic effects, which may lead to the development of new antiallergic agents. Efficient inhibition activity of the PL-catechin conjugate was found, while the monomeric catechin showed an almost negligible inhibition effect. The PL–catechin conjugate also showed good inhibition effects for XO.

Gelatin is the most widespread water-soluble protein in the body, resulting from partial degradation of water-insoluble collagen. Gelatin has widely

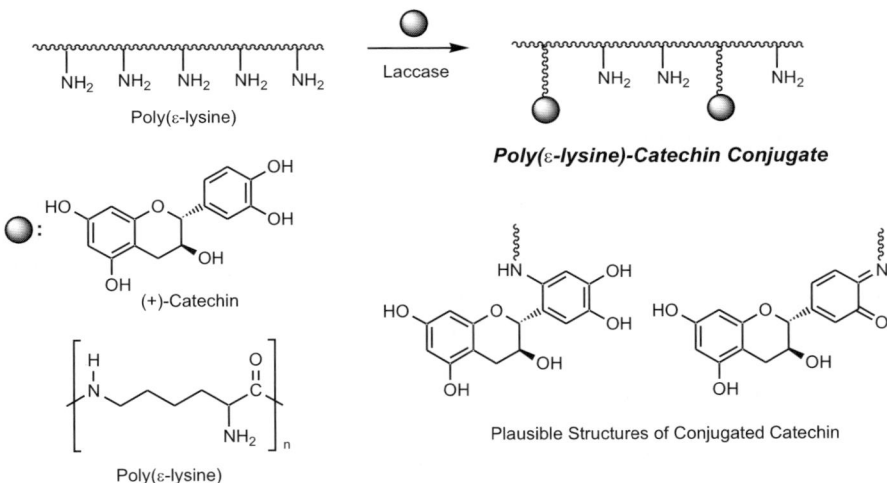

Fig. 9 Enzymatic synthesis of poly(ε-lysine)–catechin conjugate

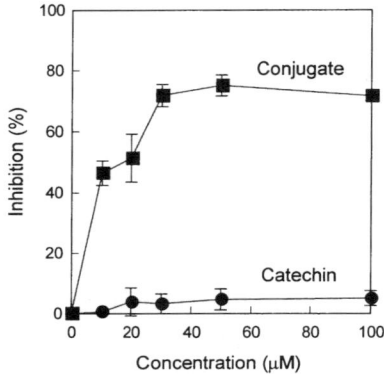

Fig. 10 Inhibition activity of catechin and poly(ε-lysine)–catechin conjugate (catechin content = 3.4%) against collagenase

been used in the food, pharmaceutical, and photographic industries. Gelatin–catechin conjugate was synthesized by the laccase-catalyzed oxidation of catechin in the presence of gelatin, in which the lysine residue of gelatin was used for the grafting of catechin [64].

Antioxidant properties of the gelatin–catechin conjugate were evaluated. The conjugate possessed scavenging properties against superoxide anion, whereas gelatin was not active for the scavenging. Furthermore, the conjugate showed greater inhibitory activity against AAPH-induced LDL oxidation in a catechin moiety-concentration-dependent manner compared to unconjugated catechin. The inhibition effect of the conjugate lasted longer than the inhibition effect of unconjugated catechin. Gelatin itself exhibited no inhibition effect on the LDL oxidation in this system. These data are indicative of the good antioxidant properties of the gelatin–catechin conjugate.

4.3
Conjugates of Synthetic Polymers

The conjugation of catechin on poly(allylamine) using ML as a catalyst was examined in air [65]. During the conjugation, the reaction mixture turned brown, and a new peak at 430 nm was observed in the UV-vis spectrum. At pH 7, the reaction rate was the highest. The conjugation hardly occurred in the absence of laccase, indicating that the reaction proceeded via enzyme catalysis. The resulting poly(allylamine)–catechin conjugate showed more lasting antioxidant activity against LDL peroxidation induced by AAPH compared with unconjugated catechin.

Polyhedral oligomeric silsesquioxane (POSS) has been extensively studied as starting substrate to construct nanocomposites with precise control of nanoarchitecture and properties. Octahedral derivatives are the most representative ones of this family. It was reported that the HRP-catalyzed conjugation of catechin on amine-substituted octahedral silsesquioxane amplified the beneficial physiological property of flavonoids [66]. The POSS-catechin conjugate exhibited great improvement in scavenging activity against superoxide anions compared with intact catechin. In addition, the conjugate showed high inhibitory effect on XO activity, while the inhibition effect of catechin was very low. These unique properties of the conjugate may be derived from the POSS structure.

Catechin-immobilizing polymer particles were prepared by laccase-catalyzed oxidation of catechin in the presence of amine-containing porous polymer particles [67]. The resulting particles showed good scavenging activity toward stable free 1,1-diphenyl-2-picryl-hydrazyl radical and 2,2′-azinobis(3-ethylbenzothiazoline-6-sulfonate) radical cations. These particles may be applied to packed column systems to remove radical species such as reactive oxygen closely related to various diseases. Poly(4-hydroxystyrene) was modified with aniline by using PPO catalyst [68]. The incorporated ratio of aniline into the polymer was very low (1.3%).

5
Concluding Remarks

This review gives an overview of enzymatic synthesis and the properties of polymers derived from polyphenols. Catechol derivatives were enzymatically oxidized to form polymers. Urushiol analogues were designed and cured by laccase catalyst to produce artificial urushi of good elasticity.

New flavonoid polymers were synthesized through enzyme catalysis. PPO, laccase, and peroxidase efficiently induced the oxidative coupling of flavonoids. Furthermore, the catechin-grafting polymers were successfully synthesized using an enzyme catalyst. In some cases, the biological and pharmacological activity of flavonoids was greatly amplified by the enzymatic treatments. These molecular designs will expand applications of natural flavonoids in the food, cosmetic, and medical fields.

References

1. Kobayashi S, Shoda S, Uyama H (1995) Adv Polym Sci 121:1
2. Kobayashi S, Shoda S, Uyama H (1996) In: Salamone JC (ed) The polymeric materials encyclopedia. CRC Press, Boca Raton, p 2102
3. Kobayashi S, Shoda S, Uyama H (1997) Enzymatic catalysis. In: Kobayashi S (ed) Catalysis in precision polymerization. Wiley, New York, p 417
4. Kobayashi S (1999) J Polym Sci: Part A: Polym Chem 37:3041
5. Kobayashi S, Uyama H, Kimura S (2001) Chem Rev 101:3793
6. Kobayashi S, Uyama H, Ohmae M (2001) Bull Chem Soc Jpn 74:613

7. Gross RA, Kumar A, Kalra B (2001) Chem Rev 101:2097
8. Kobayashi S, Uyama H (2001) Enzymatic polymerization. In: Kroschwitz JI (ed) Encyclopedia of polymer science and technology, 3rd edn. Wiley, New York
9. Uyama H, Kobayashi S (2002) J Mol Catal B Enz 19–20:117
10. Uyama H, Kobayashi S (2003) Curr Org Chem 7:1397
11. Dubey S, Singh D, Misra RA (1998) Enz Microb Technol 23:432
12. Aktaş N, Sahiner N, Kantoğlu Ö, Salih B, Tanyolaç A (2003) J Polym Environ 11:123
13. Ward G, Parales RE, Dosoretz CG (2004) Environ Sci Technol 38:4753
14. Majima R (1909) Ber Dtsch Chem Ges 42B:1418
15. Majima R (1922) Ber Dtsch Chem Ges 55B:191
16. Terada M, Oyabu H, Aso Y (1994) J Jpn Soc Colour Mater 66:681
17. Nagase K, Lu R, Miyakoshi T (2004) Chem Lett 33:90
18. Kobayashi S, Ikeda R, Oyabu H, Tanaka H, Uyama H (2000) Chem Lett 1214
19. Ikeda R, Tsujimoto T, Tanaka H, Oyabu H, Uyama H, Kobayashi S (2000) Proc Acad Jpn 76B:155
20. Ikeda R, Tanaka H, Oyabu H, Uyama H, Kobayashi S (2001) Bull Chem Soc Jpn 74:1067
21. Kobayashi S, Uyama H, Ikeda R (2001) Chem Eur J 7:4754
22. Tsujimoto T, Ikeda R, Uyama H, Kobayashi S (2000) Chem Lett 1122
23. Tsujimoto T, Ikeda R, Uyama H, Kobayashi S (2001) Macromol Chem Phys 202:3420
24. Tsujimoto T, Uyama H, Kobayashi S (2004) Macromolecules 37:1777
25. Ikeda R, Tanaka H, Uyama H, Kobayashi S (2000) Macromol Rapid Commun 21:496
26. Ikeda R, Tanaka H, Uyama H, Kobayashi S (2002) Polymer 43:3475
27. Bravo L (1998) Nutr Rev 56:317
28. Vinson JA (1998) Adv Exp Biol Med 439:151
29. Jankun J, Selman SH, Swiercz R, Skrzypczak-Jankun E (1997) Nature 387:561
30. Bordoni A, Hrelia S, Angeloni C, Giordano E, Guarnieri C, Caldarera CM, Biagi PL (2002) J Nutr Biochem 13:103
31. Nakagawa K, Ninomiya M, Okubo T, Aoi N, Juneja LR, Kim M, Yamanaka K, Miyazawa T (1999) J Agric Food Chem 47:3967
32. Mukhtar H, Ahmad N (1999) Toxic Sci 52S:111
33. Yang CS, Chung JY, Yang G-Y, Chhabra SK, Lee M-J (2000) J Nutr 130:472S
34. Mukhtar H, Ahmad N (1999) Am J Clin Nutr 71S:1698S
35. Hathway DE, Seakins JWT (1957) Biochem 67:239
36. Hathway DE, Seakins JWT (1955) Nature 176:218
37. Guyot S, Cheynier V, Souquet J-M, Moutounet M (1995) J Agric Food Chem 43:2458
38. Guyot S, Vercauteren J, Cheynier V (1996) Phytochem 42:1279
39. Janse van Rensburg W, Ferreira D, Malan E, Steenkamp JA (2000) Phytochem 53:285
40. Hamada S, Kontani M, Hosono H, Ono H, Tanaka T, Ooshima T, Mitsunaga T, Abe I (1996) FEMS Microbiol Lett 143:35
41. López-Serrano M, Ros Barceló A (1997) J Food Sci 62:676
42. López-Serrano M, Ros Barceló A (2001) J Chromatograph A 919:267
43. López-Serrano M, Ros Barceló A (2002) J Agric Food Chem 50:1218
44. Hosny M, Rosazza JP (2002) J Agric Food Chem 50:5539
45. Dordick JS, Marletta MA, Klibanov AM (1987) Biotechnol Bioeng 30:31
46. Oguchi T, Tawaki S, Uyama H, Kobayashi S (1999) Macromol Rapid Commun 20:401
47. Akkara JA, Ayyagari MSR, Bruno FF (1999) Trends Biotechnol 17:67
48. Mejias L, Reihmann MH, Sepulveda-Boza S, Ritter H (2002) Macromol Biosci 2:24
49. Kurisawa M, Chung JE, Kim YJ, Uyama H, Kobayashi S (2003) Biomacromolecules 4:469

50. Kurisawa M, Chung JE, Uyama H, Kobayashi S (2003) Macromol Biosci 3:758
51. Jiménez M, García-Carmona F (1999) J Agric Food Chem 47:56
52. Takahama U (1986) Biochim Biophys Acta 882:445
53. Kurisawa M, Chung JE, Uyama H, Kobayashi S (2003) Biomacromolecules 4:1394
54. Kurisawa M, Chung JE, Uyama H, Kobayashi S (2004) Chem Commun 294
55. Payne GF, Chaubal MV, Barbari TA (1996) Polymer 37:4643
56. Kumar G, Bristow JF, Smith PJ, Payne GF (2000) Polymer 41:2157
57. Kumar G, Smith PJ, Payne GF (1999) Biotechnol Bioeng 63:154
58. Yamada K, Chen T, Kumar G, Vesnovsky L, Topoleski LDT, Pyane GF (2000) Biomacromolecules 1:252
59. Muzzarelli RAA, Ilari P, Xia W, Pinotti M, Tomasetti M (1994) Carbohydr Polym 24:295
60. Wu LQ, Embree HD, Balgley BM, Smith PJ, Payne GF (2002) Environ Sci Technol 36:3446
61. Aberg CM, Chen T, Olumide A, Raghavan SR, Payne GF (2004) J Agric Food Chem 52:788
62. Chen T, Small DA, Wu LQ, Rubloff GW, Ghodssi R, Vazquez-Duhalt R, Bentley WE, Payne GF (2003) Langmuir 19:9382
63. Ihara N, Schmitz S, Kurisawa M, Chung JE, Uyama H, Kobayashi S (2004) Biomacromolecules 5:1633
64. Chung JE, Kurisawa M, Uyama H, Kobayashi S (2003) Biotechnol Lett 25:1993
65. Chung JE, Kurisawa M, Tachibana Y, Uyama H, Kobayashi S (2003) Chem Lett 32:620
66. Ihara N, Kurisawa M, Chung JE, Uyama H, Kobayashi S (2005) Appl Microbiol Biotechnol 66:430
67. Ihara N, Tachibana Y, Chung JE, Kurisawa M, Uyama H, Kobayashi S (2003) Chem Lett 32:816
68. Shao L, Kumar G, Lenhart JL, Smith PJ, Payne GF (1999) Enz Microb Technol 25:660

# Enzymatic Catalysis in the Synthesis of Polyanilines and Derivatives of Polyanilines

Peng Xu[1] · Amarjit Singh[2] · David L. Kaplan[1] (✉)

[1]Departments of Biomedical Engineering and Chemical & Biological Engineering, Bioengineering & Biotechnology Center, Tufts University, Medford, MA 02155, USA
peng.Xu@tufts.edu, david.kaplan@tufts.edu

[2]Department of Radiology, Harvard Medical School, Boston, MA 02155, USA
amarjit_singh@hms.harvard.edu

1	Introduction	71
2	**Polyaniline Synthesis in Aqueous Organic and Aqueous Solvents**	71
2.1	Enzymatic Synthesis of Polyanilines in Organic Solvents	71
2.2	Synthesis of Polyanilines in Aqueous Solutions	73
2.3	Application of the Enzymatic Approach	73
2.4	HRP Immobilization	74
2.5	Substrate Behavior and Substituent Sites	75
2.6	Other Enzymes in the Synthesis of Polyanilines	76
2.7	The Use of Hematin	77
3	**Template-Assisted Polymerization**	79
3.1	SPS Template	79
3.2	Other Polyelectrolyte Templates	81
3.3	Biocompatible Templates	81
3.4	Modified Hematin	82
4	**Mechanistic Aspects of Template-Assisted Polyaniline Synthesis**	82
4.1	The Role of the Template	82
4.2	The Substituent Position	84
4.3	The Application of the Fast Kinetics	85
5	**Micellar Assisted and Interfacial Polymerization**	85
5.1	Micellar Systems	85
5.2	Interfacial Polymerization	86
6	**Dip-pen Nanolithography**	86
7	**Analytical Studies on Polyanilines**	88
7.1	Solid-State NMR Studies	89
7.2	Circular Dichroism Study	91
7.3	Other Analytical Methods	92
8	Conclusions	92
	References	92

Abstract Enzyme catalysis in the synthesis of polyaniline is reviewed. Oxidoreductase enzymes and biomimetic catalysts have been used for the polymerization in aqueous, aqueous organic, and interfacial or templating environments to optimize polymer features. The results showed significant structural control in the presence of templates to produce water soluble and processable polymeric materials. Dip-pen nanolithography technology has also been used to process these materials into conducting nano-wires and thus provides new opportunities for the formation of electrical contacts among biological components for various applications such as synthetic biological interfaces.

Abbreviations

AFM	Atomic force microscopy
BOD	Bilirubin oxidase
CD	Circular dichroism
CSA	Camphorsulphonic acid
DBSA	Dodecylbenzenesulfonic acid
DMF	Dimethyl Formaldehyde
DMSO	Dimethylsulfoxide
DP	1,2-Diamino propane
DPN	Dip-pen nanolithography
EDOT	Ethylenedioxythiophene
ELBL	Electrostatic layer-by-layer
FeB	2,6-bis[1-2,6-diisopropylphenylimino-ethyl]pyridine iron(II) chloride
HRP	Horseradish peroxidase
HRPC	Horseradish peroxidase isoenzyme C
HTBA	Hexadecyltrimethylammonium bromide
LB	Langmuir-Blodgett
LGS	Lignosulfonate
LT	Langmuir trough
MALDI	Matrix-assisted laser desorption ionization
OC	Oxalyl chloride
PAA	Poly(acrylic acid)
PANI	Polyaniline
PDAC	Poly(dimethyl diallylammonium chloride)
PEO	Poly(ethylene oxide)
PVA	Poly(vinyl alcohol)
PVP	Poly(vinylphosphonic acid)
PYR	Pyrrole
SBP	Soybean peroxidase
SDBS	Sodium dodecyl benzenesulfonate
SPE	Screen-printed electrode
SPM	Scanning probe microscopy
SPS	Sulfonated polystyrene
TOF	Time of flight
Triton	Polyoxyethylene(10) isooctylphenyl ether
XPS	X-ray photoelectron spectroscopy

1
Introduction

There has been a significant increment in the use of conducting polymers in electronic and optical applications in recent years. Polyaniline (PANI), one of the most important members in the conducting polymer family, has been widely used in light-weight organic batteries [1], rechargeable batteries [2], optical displays [3], electromagnetic shielding [4], chemical sensors [5], organic LEDs [6], antistatic coatings [7], and potential molecular electronic devices because of its electric, electrochemical, and optical properties, as well as good environmental and thermal stability [8, 9]. Some other conducting polymers such as polythiophene and polypyrrole are also being extensively investigated for similar applications.

The most common methods for the synthesis of polyaniline are either chemical or electrochemical oxidation of aniline monomer resulting in polymerization through a head-to-tail coupling mechanism [10–12]. The reaction conditions are harsh with extreme pH, high temperature, strong oxidants and highly toxic solvents required. In order to improve the solubility and the processability of the polymeric products, the synthesized polyaniline is usually post-polymerization-treated with fuming sulfuric acid [13, 14]. Enzyme catalyzed polymerization of aniline, its derivatives and other aromatic compounds in the presence of hydrogen peroxide provided an alternative method toward the formation of soluble and processable conducting polymers. These enzyme-catalyzed reactions are usually carried out at room temperature, in aqueous organic environments at neutral pH. The reaction conditions were greatly improved and the purification process of the final products was simplified when compared to the traditional chemical or electrochemical oxidation methods.

2
Polyaniline Synthesis in Aqueous Organic and Aqueous Solvents

2.1
Enzymatic Synthesis of Polyanilines in Organic Solvents

Enzymatic polymerization of a series of aniline derivatives was studied in aqueous and aqueous organic solvents. Phenylenediamines and aminophenols were polymerized in ambient conditions by the catalysis of horseradish peroxidase (HRP) in dioxane without the presence of templates [15, 16]. The products were soluble in DMF and DMSO. The structures were analyzed by FT-IR and NMR. The larger molar excess of hydrogen peroxide was necessary in reaching a higher polymer yield. Poly(2-aminophenol) and poly(4-aminophenol) resulting from these reactions have been shown to have electroactive properties.

Poly(azobenzene) and its derivatives have applications in optical devices [17]. A novel polyaniline containing azo groups was synthesized by the HRP-catalyzed oxidative coupling of 4,4′-diaminoazobenzene (Scheme 1). The polymerization was carried out at pH 6.0 in tris buffer with 70% yield. The polymer analyzed by GPC (80 000, PD = 4.8) was soluble in DMSO and DMF. Azo groups were detected in the main chains as well as in the side chains. Photoexcitation studies indicated that *cis–trans* isomerization of the chromophore may be the result of structural constraints in the polymer [18]. Photodynamic properties of the azo functionalized polyaniline in their relaxed or constrained conformations were quite different.

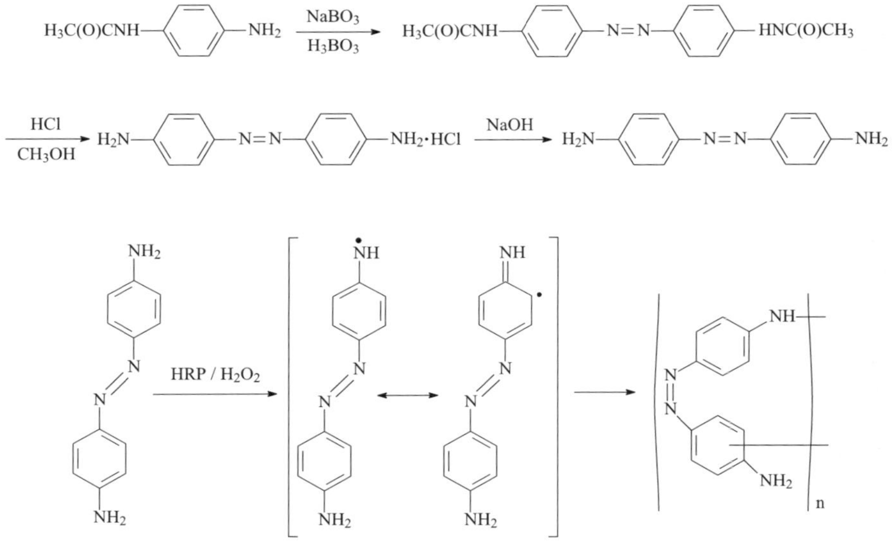

Scheme 1 Polymerization of diaminoazobenzene by HRP in the presence of hydrogen peroxide

A similar method was used for the synthesis of 4-phenylazoaniline (Scheme 2). Raman spectroscopy of the polymer showed N = N stretching, indicating the presence of the azo pendent group in the product. The photodynamic studies have single-beam or two-beam interference patterns and surface relief gratings. Further characterization detected azobenzene groups both in the main chain as well as in the side chain of the polymer. There was an equilibrium population of *trans* and *cis* states in the constrained conformations before photoexcitation, which led to the build-up of *cis* and depletion of *trans* states. Upon relaxation after the photoexcitation, a net increase in the *trans* state occurred, implying that some of the constrained predominantly *cis* segments in the original polymer relaxed to a *trans* state [17, 18].

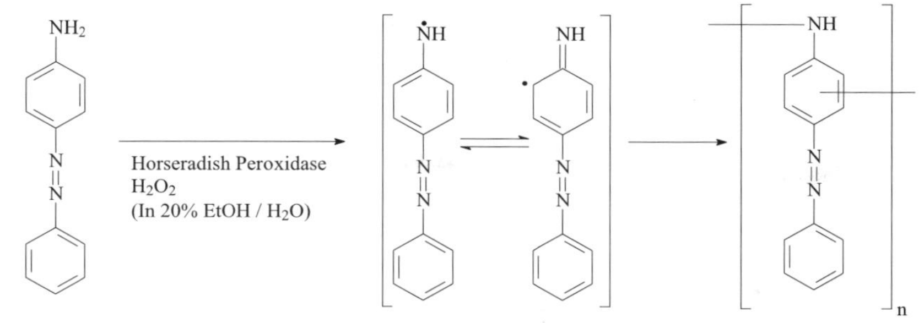

Scheme 2 Synthesis of azo functionalized polyaniline

2.2
Synthesis of Polyanilines in Aqueous Solutions

Water-soluble polyanilines were synthesized from sulfonated anilines. Previously water-soluble polyanilines were synthesized by enzymatic templating, chemical or electrochemical methods, or copolymerization of aniline with sulfonated aniline [19–22]. The presence of pendent sulfonic acid groups throughout the polyaniline made it water-soluble in all pH conditions. The product was electroactive with an average molecular weight of 18 000 Da. Conductivity was determined to be in the semiconducting region (10^{-5} S/cm) at pH 6. The conductivity of poly(2,5-diaminobenzenesulfonate) was about 3 to 4 orders of magnitude lower because of its branched structure (Scheme 3). The product was versatile for fabricating thin films by a layer-by-layer technique due to its diamine feature.

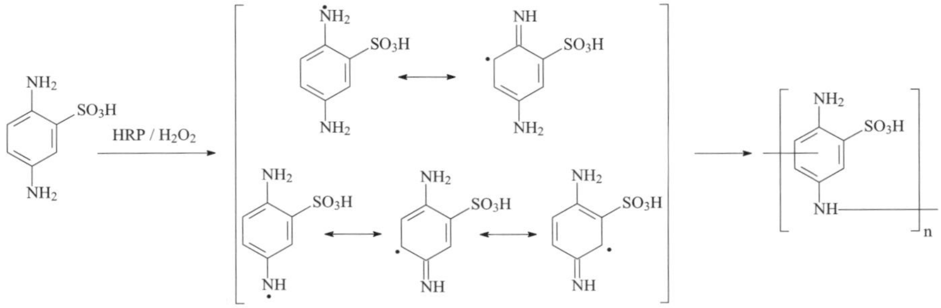

Scheme 3 Enzyme catalyzed polymerization of 2,5 diaminobenzene sulfonic acid

2.3
Application of the Enzymatic Approach

A catalytic coupling reaction between 4-amino antipyrine and an N,N-disubstituted aniline derivative had been exploited as a less hazardous chromogen system in the detection of HRP and the iron(III) sulfonated tetraphenyl porphyrin, which was a biomimetic catalyst [23]. A dye P^+, which was intrinsically electroactive due to the presence of a quinone-iminium feature, was generated through an oxidative coupling reaction (Scheme 4), where 3-methyl-N-ethyl-N-(β-hydroxyethyl)aniline was used as the disubstituted aniline derivative. The detection of HRP was carried out by linear scan voltammetry equipped with a Nafion-film-modified screen-printed electrode (SPE). The cation exchange properties of the Nafion film coating were beneficially exploited to retain and concentrate the P^+ cation formed during the enzymatic oxidation for quantitative electrochemical or colorimetric analysis. This method allowed the determination of HRP with a detection limit of 10^{-12} M.

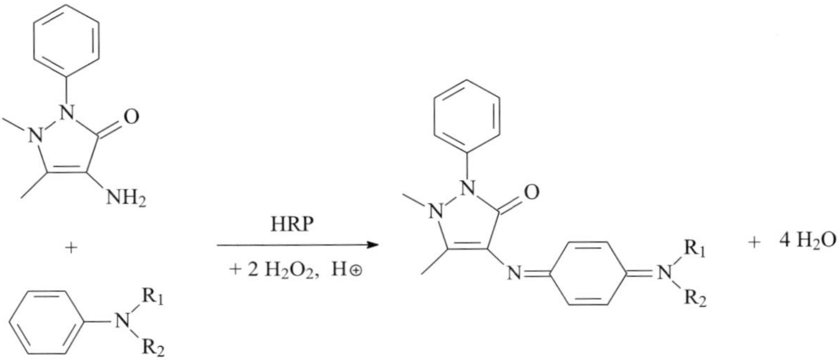

Scheme 4 Quinone-iminium dye generated through an oxidative coupling reaction

A device called an "enzyme switch" based on polyaniline and HRP was developed. The enzyme HRP was either adsorbed onto the PANI film or immobilized on an insulating poly(1,2-diaminobenzene) polymer grown electrochemically on top of the PANI film [24]. The conductivity of the PANI film changed from a conducting state to an insulating state with the addition of hydrogen peroxide. The changes in the conductivity were confirmed by measuring the potential of the PANI film as the conductivity was switched. The experimental results showed that the hydrogen peroxide on its own had no effect on the conductivity of the PANI film at pH 5.0 without the adsorbed or immobilized HRP. The electrochemical reduction of the oxidized PANI film allowed for the recovery of the enzyme switch responsive to the hydrogen peroxide, which operated in the "on" to "off" direction. It was also functionalized as a sensitive hydrogen peroxide detector.

2.4
HRP Immobilization

Controlled-thickness polyaniline layers on various substrate surfaces were synthesized using immobilized HRP as a catalyst in an aqueous solution [25].

The polyethylene (PE) surfaces were treated with argon/dichlorosilane (Ar/DS) plasma to assist the implantation of $Si-(Cl)_x$ functionalities. Further, the PE surfaces were functionalized in situ in the gas phase with 1,2-diamino propane (DP) and oxalyl chloride (OC). The HRP was covalently immobilized on the modified PE surfaces using sodium cyanoborohydride. XPS studies verified the surface chemistry of each of these three steps for the HRP immobilization. Aniline was polymerized in an aqueous solution by catalysis with immobilized HRP. The molecular weight was evaluated by GPC. Similar distributions were observed regardless of whether the enzyme was in free form or immobilized on PE.

2.5
Substrate Behavior and Substituent Sites

The substrate behavior and the influence of the substituent groups on the properties of the polymerized aniline derivatives were investigated by comparing the effects of the Hammett constants of the substituents. The Hammett equation (Hammett relation) is given as:

$$\lg(K/K_0) = \rho\sigma \quad (1)$$

$$\lg(k/k_0) = \rho\sigma \quad (2)$$

where K or k is the equilibrium or rate constant, respectively, for the given reaction of m- or p-XC_6H_4Y. K_0 or k_0 refers to the reaction of C_6H_5Y, i.e., $X = H$. σ is the substituent constant characteristic of m- or p-X. ρ is the reaction constant characteristic of the given reaction of Y. This equation was applied to the influence of *meta-* or *para*-substituents X on the reactivity of the functional group Y in the benzene derivative m- or p-XC_6H_4Y.

In the luminol-H_2O_2-HRP chemiluminescence system, some of the aniline derivatives were functionalized as enhancers, while some were inhibitors of the fluorescence [26, 27]. A variety of *meta-*, *para-*, and *ortho*-substituted aniline were studied. The structures of some of these aniline derivatives are listed in Scheme 5. The reaction constants with HRP-I and HRP-II were measured, and the Hammett constants (σ) of their substituents were determined. The results indicated that the aniline substituents with σ less than -0.27 exhibited inhibition behavior. The aniline derivatives acted as enhancers when σ was between -0.27 and $+0.18$. A slight enhancement or inhibition was observed if σ is greater than $+0.18$. The factors of enhancement were determined by the potential of reduction of its aniline radicals, the reaction rates of the aniline with HRP-I and HRP-II, and the maximum concentration of the aniline radicals allowed during the oxidation [28].

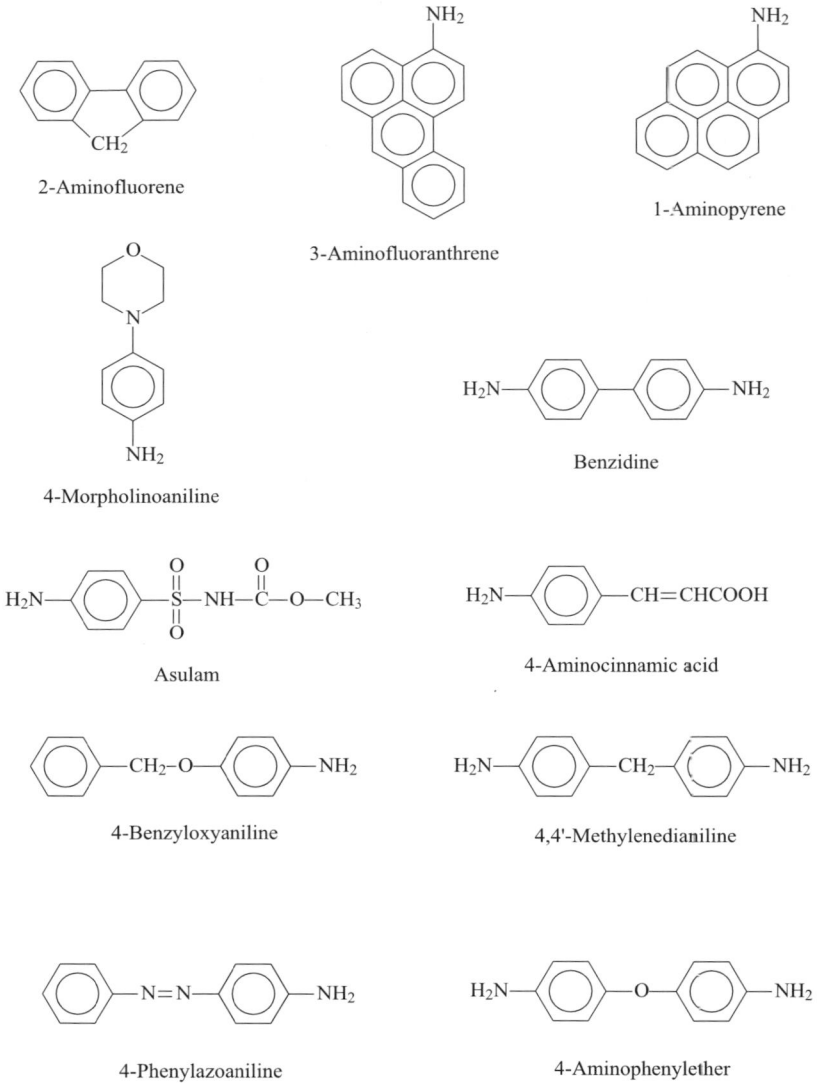

Scheme 5 Structure of some aromatic amines studied for the influence of the substituent groups

2.6
Other Enzymes in the Synthesis of Polyanilines

Polythiophenes and polypyrroles have been reported to have high electrical conductivity and are of significant technological importance [29]. These polymers cannot be directly synthesized by HRP. According to the mechanism of

HRP catalysis, an oxidized iron–heme complex is formed in the presence of hydrogen peroxide. It reacts with the substrate in a one-electron transfer process, which leads to the formation of substrate radicals and the recovery of the iron–heme complex [30]. Monomers such as 3,4-ethylenedioxythiophene (EDOT) and pyrrole (PYR) coupled to the active site of the enzyme result in the deactivation of the catalyst.

Apart from HRP, other peroxidases and biomimetic catalysts are used in polyaniline synthesis. The oxidation of a series of phenols, anilines, and benzenethiols were performed with several fungal laccases [31]. Soybean peroxidase (SBP) is an attractive biocatalyst because of its thermal and pH stability. The use of soybean peroxidase for the synthesis of conducting polyaniline with controlled molecular weight was reported [32]. This method allowed the synthesis low-molecular-weight PANI oligomers. The pH was maintained above 3.5, but reactions can be run at a pH as low as 1.0. The soybean peroxidase was preferable in these types of reactions due to its low pH stability, compared to such enzymes as HRP, tobacco peroxidase, tomato peroxidase, legume peroxidase, bacterial peroxidase, and chloroperoxidase. PANI molecular weight as high as 200 000 Da was obtained. The conductivity of the polymer was measured as 10^{-3} to 10^{-1} S/cm, which allowed blending with the thermoplastic or thermosetting polymers for antistatic purposes.

Electroactive polyaniline films were synthesized by the catalysis of bilirubin oxidase (BOD, a copper-containing oxidoreductase). The polymerization of aniline was carried out on the surface of a solid matrix such as glass slide, plastic plate, or platinum electrode to form homogeneous films [33]. The BOD was immobilized on the surface by physical absorption. The optimum pH was around 5.5. Some aniline derivatives such as *p*-aminophenol and *p*-phenylenediamine were good substrates for BOD. Structural analysis suggested the BOD synthesized polyaniline possessed partially 1,2-substititued structures. Cyclic voltammetric studies demonstrated that the PANI films were electrochemically reversible in redox properties, but differed from that of chemically or electrochemically synthesized PANI. The difference was attributed to the partial 1,2-substitution. Laccases are known to oxidize phenolic compounds in nature in the presence of oxygen and are capable in polyaniline synthesis in vitro [34–36].

2.7
The Use of Hematin

Synthetic porphyrin complexes were extensively studied for their potential catalytic applications. The hydroxy ferriprotoporphyrin compound hematin (Scheme 6) was used as a catalyst in the polymerization of phenols [37].

The catalytic application of hematin in the synthesis of polyaniline was further studied. The electrostatic layer-by-layer (ELBL) self-assembly of a polyelectrolyte poly(dimethyl diallylammonium chloride) (PDAC) and

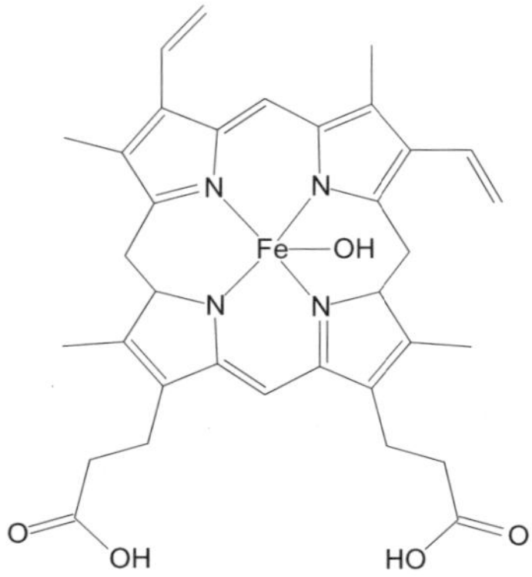

Scheme 6 Structure of hematin

hematin was utilized to construct a nanocomposite film catalyst [38]. The conductive form of PANI was formed not only on the surface of ELBL as a coating, but in the bulk solution. In contrast to the immobilization method, where enzyme was covalently coupled onto the substrate, hematin desorption was required. The polymerization rate was strongly dependent on the degree of hematin desorption, which was controllable. The reaction rate was also affected by the pH value, the ionic strength of the medium, the amount of hematin, and the type of PANI. Hematin had been demonstrated as the best-suited biomimetic compound for these types of reactions in comparison to catalase, horseradish peroxidase isoenzyme C (HRPC), and FeB-2,6-bis[1-2,6-diisopropylphenylimino-ethyl]pyridine iron(II) chloride (FeB) in the UV/Visible study of their catalytic activities [39].

Some aromatic organometallics, such as p-ferrocenylaniline and p-ferrocenylphenol, are reported as good substrates for HRP- or laccase-catalyzed polymerization [40]. These compounds are advantageous because of their high reactivity, easy control of redox transformation, water-soluble products and avoidance of deactivation of the enzyme. The unique properties of ferrocene were introduced into the polymer. The reaction process is shown in Scheme 7. The p-ferrocenylaniline was electrochemically active when adsorbed to a carbon electrode.

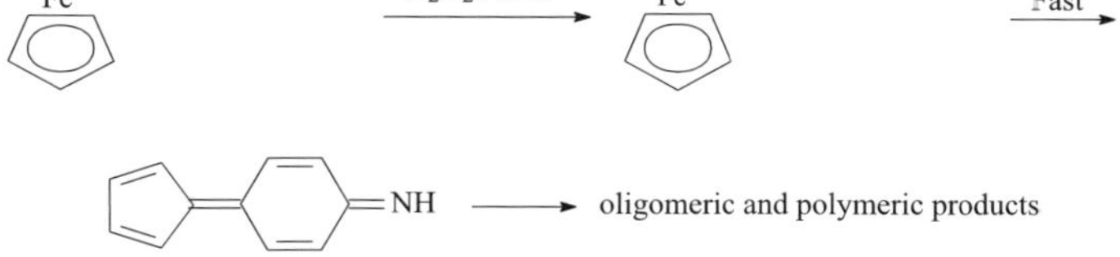

Scheme 7 The HRP catalyzed reaction of p-ferrocenylaniline

3
Template-Assisted Polymerization

3.1
SPS Template

Template-assisted synthesis of water-soluble conducting polyaniline was achieved in an aqueous medium. The aniline complex was formed in the system in the presence of sulfonated polystyrene (SPS), which functioned as a polyanionic template. HRP-catalyzed polymerization was carried out at pH 4.3 in a buffered aqueous solution with a stoichiometric amount of hydrogen peroxide. The resulting polymer was complexed to SPS and exhibited electroactivity. The pH of the reactant solution is critical in controlling the aniline-SPS complex formation and hence the electrical activity of the resulting backbone. The effect of the pH was studied. The optimized ratio of the SPS to aniline was obtained at optimum pH. The reduction/oxidation reversibility of the PANI/SPS complex was demonstrated [41]. Spectroscopic studies indicated that the conductivity of the PANI/SPS complex was pH-dependent due to the self-doping mechanism during synthesis. The molecular weight of the complex was measured by GPC during the course of the reaction and compared with pure SPS as a control to demonstrate the polymerization [42]. The low pKa (benzenesulfonic group has a pKa of 0.7) provided the necessary cationic and anionic charges for preferential monomer alignment and salt formation along the SPS backbone [43]. The desired structure of the conducting polyaniline was synthesized by enzymatic oxidation of the macromolecular complex as shown in Scheme 8.

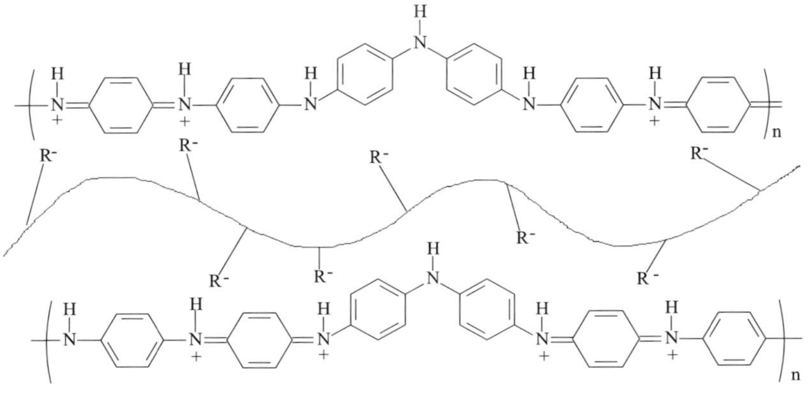

$R=SO_3^-, PO_4^-$

Scheme 8 Representation of enzymatic polymerization of aniline (at pH 4.3) in the presence of SPS

Cyclic voltammetric studies showed only one set of redox peaks over the potential range of − 0.2 V to 1.2 V, which suggested that the complexes are more oxidatively stable. The conductivity of the PANI/SPS complex was measured and was found to be 0.005 S/cm. The values increased to 0.15 S/cm after HCl doping and could be increased further with an increase in the aniline-to-SPS molar ratio [43].

Laccase from *Coriolus hirsutus* was used in the synthesis of water-soluble conducting polyaniline in the presence of an SPS template [44]. This enzyme showed remarkable advantages during the polymerization compared to commonly used HRP due to its high activity and stability under acidic conditions. HRP exhibited low activity and poor stability at a pH below 4.5 [45, 46]. The HRP catalyzed reactions resulted in the consumption of a large amount of enzyme in the polymerization at low pH. Usually, the HRP-catalyzed polymerization of aniline is carried out at pH 4.3 [43]. The laccase is active at pH 3.5 for about 4 to 5 days. The polyaniline synthesized by HRP and laccase showed similar structures. The conductivity of the PANI/SPS complex was measured as 2×10^{-4} S/cm. Cyclic voltammetry studies showed that the C–V curve generated with laccase was not identical to the HRP-catalyzed polyaniline. A possible reason was that different structures resulted due to the different catalytic mechanisms of HRP and laccase. The catalysis of HRP took place in a hydrogen-peroxide-oxidized ferric center through two sequential one-electron reduction steps. In the case of laccase, the one-electron oxidation was catalyzed via the four copper centers in the presence of oxygen.

Since HRP exhibited low activity and stability at a pH less than 4.5 [43, 45] palm tree peroxidase was reported as a useful biocatalyst in the synthesis of polyaniline in the presence of an SPS template [47]. This peroxidase was isolated from leaves of the African oil palm tree and royal palm tree and exhibited an unusual stability at high temperatures and over a broad range of pH [48]. The polyelectrolyte complex of PANI/SPS was studied by computer modeling. The SPS presented in a syndiotactic configuration may interact with two PANI polymers forming a regular pattern of intermolecular contact, rather than forming a less stable complex with one SPS to one PANI molecule. The polymerization at low pH resulted in the aggregation state of the complex. Optical microscopy showed that complex was not in soluble form, but in dispersion form with a maximum particle size of 1.5 to 2 μm.

In order to recycle the enzyme, since it is quite expensive, a new method of immobilizing the enzyme was introduced. The HRP was immobilized on chitosan powder by a simple glutaraldehyde bridge method [49]. The immobilized HRP had the same catalytic function as the enzyme solution. The spectroscopic characterization indicated that the polyaniline obtained by this method was similar to that formed using HRP in solution as a catalyst [50].

The synthesis of water-soluble polyaniline greatly improved processability. Conducting fibers were fabricated from polyaniline–SPS complex by a dry-spinning technique, which took advantage of water-soluble products attributed to the enzymatic synthesis method [51]. The conductivity of the dry-spun fibers were measured to be approximately 10^{-3} S/cm, one order of magnitude higher than that of films fabricated by the same PANI/SPS complex process as for the fibers. The higher conductivity of the fibers was attributed to the orientation of the polymer chains. Compared with the chemically synthesized PANI (1–10 S/cm), the lower conductivity was due to the interchain diffusion of the charge carriers caused by the separation of the PANI chains from the SPS backbones. The conductivity of the fibers increased to 10^{-2} S/cm by exposure to HCl vapor for two hours. The PANI conducting fibers exhibited improved tensile strength and stability in thermal analysis.

3.2
Other Polyelectrolyte Templates

Some other templates were studied and proved to have similar functionality as SPS. These templates are all polyelectrolytes. Poly(vinylphosphonic acid) (PVP) was reported to be a good template to form complexes with aniline. The conducting macromolecular complex was synthesized in aqueous buffer at pH 4.0 by the catalysis of HRP in the presence of hydrogen peroxide [52]. The thermal stability of the resultant PANI/PVP complex was improved when compared to the chemically synthesized complex. A constant ratio of 1.5 to 1 between the phosphonic acid repeat unit and the aniline unit prior to precipitation was found when PVP was in excess. Appropriate control of the polymerization could be used to tune the water solubility and conductivity of the complex. It was possible to separate the template from the polymer by dedoping the PANI/PVP complex with an aqueous NH_4OH solution. The ^{13}C and ^{15}N cross-polarization with magic angle spinning (CP/MAS) NMR distinguished the charged and uncharged domains presented in the various forms of PANI [53]. The conductivity of PANI/PVP was measured as 1.2×10^{-2} S/cm to 5.5×10^{-2} S/cm and was found to be about one magnitude lower than that of PANI purchased from commercial sources or generated by nonenzymatic methods.

3.3
Biocompatible Templates

Biomacromolecules were exploited as templates in the enzymatic synthesis of electronic and photonic materials for the development of biofunctional complexes and selective biosensors. The aniline was polymerized by the catalysis of HRP on a calf thymus DNA matrix consisting of a synthetic oligonucleotide (poly[dA-dC].poly[dG-dT]) [54]. The complexation of PANI with DNA and the oligonucleotide was found to induce reversible changes in the secondary structure of the nucleic acid template, leading to the formation of an overwound polymorph. The melting behavior of the PANI/DNA complex demonstrated stabilization of the DNA by the complexation with the PANI. Cyclic voltammetry studies confirmed the presence of an electrochemically redox reversible form of the complex. The conductivity of the PANI/DNA complex was measured as 40.9×10^{-7} S/cm after doping, which was almost an 800% increase compared to prior to doping. The conductivity of the DNA template, which increased only 56.8% after doping, was measured as the control. Scanning probe microscopy (SPM) studies indicated that the agglomerate of the PANI/DNA strands was caused by the formation of PANI. But the secondary structure of the individual DNA was retained. This study also resulted in the development of a new route to the fabrication of electro-responsive biomaterials. The secondary structure of the DNA may be reversibly controlled from the native form to an overwound polymorph by simply changing the redox state of the polyaniline [55].

Lignosulfonate (LGS), an inexpensive byproduct from pulp processing industries, functioned similarly as SPS during the polymerization. The product presented thermal stability and electrical conductivity. The conductivity of PNAN/LGS complex synthesized at pH 1.0 was determined to be about 10^{-3} S/cm without further external doping.

3.4
Modified Hematin

Pegylated hematin was developed as an effective biomimetic catalyst for polymerization of various anilines [56]. PEG modified hematin, a biomimetic catalyst, was used in the synthesis of PEDOT and PPYR at the presence of SPS template [57]. The polymerization was carried out in an aqueous solution at pH 1.0. The polymers were electrochemically active. Conductivity was measured as 10^{-5} S/cm for PEDOT and 10^{-4} S/cm for PPYR. The conductivity was substantially improved by the copolymerization of PEDOT and PPYR and was found to be in the range of 0.1–1.0 S/cm.

4
Mechanistic Aspects of Template-Assisted Polyaniline Synthesis

4.1
The Role of the Template

HRP is a single-chain β-type hemoprotein that catalyzes the decomposition of hydrogen peroxide at the expense of aromatic proton donors [58]. HRP is an iron-containing porphyrin- type structure and is well-known to catalyze the oxidative coupling of aniline and its derivatives in the presence of hydrogen peroxide. Usually, the HRP catalyzed polymerization of aniline is carried out in aqueous organic solvents, and the resultant polymers are generated in low yields and have complex structures [59]. Typically, a mixture of at least two structurally different types of PANIs is obtained as shown in Scheme 9 [43]. The first one is that of mixed carbon–carbon or carbon–nitrogen coupling at the *ortho*- or *para*-positions and the second is that of the desired benzenoid-quinoid (head-to-tail) structure, which is formed in the traditional chemical or electrochemical synthesis of polyaniline.

In order to understand the mechanistic role of the template in the enzymatic synthesis of water- soluble conducting polyaniline, a variety of macromolecular and surfactant templates were systematically investigated [60]. Polyelectrolytes were specifically selected to investigate the effects of ionizability (pK_a), ionic charge type (anionic, cationic, or neutral), and surface charge density of the templates in the reaction. The structures of these templates are given in Scheme 10. It is generally accepted that a highly acidic medium is necessary for linear chain growth of conducting polyaniline [61–63]. The enzymatic template polymerization of aniline was also pH-dependent. The template provided a necessary "local" environment where the pH and charge density differed from that of the bulk solution and served as a nanoreactor that was critical in anchoring, aligning, and reacting with the aniline monomers to control the form of polyaniline generated in the reaction. Strong acidic polyelectrolytes such as SPS were the most favorable because of the lower local pH provided. The formation and control of lower local pH allowed the reaction to be carried out at a higher bulk pH which

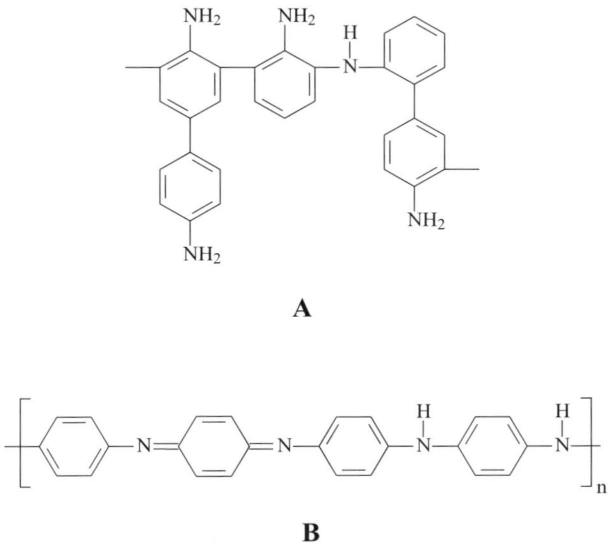

Scheme 9 Two structurally different types of PANIs obtained in typical enzymatic polymerization without template: **A** ortho-, para-substituted branched structure and **B** benzenoid-quinoid linear structure

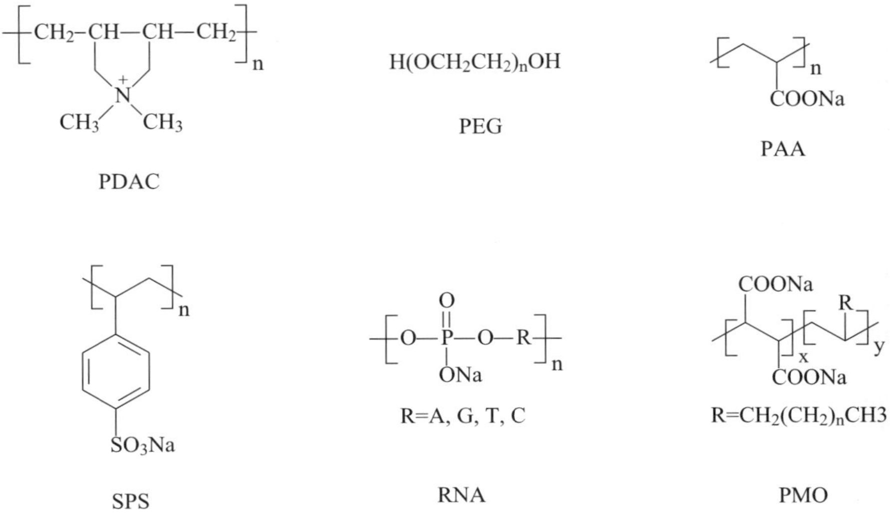

Scheme 10 Molecular structures of the macromolecular templates used in the enzymatic polymerization

was favorable to the enzyme and prolonged the catalytic activity. The template also provided hydrophobic domains that served to solubilize and orient the monomer prior to reaction. There is a direct dependence on the template structure and the type of polyaniline that is formed.

4.2
The Substituent Position

The kinetics of these reactions was studied for a two-step mechanism using the cationic HRP isoenzyme C (HRPC) as a catalyst. Similar conclusions were drawn based on the Hammett constants (σ), which were obtained according to kinetic studies of the *meta-* or *para*-mono-substituted aniline that included: 4-methoxyaniline, *p*-toluidine, *m*-toluidine, and 4-amino-*N,N*-dimethylaniline. The kinetics indicated that a two-step mechanism operated, in which the first step corresponded to the formation of an enzyme–substrate complex, and the second step corresponded to the electron transfer from the substrate to the iron atom. The size and hydrophobicity of the substrates controlled their access to the hydrophobic binding site of HRPC [64–66]. In general, the electron donating substituents on the *para*-site resulted in higher values of reaction rate constants. Some enhancers of the chemiluminescent luminol-HRP-H_2O_2 system were studied for their effects with respect to pH and concentration. The results demonstrated a previous conclusion that the active substituents in the *para*-position produced a greater increase in chemiluminescent properties at low concentration than those with deactivating substitutions [67].

4.3
The Application of the Fast Kinetics

The HRP, aniline (or PANI), and hydrogen peroxide made up a specific oxidative system based on fast reaction kinetics. Devices were developed according to this system in which any two of the three components can be used as a sensor to detect the third component. The electronic signals were collected and analyzed for changes in the electrochemical properties of the polyaniline or its derivatives when the oxidative coupling reactions occurred. Polyaniline-coated polypropylene membranes exhibited strong tendency to adsorb and immobilize HRP [68].

5
Micellar Assisted and Interfacial Polymerization

5.1
Micellar Systems

Aqueous micellar systems were similar to enzymatic templating polymerization in the reactions. The local environment in the vicinity of the template moieties had a charge density and pH that were different from those of the bulk solution [60]. Micelles served as nanoreactors, which complexed to the previously charged aniline monomers prior to the reaction and provided the necessary low-pH local environment for the growth of conducting polyaniline. This local environment was critical in anchoring and aligning the aniline monomers for reaction and ultimately controlled the forms of polyaniline (conducting or insulating) obtained during the reaction. Different types of surfactants—sodium dodecyl benzenesulfonate (SDBS), hexadecyltrimethylammonium bromide (HTAB), and polyoxyethylene(10) isooctylphenyl ether (Triton X-100)—were selected to form the micelle templates for the enzymatic polymerization of aniline [69]. The polyanilines obtained were soluble in organic solvents such as DMF and DMSO in emeraldine base form and were electrochemically active and thermally stable. The properties were strongly dependent on the structure of the surfactants selected. The SDBS was considered as a suitable micellar template in this system. The micellar system was also considered as nanoreactors for the synthesis of conducting PANI. A jump in pH from 4.3 to 5.1 was observed with the addition of aniline to the SDBS micelle solution [70]. No chemical reaction was involved in this mixing process. The bulk pH change was ascribed to the inhomogeneous distribution of protons in the media, which were trapped in the local molecular environments formed by the aqueous micelles. Besides SDBS, it was found that DNA could also provide this type of nanoreactor environment.

5.2
Interfacial Polymerization

A new approach for the in situ synthesis of conducting polyaniline and its derivatives in two dimensions was achieved at a Langmuir trough (LT) air-water interface. The LT was used to orient and order monomers prior to enzyme-based polymerization [71]. The 4-(tetradecyloxy)phenol was synthesized by o-alkylation of hydroquinone with 1-bromotetradecane and functionalized as amphiphilic surfactant at the interface. The monomer aniline or 4-hexadecylaniline was assembled on the surfactant monolayer and compressed. The polymerization was carried out by injecting hydrogen peroxide into the aqueous subphase, which contained HRP. The polymer monolayer was transferred onto an appropriate slide using the Langmuir-Blodgett (LB) technique. The use of LT to organize and orient the reactants resulted in improved control of monomer reactivity and orientation in the resulting polymer, compared to bulk synthesis. By using this method, the properties of the resulting polyaniline were improved in terms of heat stability, nonlinear optical response, and electrical conductivity. Upon doping with iodine, the conductivity for the bulk polymer was measured as 10^{-5} S/cm. However the polymer monolayer exhibited a conductivity of 1.07×10^{-2} S/cm. The conductivity was stable for more than four weeks at ambient conditions [72]. The improvement of conductivity was attributed to the 2D surface orientation of monomers to achieve an ordered lattice, which reduced the possibility of branching during the synthesis.

Polyaniline particles were synthesized by the HRP-catalyzed template-guided polymerization. Poly(sodium 4-styrenesulfonate) (SPS), poly(vinyl pyrrolidone) (PVP), poly(vinyl alcohol) (PVA), and poly(ethylene oxide) (PEO) were studied and used as the steric stabilizers [73]. The SPS and PVP were found to have strong interaction with PANI and remained in the particles formed in the reactions. Large PANI particles were obtained when using PVP and PEO as matrix polymers due to the ketone or ether groups, which had strong proton-accepting ability to aggregate monomers. Uniform microspheres of PANI particles were observed when low-molecular-weight (M_W 100 000) PEO was applied. The molecular weight of the particles was not influenced by the molecular weight of the matrix polymers. The microsphere sizes were also controlled by the chemical structures and properties of the matrix polymers.

6
Dip-pen Nanolithography

Dip-pen nanolithography (DPN) technology is a new tool for surface patterning at the nanometer scale. 4-Aminothiophenol was processed into nanowires via DPN-assisted enzymatic polymerization on gold surfaces

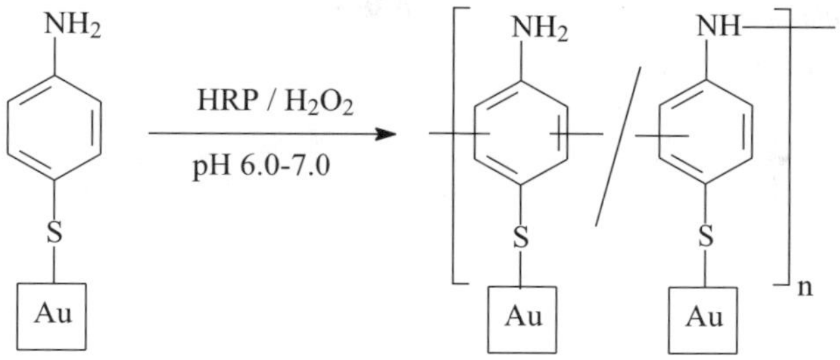

Enzymatic Catalysis in the Synthesis of Polyanilines

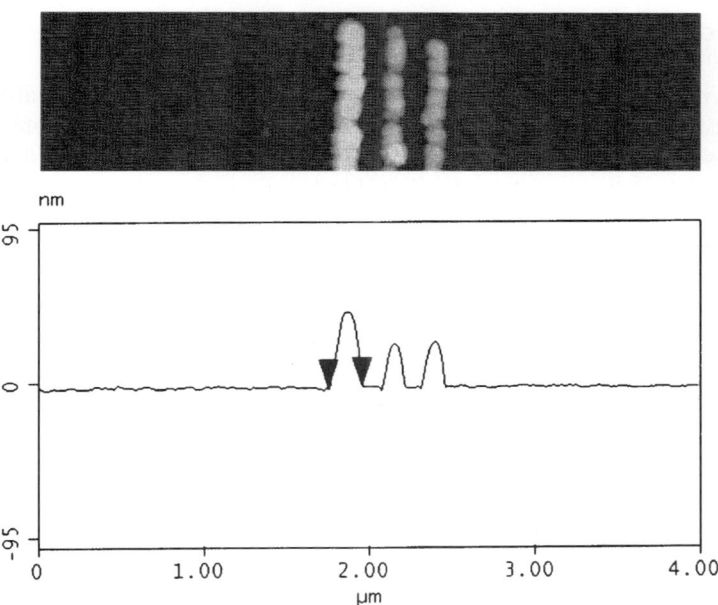

Scheme 11 Schematic representation of enzymatic polymerized 4-aminothiophenol on gold surface

(Scheme 11) [74, 75]. Reactions were performed in methanol/water (1 : 1 v/v) with aminothiophenol at 0.01 M. The monomer was patterned onto gold via DPN by scratching the surface at the rate of 0.1 Hz with a force varying from 0.35 nN with a step point of 0.05 nN. A series of parallel lines was patterned via the AFM tip and imaged in tapping mode at a scan rate of 1 Hz (Figs. 1 and 2). The lines are approximately 4 µm long with an average line width of 170 nm and average height of 35 nm. The measured topography of the polymer of aminothiophenol indicated lines 4 mm long with an average width of 210 nm and average height of 25 nm. The line width increased and the line height decreased when compared to the monomer patterns.

Fig. 1 AFM topography image of monomer (4-aminothiophenol) nanopatterns via DPN on gold surfaces

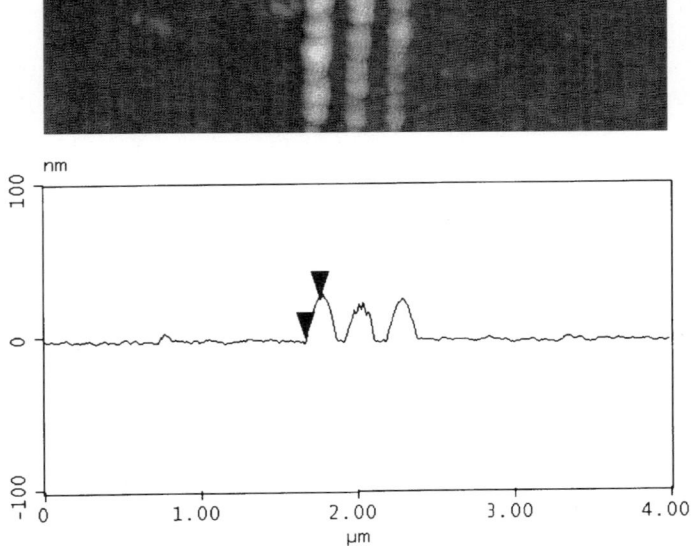

Fig. 2 AFM topography image of polymer (4-aminothiophenol after peroxidase-catalyzed polymerization) nanopatterns via DPN on gold surfaces

7
Analytical Studies on Polyanilines

In general, enzymatically synthesized polyaniline is in a protonated form, which can be converted to an unprotonated base form via treatment with an aqueous ammonia solution or other suitable bases. The unprotonated base form of polyaniline consists of reduced base units "A" and oxidized base units

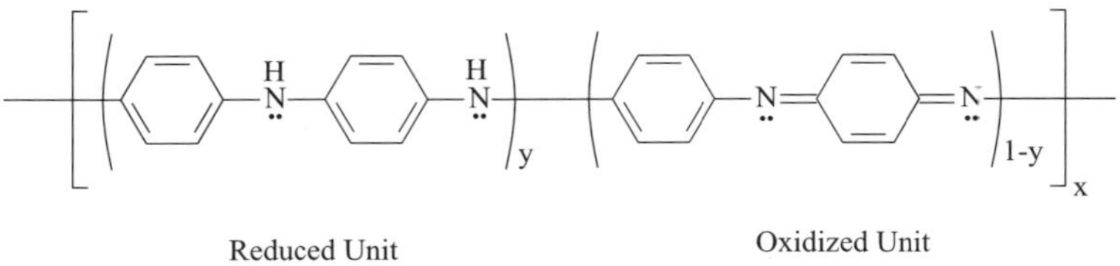

y=1   Leucoemeraldine base

y=0.5   Emeraldine base

y=0   Pernigraniline base

Scheme 12 Various oxidation states of the base form of polyaniline

"B" as repeat units, where the oxidation state of the polymer increases with decreasing values of y ($0 \leq y \leq 1$) (Scheme 12). The three extreme possibilities for value of y are 0, 0.5, and 1, corresponding to fully oxidized polyaniline (pernigraniline), half oxidized polyaniline (emeraldine), and fully reduced polyaniline (leucoemeraldine), respectively [53, 76].

7.1
Solid-State NMR Studies

Solid-state NMR spectroscopy was exploited in the structural analysis of enzymatically synthesized polyaniline. The variation of the molecular and electronic structure of doped (as synthesized) conducting, dedoped base and redoped conducting forms of polyaniline prepared by enzymatic polymerization of aniline with and without the presence of the polyelectrolyte templates was studied by solid-state ^{13}C and ^{15}N NMR. The different templates—SPS, PVP, and dodecylbenzenesulfonic acid (DBSA)—were used for the synthesis of conducting polyaniline complexes [53, 77]. The highest conductivity, on the order of approximately 10 S/cm, was observed in the case of the PANI/DBSA complex. The conductivity-related structural differences were analyzed at different doping states. The doping forms of conducting polyaniline are shown in Scheme 13. The CP/MAS NMR spectra indicated that the redoped PANI showed entirely different features than the doped (as synthesized) form, although the redoping process can restore the conductivity. The charge distribution along the PANI backbone was more inhomogeneous compared to that of the doped (as synthesized) form. The sequential redoping of the dedoped PANI showed a transformation of amine to imine and a charge delocalization effect on the spectral features [78]. The charge carriers during conduction are believed to be bipolarons with dicationic lattices and polarons having semiquinone radical structures.

The enzymatically synthesized polyanilines were studied in self-doped conducting, dedoped base and redoped conducting forms by solid-state ^{13}C and ^{15}N CP/MAS NMR to understand the structural differences in cases where the synthesis was conducted with and without the presence of template and oxidative coupling positions [79]. This approach revealed a variation in the structural features and verified that a 1,4-coupling (head-to-tail) mechanism occurred in the presence of a template. The resulting polyaniline contained alternate benzenoid–quinoid repeating units in the backbone, which was favorable in the conducting form. A highly branched structure was proposed in the absence of a template. The dominant coupling mechanism of either C – C or C – N – C at the 1,2- and 1,4-positions was confirmed by NMR spectra. The branched structure model was shown in Scheme 14. The required benzenoid-quinoid repeating units were minor constituents of such polyaniline. The ^{15}N NMR indicated that the imine nitrogen species exhibited intramolecular hydrogen bonding with the neighboring benzenoid amine.

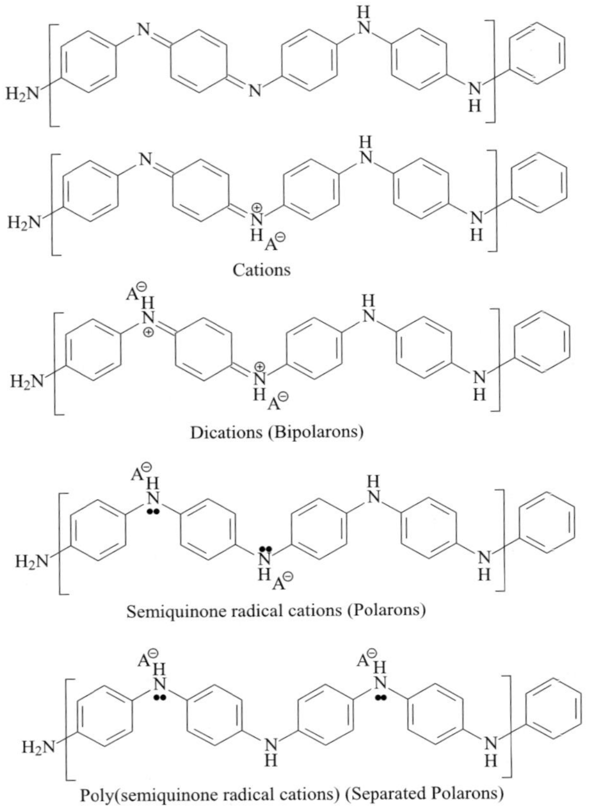

Scheme 13 Possible structures of polyaniline in base and conducting form

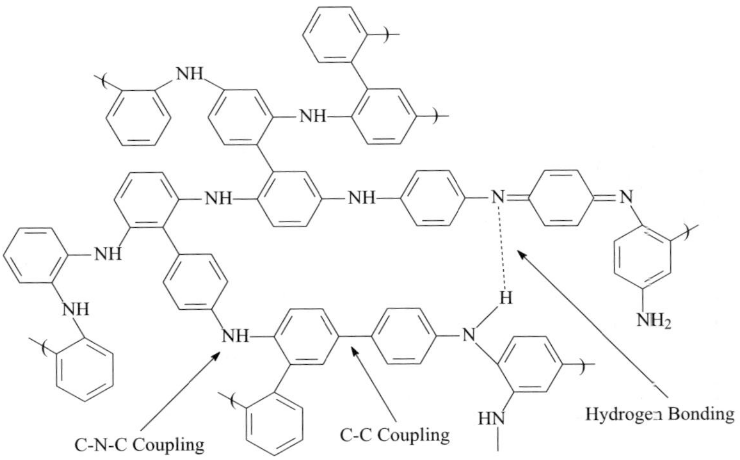

Scheme 14 Model for branched polyaniline structure showing C – C and C – N – C coupling at 1,2- and 1,4-positions as well as possible hydrogen bondings

7.2
Circular Dichroism Study

Chiral conducting polyaniline nanocomposites were synthesized by enzymatic polymerization of aniline/poly(acrylic acid) (PAA)/camphorsulphonic acid (CSA) complexes. Solid-state ^{13}C NMR studies found that the enzyme HRP, apart from being a biocatalyst, played an important role during the polymerization, which allowed PANI to prefer a specific helical conformation whether the induced chirality in the monomer–CSA complex was formed with (+) CSA or (–) CSA [80]. Poly(ethylene glycol)-modified hematin was used for a comparative study to examine the unique enzymatic structural specificity of HRP. Circular dichroism (CD) results without the helical conformation specificity found in the resulting PANI suggested that biomimetic catalyst hematin had no capacity to direct enantiospecificity for the PANI chains, which was a unique property for the enzyme HRP in these reactions [81]. The PAA and CSA templates were removed upon NH$_4$OH dedoping. Low conductivity of the nanocomposite was observed due to the presence of non-conducting domains which was supported by ^{13}C NMR analysis.

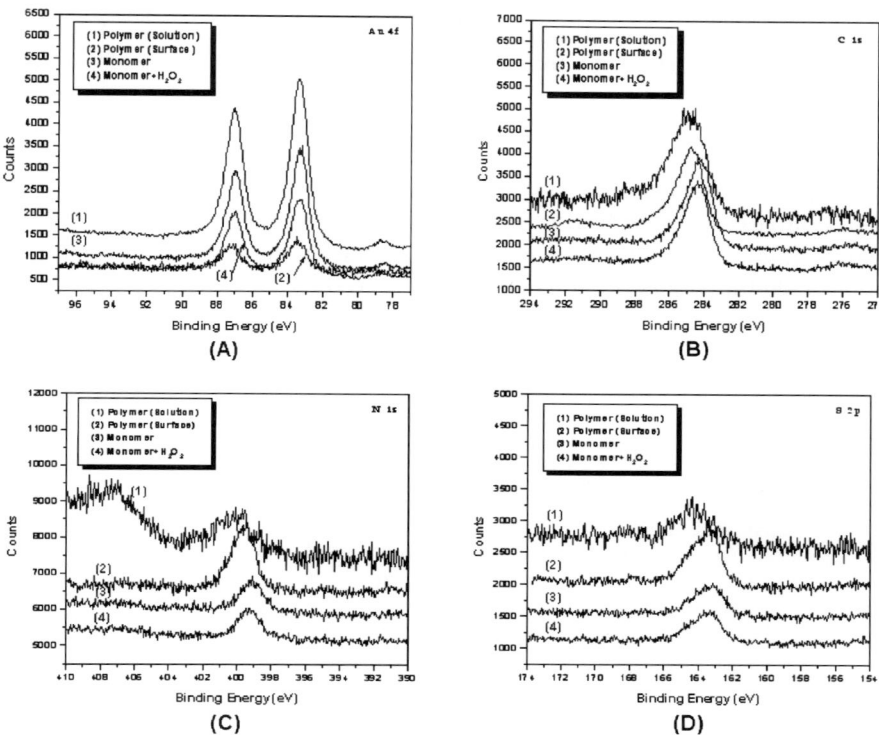

Fig. 3 XPS spectra for the 4-aminothiophenol monomer and its series of reactions on gold surfaces: **A** Au 4f, **B** C 1s, **C** N 1s, **D** S 2p

7.3
Other Analytical Methods

MALDI-TOF analysis of the HRP catalyzed *in situ* polymerization of 4-aminothiophenol measured a molecular weight of 1000 Da. XPS data reflected prominent shifts in the binding energy of N, C, and S with the polymer and supported the polymerization on the gold surface (Fig. 3).

8
Conclusions

During the past two decades, biocatalysis in polymer synthesis has sparked new insights into the synthesis and functional features of materials formed in aqueous ambient environments with better structural control. Template-assisted polyaniline synthesis results in water-soluble materials that are relatively easy to process, unlike similar materials produced by traditional synthetic means. More studies are needed to further elucidate the potential interactions between enzyme, template, substrate, and environmental surroundings. Dip-pen nanolithography in combination with biocatalysis has opened new opportunities to integrate polymers at the nanoscale, a step forward in making connections with biological components. Continuing improvements in biocatalysts and a better understanding of the reactions will continue to lead to new scientific approaches in the field with eventual technological impacts. Considering the current advantages of biocatalysis in polymer synthesis, it is likely that future efforts will focus on enzyme-assisted nanotechnologies to produce polymeric materials for a variety of applications, including medical and energy-related needs.

References

1. Genies EM, Hany P, Sanier CJ (1988) Appl J Electrochem 18:285
2. Naoi K, Ogano S, Osaka T (1988) Electrochem J Soc 135:C119
3. Jelle BP, Hagen G (1993) Electrochem J Soc 140:3560
4. Joo J, Epstein A (1994) J Appl Phys Lett 65:2278
5. Agbor NE, Cresswell JP, Petty MC, Monkman AP (1997) Sens Actuat B 41:137
6. Gustafsson G, Cao Y, Treacy GM, Colaneri N, Heeger AJ (1992) Nature 357:477
7. Wood AS (1991) Mod Plast August 68:47
8. MacDiarmid AG (1997) Synth Met 84:27
9. Westerweele W, Smith P, Heeger AJ (1995) Adv Mater 7:788
10. Verghese MM, Ramanathan K, Ashraf SM, Kamalasanan MN, Malhotra BD (1996) Chem Mater 8:822
11. Liu G, Freund MS (1997) Macromolecules 30:5660
12. Wei X, Epstein AJ (1995) Synth Met 74:123

13. Chen SA, Hwang GW (1994) J Am Chem Soc 116:7939
14. Chen SA, Hwang GW (1996) Macromolecules 29:3950
15. Shan J, Cao S (2000) Polym Adv Technol 11:288
16. Shan J, Han L, Bai F, Cao S (2003) Polym Adv Technol 14:330
17. Tripathy SK, Kim DY, Li L, Viswanathan NK, Balasubramanian S, Liu W, Wu P, Bian S, Samuelson L, Kumar J (1999) Synth Met 102:893
18. Alva KS, Lee TS, Kumar J, Tripathy SK (1998) Chem Mater 10:1270
19. Chen SA, Hwang GW (1995) J Am Chem Soc 117:10055
20. Nguyen MT, Kasai P, Miller JL (1994) Macromolecules 27:3625
21. Chan HSO, Ho HPK, Ng SC, Tan BTG, Tan KL (1995) J Am Chem Soc 117:8517
22. Shridhara K, Kumar J, Marx KA, Tripathy SK (1997) Macromolecules 30:4024
23. Degrand C, Limoges B, Martre AM, Schollhorn B (2001) Analyst 126:887
24. Bartlett PN, Birkin PR, Wang JH (1998) Anal Chem 70:3685
25. Alvarez S, Manolache S, Denes F (2003) J Appl Polym Sci 88:369
26. Hodgson M, Jones P (1989) J Biolumin Chemilumin 3:21
27. Navas Diaz A, Garcia Sanchez F, Gonzalez Garcia JA (1995) J Photochem Photobiol A: Chem 87:99
28. Navas Diaz A, Garcia Sanchez F, Gonzalez Garcia JA (1998) J Photochem Photobiol A: Chem 113:27
29. Skotheim TA, Elsenbaumer RL, Reynolds JR (1998) Handbook of Conducting Polymer. Marcel Dekker, New York, NY
30. Ryu K, McEldoon JP, Pokora AR, Cyrus W, Dordick JS (1993) Biotech Bioeng 42:807
31. Xu F (1996) Biochemistry 35:7608
32. Jakowski JD, Kaplan GM (2001) Int App PCT, WO01/49733A1
33. Aizawa M, Wang L, Shinohara H, Ikariyama Y (1990) J Biotechnol 14:301
34. Solomon EI, Lowery MD (1993) Science 259:1575
35. Yaropolov A, Skorobogat'ko OV, Vartanov SS, Varfolomeyev SD (1994) Appl Biochem Biotechnol 49:257
36. Solomon EI, Sundaram UM, Machonkin TE (1996) Chem Rev 96:2563
37. Akkara JA, Wang J, Yang DP, Gonsalves KE (2000) Macromolecules 33:2377
38. Ku BC, Lee SY, Liu W, He JA, Kumar J, Bruno FF, Samuelson LA (2003) Macromol J Sci Pure Appl Chem A40:1335
39. Ferreira ML (2003) Macromol Biosci 3:179
40. Ryabov AD, Kurova VS, Goral VN, Reshetova MD, Razumiene J, Simkus R, Laurinavicius V (1999) Chem Mater 11:600
41. Samuelson LA, Anagnostopoulos A, Alva KS, Kumar J, Tripathy SK (1998) Macromolecules 31:4376
42. Liu W, Kumara J, Tripathy S, Bruno FF, Senecal K, Samuelson L, Anagnostopoulos A (1999) Synth Met 101:738
43. Liu W, Kumar J, Tripathy S, Senecal KJ, Samuelson L (1999) J Am Chem Soc 121:71
44. Karamyshev AV, Shleev SV, Koroleva OV, Yaropolov AI, Sakharov IY (2003) Enzyme Microb Technol 33:556
45. Pina DG, Shnyrova AV, Gavilanes F, Rodriguez A, Leal F, Roig MG, Sakharov IY, Zhadan GG, Villar E, Shnyrov VL (2001) Eur J Biochem 268:120
46. Chattopadhyay K, Mazumdar S (2000) Biochemistry 39:263
47. Sakharov IY, Ouporov IV, Vorobiev AK, Roig MG, Pletjushkina OY (2004) Synth Met 142:127
48. Sakharov IY, Sakharova IV (2002) Biochim Biophys Acta 1598:115
49. Sakuragawa A, Taniai T, Okutani T (1998) Anal Chim Acta 374:191
50. Jin Z, Su Y, Duan Y (2001) Synth Met 122:237

51. Wang X, Schreuder-Gibson H, Downey M, Tripathy S, Samuelson L (1999) Synth Met 107:117
52. Nagarajan R, Tripathy S, Kumar J, Bruno FF, Samuelson L (2000) Macromolecules 33:9542
53. Sahoo SK, Nagarajan R, Samuelson LA, Kumar J, Cholli AL, Tripathy SK (2001) J Macromol Sci Pure Appl Che A38:1315
54. Nagarajan R, Roy S, Kumar J, Tripathy SK, Dolukhanyan T, Sung C, Bruno F, Samuelson LA (2001) Macromol J Sci Pure Appl Chem A38:1519
55. Nagarajan R, Liu W, Kumar J, Tripathy SK (2001) Macromolecules 34:3921
56. Roy S, Fortier JM, Nagarajan R, Tripathy S, Kumar J, Samuelson LA, Bruno FF (2002) Biomacromolecules 3:937
57. Bruno FF, Nagarajan R, Roy S, Kumar J, Samuelson LA (2003) J Macromol Sci Pure Appl Chem A40:1327
58. Kobayashi S, Uyama H, Kimura S (2001) Chem Rev 101:3793
59. Akkara JA, Salapu P, Kaplan DL (1992) Ind J Chem 31B:855
60. Liu W, Cholli AL, Nagarajan R, Kumar J, Tripathy S, Bruno FF, Samuelson L (1999) J Am Chem Soc 121:11345
61. Gospodinova N, Mokreva P, Terlemezyan L (1993) Polymer 34:1330
62. Gospodinova N, Terlemezyan L, Mokreva P, Kossev K (1993) Polymer 34:2434
63. Gospodinova N, Mokreva P, Terlemezyan L (1994) Polymer 35:3102
64. Gilabert MA, Hiner ANP, Garcia-Ruiz PA, Tudela J, Garcia-Molina F, Acosta M, Garcia-Canovas F, Rodriguez-Lopez JN (2004) Biochim Biophys Acta 1699:235
65. Rodriguez-Lopez JN, Gilabert MA, Tudela J, Thorneley RN, Garcia-Canovas F (2000) Biochemistry 39:13201
66. Regelsberger G, Jakopitsch C, Engleder M, Ruker F, Peschek GA, Obinger C (1999) Biochemistry 38:10480
67. Sanchez FG, Diaz AN, Garcia JAG (1995) J Lumin 65:33
68. Piletsky S, Piletska E, Bossi A, Turner N, Turner A (2003) Biotechnol Bioeng 82:86
69. Liu W, Kumar J, Tripathy S (2002) Langmuir 18:9696
70. Samuelson L, Liu W, Nagarajan R, Kumar J, Bruno FF, Cholli A, Tripathy S (2001) Synth Met 119:271
71. Bruno FF, Akkara JA, Kaplan DL, Sekher P, Marx KA, Tripathy SK (1995) Ind Eng Chem Res 34:4009
72. Bruno FF, Akkara JA, Samuelson LA, Kaplan DL, Mandal BK, Marx KA, Kumar J, Tripathy SK (1995) Langmuir 11:889
73. Takamuku S, Takeoka Y, Rikukawa M (2003) Synth Met 135–136:331
74. Xu P, Kaplan DL (2004) Advanced Materials 16:628
75. Xu P, Kaplan DL (2003) Polym Prep 44:948
76. Macdiarmid AG, Chiang JC, Richter AF, Epstein AJ (1987) Synthetic Metals 18:285
77. Sahoo SK, Nagarajan R, Roy S, Samuelson LA, Kumar J, Cholli AL (2002) Polym Mater Sci Eng 87:394
78. Sahoo SK, Nagarajan R, Roy S, Samuelson LA, Kumar J, Cholli AL (2004) Macromolecules 37:4130
79. Sahoo SK, Nagarajan R, Chakraborty S, Samuelson LA, Kumar J, Cholli AL (2002) J Macromol Sci Pure Appl Chem A39:1223
80. Thiyagarajan M, Kumar J, Samuelson LA, Cholli AL (2003) J Macromol Sci Pure Appl Chem A40:1347
81. Thiyagarajan M, Samuelson LA, Kumar J, Cholli AL (2003) J Am Chem Soc 125:11502

Enzymatic Synthesis of Polyesters via Ring-Opening Polymerization

Shuichi Matsumura

Faculty of Science and Technology, Keio University, 3-14-1, Hiyoshi, Kohoku-ku, 223-8522 Yokohama, Japan
matumura@applc.keio.ac.jp

1	Introduction	97
2	Ring-Opening Polymerization of Lactones into Polyesters	98
2.1	Ring-Opening Polymerization of Small-Sized Lactones to Macrolides and Cyclic Oligomers	98
2.2	Copolymerization of Lactones	104
2.3	Immobilization of Enzyme for Polymerization	105
2.4	Mechanism of Lipase-Catalyzed Ring-Opening Polymerization of Lactones	105
2.5	Responsibility of Catalytic Amino Acid Residues of the Enzyme	107
2.6	Polymerization of Lactones by PHB Depolymerase	108
3	Regioselective Ring-Opening Polymerization of Lactones	109
3.1	End-Functionalized Polyesters	109
3.2	Polymerization of Naturally Derived Lactones	110
3.3	Structurally Characteristic Polyesters	111
4	Enantioselective Ring-Opening Polymerization of Lactones	112
5	Polymerization of Cyclic Diacid Anhydrides with Diols or Oxiranes	115
6	Polycarbonate	116
6.1	Ring-Opening Polymerization of Trimethylene Carbonate	116
6.2	Ring-Opening Polymerization of Substituted Trimethylene Carbonate	118
6.3	Cyclic Carbonate Oligomers for Ring-Opening Polymerization	120
7	Polythioester and Polyphosphate via Ring-Opening Polymerization	120
7.1	Polythioester	120
7.2	Polyphosphate	121
8	Degradative Transformation into Cyclic Oligomers as Repetitive Chemical Recycling	122
8.1	Lipase-Catalyzed Transformation of Biodegradable Polymers into Cyclic Oligomers	122
8.2	Continuous Degradation of Polyesters Using the Enzyme-Packed Column	124
8.3	Poly(ester/Carbonate) Urethane	125
9	Green Solvent for Lipase-Catalyzed Polymerization and Degradation	127
10	Concluding Remarks	127
References		128

Abstract An enzyme can act as a powerful catalyst for the production and chemical recycling of green and sustainable polymers via a ring-opening polymerization. Also, enzymes will provide versatile synthetic tools for regio- and enantioselective polymerizations. In this article, lipase-catalyzed ring-opening polymerization and copolymerization of small-sized four-membered lactones to macrolides and cyclic oligomers having a molecular weight of a few hundred into high-molecular-weight polyesters, ring-opening polymerization of substituted/unsubstituted cyclic carbonates and cyclic phosphates into polycarbonates and polyphosphates, and the mechanism of lipase-catalyzed ring-opening polymerization of lactones and cyclic carbonates are reviewed. Also degradative transformation of polyesters and polycarbonates into the corresponding cyclic oligomers as repetitive chemical recycling and trends for enzymatic ring-opening polymerization and degradation are described.

Keywords Chemical recycling · Lactone · Lipase · Polyesters ·
Ring-opening polymerization

Abbreviations

β-BL	β-butyrolactone
BTMC	5-benzyloxy-1,3-dioxan-2-one
CCL	lipase from *Candida cylindracea*
CL	6-hydroxyhexanoate
ε-CL	ε-caprolactone
scCO$_2$	supercritical carbon dioxide
DCL	dicaprolactone
DDL	12-dodecanolide
DP	degree of polymerization
DTC	5,5-dimethyl-trimethylene carbonate
EM	enzyme-activated monomer
EEP	ethyl ethylene phosphate
(R, S)-4-EtCL	4-ethyl-caprolactone
3HB	3-hydroxybutanoate
K_m	Michaelis constant
lipase CA	immobilized lipase from *Candida antarctica*
MBC	5-methyl-5-benzyloxycarbonyl-1,3-dioxan-2-one
(R, S)-4-MeCL	racemic 4-methyl-caprolactone
M_n	number-average molecular weight
MOHEL	3-methyl-4-oxa-6-hexanolides
3MP	3-mercaptopropionic acid
MPL	α-methyl-β-propiolactone
M_w	weight-average molecular weight
11MU	11-mercaptoundecanoic acid
PBA	poly(butylene adipate)
PCL	poly(ε-caprolactone)
PDL	15-pentadecanolide
PDLLA	poly(DL-lactic acid)
PEG	poly(ethylene glycol)
PHB	poly(3-hydroxybutanoate)
PHA	poly(R-3-hydroxyalkanoate)
PhaZ*Afa*	PHB depolymerase from *Alcaligenes faecalis* T1
PLA	poly(lactic acid)

PLLA	poly(L-lactic acid)
β-PL	β-propiolactone
PPL	porcine pancrease lipase
SBD	substrate binding domain
SL	sophorolipid
T_m	glass transition temperature
TMC	trimethylene carbonate (1,3-dioxan-2-one)
UDL	11-undecanolide
V_{max}	maximal velocity

1
Introduction

An enzyme can act as a powerful catalyst for the production and chemical recycling of green and sustainable polymers via a ring-opening polymerization. Also, enzymes will provide versatile synthetic tools for regio- and enantioselective polymerizations. Lipase-catalyzed polymerization emerged about 20 years ago using the hydrolase enzyme. The early studies were generally carried out at ambient temperature. One of the events that has led to developing this new field of enzyme-catalyzed polymerization may be ascribed to the invention of an enzyme acting in organic medium at a high temperature. The pioneering work of Klibanov and coworkers implied that the enzyme-catalyzed reactions become novel and valuable tools in the field of both synthetic organic chemistry and polymer chemistry [1, 2]. Not only can the dry lipase withstand heating at 90–120° for many hours, but it exhibits a high catalytic activity at that temperature [1–6].

Enzyme-catalyzed polymerization has quickly developed as a novel methodology of polymer synthesis since the invention of the lipase-catalyzed ring-opening polymerization of lactones presented by two independent groups, Uyama and Kobayashi et al. [7, 8] and Knani et al. in 1993 [9]. Small- to large-sized (4- to 16-membered) lactones were found to be polymerized in a lipase-catalyzed ring-opening fashion, though their polymerizability varied according to ring size as well as enzyme origin (Scheme 1) [10]. These lactones could be polymerized by conventional chemical catalysts; however, there are some characteristic features in the lipase-catalyzed polymerization profiles. The large-sized lactones were efficiently polymerized by lipase. On the other hand, when using conventional anionic catalysts, the polymerizability of these large-sized lactones was much lower than that of the medium-sized ε-caprolactone (ε-CL) due to the lower ring strain. In general, similar to chemical catalysts, the lipase-catalyzed ring-opening polymerization of lactones gave both higher molecular weights and higher monomer conversions than the condensation polymerization of hydroxyacid/esters [11]. In organic solvents, cyclic oligomers were mainly produced [12]. When such a polymer was cleaved by lipase under water-restricted and diluted conditions, oligomer forms a cyclic

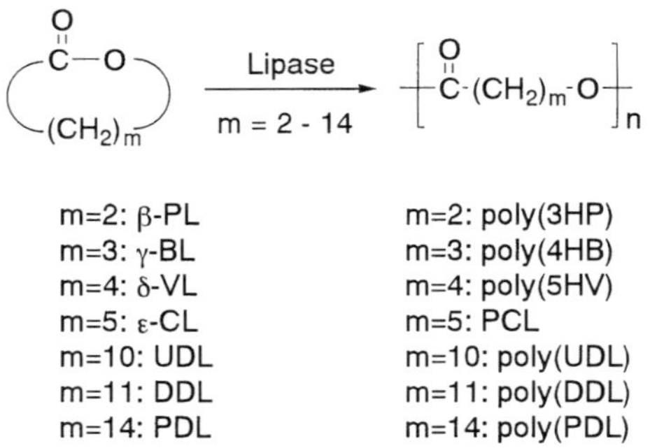

$$\text{cyclic } (CH_2)_m\text{-C(=O)-O} \xrightarrow{\text{Lipase} \atop m = 2 - 14} \text{–[C(=O)-(CH}_2)_m\text{-O]–}_n$$

m=2: β-PL	m=2: poly(3HP)
m=3: γ-BL	m=3: poly(4HB)
m=4: δ-VL	m=4: poly(5HV)
m=5: ε-CL	m=5: PCL
m=10: UDL	m=10: poly(UDL)
m=11: DDL	m=11: poly(DDL)
m=14: PDL	m=14: poly(PDL)

Scheme 1

structure were produced. The cyclic oligomer may have promising properties as monomers for the ring-opening polymerization and as versatile intermediates for the production of green polymers. Therefore, the enzymatic method may have significant possibilities for establishing a sustainable polymer recycling system. A rapidly increasing number of publications now exist that showcase the potential of in vitro enzyme catalysis to provide a wide range of polymer structures [13–21].

This article reviews the enzyme-catalyzed polymerization and chemical recycling of biodegradable aliphatic polyesters, polycarbonates, polythioesters, polyphosphates, and polysiloxane, particularly highlighted for the nascent green polymer chemistry.

2
Ring-Opening Polymerization of Lactones into Polyesters

Small- and medium-sized (4-, 6-, and 7-membered) and large-sized (12- to 16-membered) lactones as well as macrolides and cyclic oligomers were found to be efficiently polymerized in a lipase-catalyzed ring-opening fashion. In this section, enzyme-catalyzed ring-opening polymerization and copolymerization of lactones and cyclic oligomers are reviewed and mechanisms of lipase-catalyzed ring-opening polymerization and the responsibility of catalytic amino acid residues of the enzyme are also described.

2.1
Ring-Opening Polymerization of Small-Sized Lactones to Macrolides and Cyclic Oligomers

(1) Four-membered lactone (Scheme 2)
The unsubstituted four-membered β-propiolactone (β-PL), the smallest-sized lactone, was readily polymerized by lipase producing a polyester with

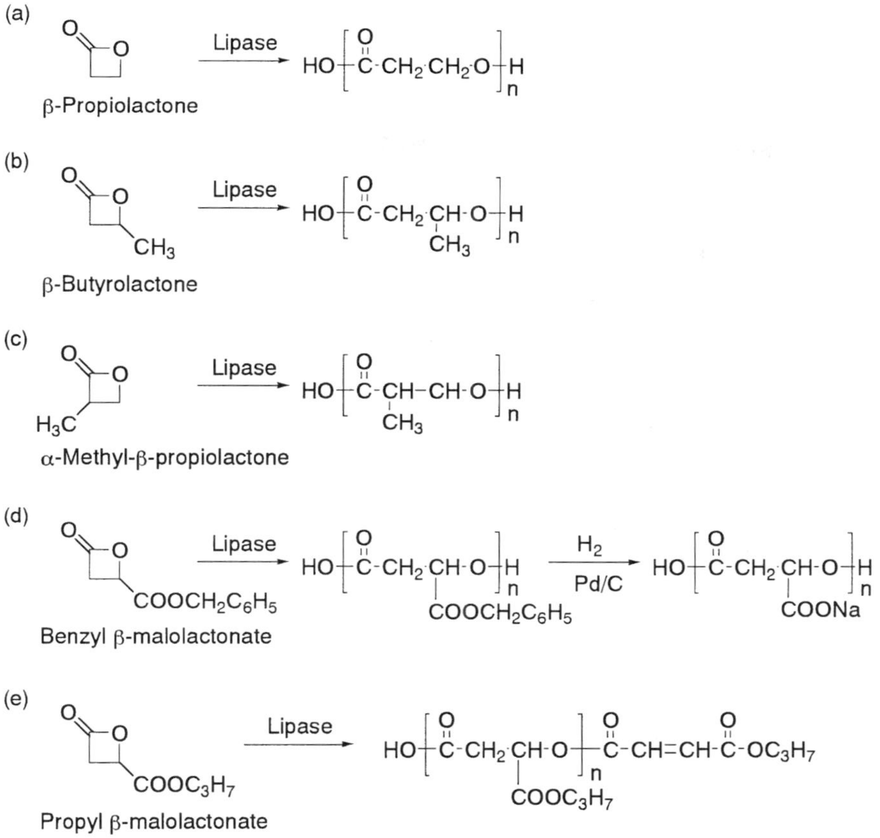

Scheme 2

(a) β-Propiolactone + Lipase → $HO\text{-}[C(O)\text{-}CH_2CH_2\text{-}O]_n\text{-}H$

(b) β-Butyrolactone + Lipase → $HO\text{-}[C(O)\text{-}CH_2CH(CH_3)\text{-}O]_n\text{-}H$

(c) α-Methyl-β-propiolactone + Lipase → $HO\text{-}[C(O)\text{-}CH(CH_3)\text{-}CH\text{-}O]_n\text{-}H$

(d) Benzyl β-malolactonate + Lipase → $HO\text{-}[C(O)\text{-}CH_2CH(COOCH_2C_6H_5)\text{-}O]_n\text{-}H$ → (H$_2$/Pd/C) → $HO\text{-}[C(O)\text{-}CH_2CH(COONa)\text{-}O]_n\text{-}H$

(e) Propyl β-malolactonate + Lipase → $HO\text{-}[C(O)\text{-}CH_2CH(COOC_3H_7)\text{-}O]_n\text{-}C(O)\text{-}CH=CH\text{-}C(O)\text{-}OC_3H_7$

a molecular weight of greater than 20 000 [22–25]. The four-membered β-lactones with various substituents, such as α-methyl-β-propiolactone [24], β-butyrolactone (β-BL) [25–29], benzyl β-malolactonate [30–34], and propyl β-malolactonate, [35] were polymerized by lipases to yield the corresponding polyesters. Poly(3-hydroxybutanoate) (PHB) obtained from the lipase-catalyzed ring-opening polymerization of (R, S) – β-BL presented three kinds of structural isomers, i.e., cyclic PHB, linear PHB with a hydroxyl group at one end and a carboxyl group at the other end, and a linear PHB with a crotonate group at one end and a carboxyl group at the other end, were identified. A small amount of the crotonate terminal was spontaneously produced by the partial elimination reaction at the labile acyl-enzyme intermediate as a side reaction of the ring-opening polymerization as shown in Scheme 3. Cyclic and hydroxy terminated linear PHBs were produced by the enzymatic intra- and intermolecular transesterification reactions of the produced polymer species [28, 29].

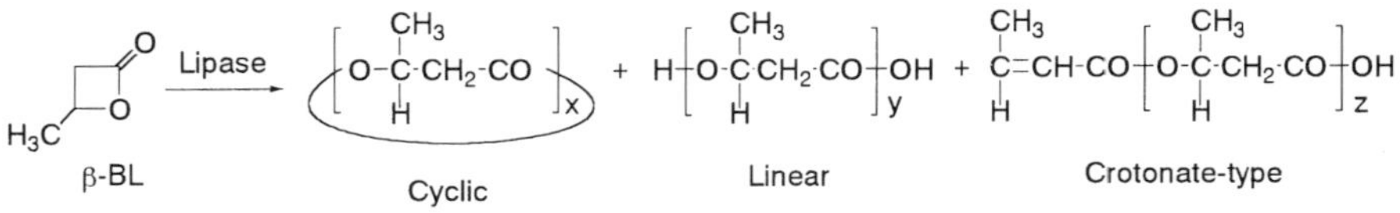

Scheme 3

Polymalate is attractive as a water-soluble biodegradable polycarboxylate in the biomedical and industrial fields. Polymalate could be produced by the ring-opening polymerization of benzyl β-malolactonate with subsequent dehydrogenation. The benzyl β-malolactonate was readily polymerized by the lipase to yield poly(benzyl β-malolactonate) with an M_w of greater than 7000. The benzyl group was removed to quantitatively produce poly(β-malic acid) and then neutralized to obtain the water-soluble and biodegradable poly(sodium β-malate) [33, 34]. The lipase-catalyzed polymerization of propyl β-malolactonate occurred in a characteristic fashion, such that the polymer chain length was controlled by the formation of double bonds due to the elimination of an α-hydrogen from the propyl β-malolactonate with the formation of a new initiator (Scheme 2e) [35].

(2) Five-Membered Lactone
The five-membered unsubstituted lactone γ-butyrolactone (γ-BL) may not polymerize when using a conventional chemical catalyst. However, it was reported that only an oligomer was produced by the ring-opening polymerization of γ-BL using PPL or lipase from *Pseudomonas* sp. [11, 36].

(3) Six-Membered Lactone (Schemes 4, 5)
The six-membered unsubstituted lactone δ-valerolactone was polymerized by lipase to produce polyesters with the relatively low molecular weight of less than 2000 [7, 37]. Also, α-methyl-δ-valerolactone was polymerized by lipase to produce the corresponding polyesters (Scheme 4b) [38]. Similarly, the six-membered 1,4-dioxan-2-one (Scheme 4c) [39] and six-membered cyclic depsipeptides, such as 3(S)-isopropylmorpholine-2,5-dione, 3-isopropyl-6-methyl-morpholine-2,5-dione, and 3(S)-sec-butylmorpholine-2,5-dione, were polymerized by lipase to produce the corresponding polymers (Scheme 5) [40–43]. Of these, poly(1,4-dioxan-2-one) is a biocompatible polymer with good flexibility and tensile strength (Scheme 4a). For medical applications, the metal-free polymerization of 1,4-dioxan-2-one by lipase may become the preferred method. Poly(1,4-dioxan-2-one) with an M_w of 41 400 was produced by an immobilized lipase from the *Candida antarctica* (lipase CA)-catalyzed ring-opening polymerization [39]. By the copolymerization of 3(S)-isopropylmorpholine-2,5-dione and lactide, the 3(S)-isopropylmorpholine-2,5-dione unit was introduced into the polylactide polymer chain in order to improve the physicochemical properties of the

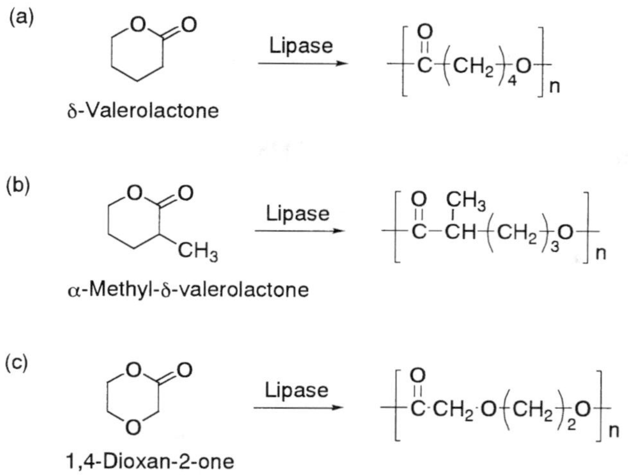

Enzymatic Synthesis of Polyesters via Ring-Opening Polymerization

Scheme 4

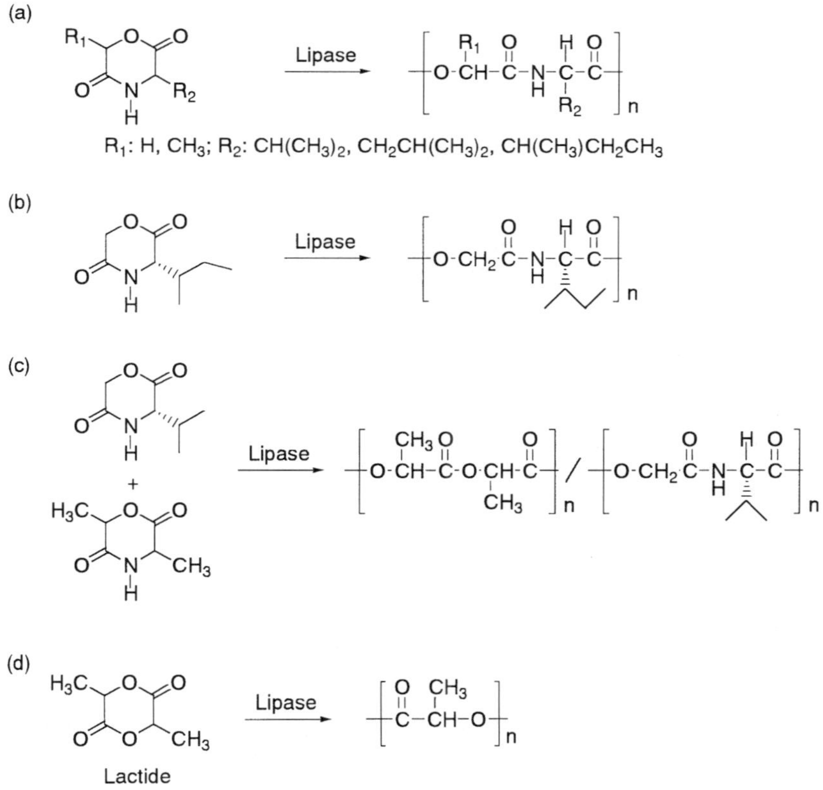

(a) δ-Valerolactone

(b) α-Methyl-δ-valerolactone

(c) 1,4-Dioxan-2-one

Scheme 5

(a) R_1: H, CH_3; R_2: $CH(CH_3)_2$, $CH_2CH(CH_3)_2$, $CH(CH_3)CH_2CH_3$

(d) Lactide

polylactides. The glass transition temperature (T_m) of the copolymers decreased with the increasing mole fraction of the DL-lactide residue in the copolymers from 74 to 40 °C [44].

The six-membered lactide, a cyclic dimer of lactic acid, was polymerized by lipase as a catalyst in bulk to yield the corresponding polylactide with an M_w of up to 126 000 with a relatively low yield. The D,L-lactide gave a higher molecular weight compared to the D,D- and L,L-lactides [45, 46]. The L,L-, D,D-, and D,L-lactides were copolymerized with trimethylene carbonate by porcine pancreatic lipase (PPL) to produce random copolymers having molecular weights of up to 21 000 [47].

(4) Seven-Membered Lactone

The seven-membered unsubstituted lactone ε-CL was the most extensively studied with respect to lipase-catalyzed ring-opening polymerization [7, 9, 37, 48–60]. ε-CL was quickly polymerized by various lipases of different origin. Of these, lipase CA was the most effective for the polymerization of ε-CL [61, 62], and under appropriate conditions PCL with a molecular weight (M_n) of greater than 89 000 was produced [50, 63]. During the polymerization of ε-CL, extensive degradation and polymerization occurred simultaneously [62]. Also, no termination and chain transfer occurred, and the molecular weight of PCL was a function of the monomer to initiator stoichiometry. Based on kinetic studies, the monomer consumption followed a first order rate law [48, 49, 64].

The α- and γ-methyl substituted seven-membered lactones α-methyl-ε-caprolactone and γ-methyl-ε-caprolactone were polymerized by lipase similar to the unsubstituted ε-CL. On the other hand, the polymerizability of ω-methyl substituted ε-caprolactone significantly decreased compared to the unsubstituted ε-caprolactone [65] (Scheme 6).

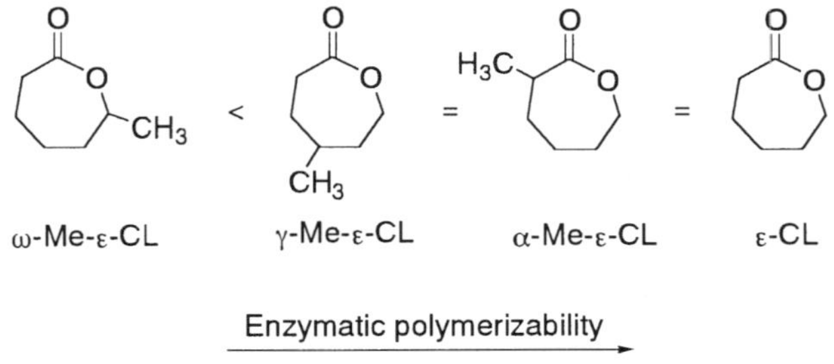

ω-Me-ε-CL < γ-Me-ε-CL = α-Me-ε-CL = ε-CL

Enzymatic polymerizability →

Scheme 6

(5) Macrolides

As the characteristic features of lipase-catalyzed ring-opening polymerization, macrolides were readily polymerized to produce high-molecular-weight polyesters that were not easily obtained by the conventional chemical catalysts. Uyama et al. first reported the lipase-catalyzed polymerization of macrolides, such as 11-undecanolide (UDL) and 12-dodecanolide (DDL) [66, 67]. The lipase-catalyzed ring-opening polymerization of macrolides was then extensively studied using 9- to 17-membered macrolides [63, 66, 68–73]. The polymerization behaviors of the macrolides were dependent on the position of the substituent as well as the ring size. The polymerizability of the α-methyl-substituted macrolides (13- and 16-membered) decreased by the introduction of the methyl substituent [65]. The bulk polymerization of the macrolide produced relatively high-molecular-weight polyesters, i.e., the polymerization of 15-pentadecanolide (PDL) using lipase CA produced the corresponding polyester with a molecular weight (M_n) of 34 400. At low reaction water levels, the immobilized lipase PS-30 on Celite catalyzed the polymerization of PDL to give poly(PDL) with an M_n of 62 000 [70]. The produced polyesters by the bulk polymerization had a carboxylic acid at one end and a hydroxy group at the other. Though a high-molecular-weight poly(PDL) was produced by the lipase-catalyzed ring-opening polymerization of PDL, its mechanical properties were evaluated. Poly(PDL) is a crystalline polymer that melts close to 100 °C and has a glass transition below room temperature. The physical properties of the polyester are somehow intermediate between those of PCL and polyethylene [74].

It is interesting that macrolides, such as UDL and DDL, are polymerized by lipase in an aqueous medium. That is, Kobayashi and Uyama et al. presented the polyester synthesis by dehydration polymerization in aqueous medium using lipase. This is due to the hydrophobic nature of the catalytic domain of the lipase as well as hydrophobic nature of the macrolide [10].

High-molecular-weight biodegradable polyester nanoparticles were prepared by the direct enzymatic polymerization of miniemulsions consisting of lactone nanoparticles. Stable miniemulsions consisting of PDL in water were obtained using a nonionic surfactant and ultrasonication. Polyester molecular weights of more than 200 000 were obtained using 0.125% lipase PS at 45 °C [75].

(6) Cyclic Oligomers

One of the characteristic features of lipase-catalyzed ring-opening polymerization is the formation of a cyclic oligomer. The formation of cyclic oligomers by the lipase-catalyzed polymerization of β-PL was first reported by Namekawa et al. [23]. The cyclic oligomers may be more readily polymerized by lipase to produce higher-molecular-weight polyesters. The ring-opening polymerization of cyclic diesters, ethylene dodecanedioate, and ethylene tridecanedioate was carried out using lipases under mild reaction conditions to give the corresponding polyesters with an M_n of several thousand (Scheme 7a) [76]. A cyclic ε-CL dimer with 14-membered ring was polymerized by lipase CA to quantitatively produce PCL with an M_n of 89 000 (Scheme 7b) [63]. Also, cyclic butylene adipate oligomers mainly consisting of a dimer showed excellent polymerizability by lipase CA to produce a high-

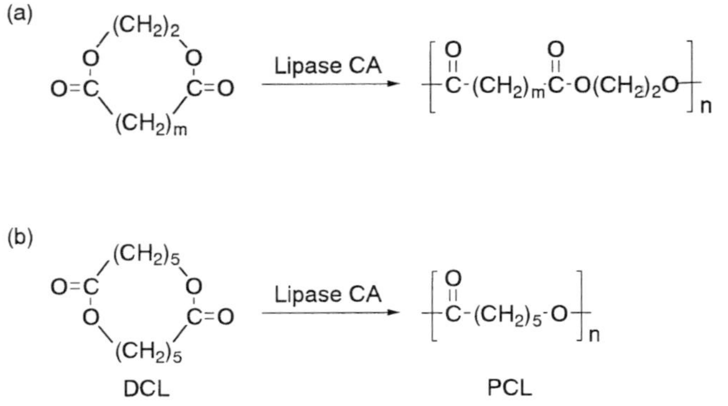

Scheme 7

molecular-weight polyester with an M_w of 52 000. Under the same conditions, the polycondensation of diethyl adipate and 1,4-butanediol produced a lower-molecular-weight polyester [77]. A larger-membered cyclic ester-urethane oligomer was also readily polymerized by lipase CA to produce a higher-molecular-weight poly(ester-urethane) with the highest molecular weight of 103 000 [78]. Based on these results, ring strain may not be primarily responsible for the ring-opening polymerization of cyclic macrolides.

2.2
Copolymerization of Lactones

Copolymerization is a convenient method of synthesizing copolymers with various properties for versatile applications. The enzymatic ring-opening copolymerization was carried out to produce some biodegradable polymers, such as poly(ester-co-carbonate)s [47], copolycarbonates [79] and polyethylene glycol-polyester block copolymer [80]. A series of PCL-PEG and PCL-PEG-PCL block copolymers were successfully synthesized for the first time using lipase CA. The block copolymers were all semicrystalline polymers with the PCL type crystalline structure. There was almost no composition and molecular-weight changes in the copolymers produced during the degradation [80].

Diblock copolymers, polybutadiene-*block*-polypentadecalactone, and polybutadiene-*block*-polycaprolactone were prepared by the monohydroxyl-terminated polybutadiene-initiated PDL and ε-CL, respectively, using immobilized *Candida antarctica* lipase B (Scheme 8) [81].

Microstructure analyses of the copolymers revealed that a statistically binary copolymer had four different diads, and the formation of random copolymers by lipase-catalyzed copolymerization was ascribed to the intermolecular transesterification of the polyesters during the copolymerization reaction [8, 69].

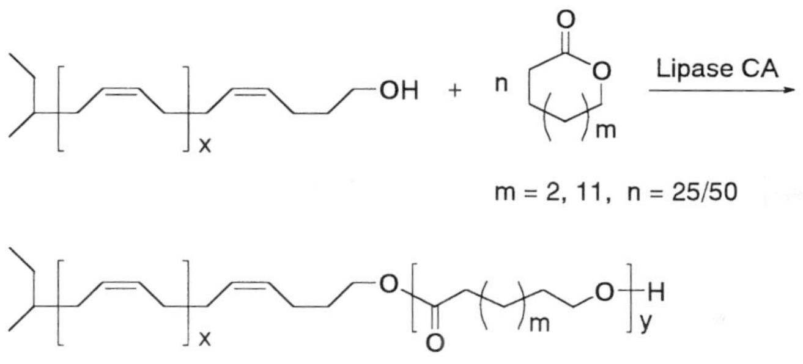

Scheme 8

m = 2, 11, n = 25/50

The ring-opening polymerization of lactones in the presence of aliphatic polyesters using lipase PF produced copolyesters with molecular weights of several thousands. The resulting polymer was a random copolymer consisting of both units [82]. Lactones were also copolymerized with divinyl esters and glycols using lipase to produce the copolyester, not a mixture of homopolymers. Two different modes of polymerization, ring-opening polymerization and polycondensation, simultaneously took place through enzymatic polymerization in one-pot to produce the copolyesters [83].

2.3
Immobilization of Enzyme for Polymerization

In order to improve the catalytic efficiency of the lipase for ring-opening polymerization, immobilized lipase on Celite [61, 84], and the surfactant-coated lipase were prepared and produced some characteristic results [85, 86]. The catalytic activity was further enhanced by the presence of a sugar or PEG during the immobilization. The surfactant-coated lipase efficiently catalyzed the ring-opening polymerization of lactones in organic solvents in which the modified enzyme was soluble [85, 86]. The *Candida antarctica* lipase immobilized on porous polypropylene more efficiently catalyzed the ring-opening polymerization of PDL as well as the polycondensation than lipase CA (*Candida antarctica* lipase immobilized on acrylic resin) [87].

2.4
Mechanism of Lipase-Catalyzed Ring-Opening Polymerization of Lactones

The mechanism for the lipase-catalyzed polymerization of lactones was first presented by Uyama et al. [67]. The proposed mechanisms for the lipase-catalyzed polymerization of lactones proceeded via an acyl-enzyme intermediate (enzyme-activated monomer, EM) at a serine residue of the catalytic site of lipase as the principal reaction course (Scheme 9) [67, 88–90]. The key step is the reaction of the lactone with lipase involving the ring opening of

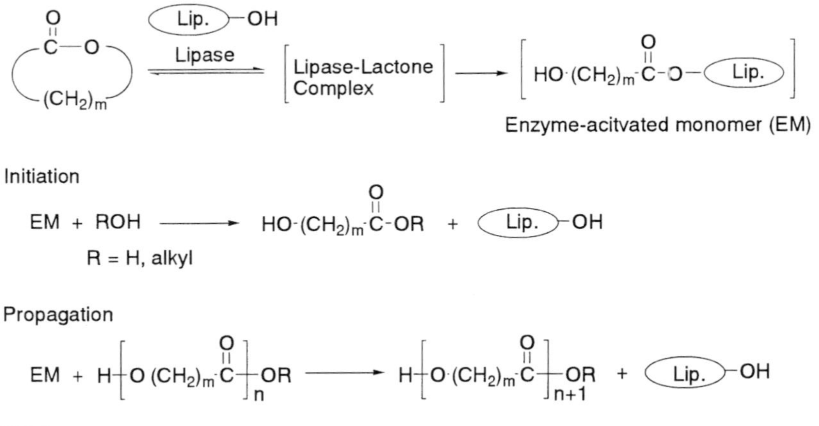

Scheme 9

[Scheme 9 shows lipase-catalyzed ring-opening polymerization mechanism: lactone + Lipase-OH ⇌ Lipase-Lactone Complex → HO-(CH₂)ₘ-C(O)-O-Lip. (Enzyme-activated monomer, EM)

Initiation: EM + ROH → HO-(CH₂)ₘ-C(O)-OR + Lip.-OH (R = H, alkyl)

Propagation: EM + H-[O(CH₂)ₘC(O)]ₙ-OR → H-[O(CH₂)ₘC(O)]ₙ₊₁-OR + Lip.-OH]

the lactone to produce the acyl-enzyme intermediate (EM). The initiation is the nucleophilic attack by water, which is probably contained in the enzyme, on the acyl carbon of the intermediate, yielding ω-hydroxycarboxylic acid, which is regarded as the basic propagation species. The EM is nucleophilically attacked by the terminal hydroxyl group of the growing polymer species during the propagation stage, leading to the formation of the additional one-unit elongated polymer chain. That is, the lipase-catalyzed polymerization proceeds via a monomer-activated mechanism. The rate-determining step of the overall polymerization is the formation of the EM. Similar mechanisms were also examined and verified [48, 91, 92].

Further studies on the mechanistic limitations in the lipase-catalyzed ring-opening polymerization of ε-CL in toluene with monomethoxy-poly(ethylene glycol) and water as initiators were carried out [93]. The apparent activation energy for the lipase-B-catalyzed ε-CL polymerization in toluene is estimated to be 2.88 kcal/mol. On the other hand, the activation energy for the aluminum-alkoxide-catalyzed ε-CL polymerization in toluene is 10.3 kcal/mol [89].

Polymerization of the macrolides proceeded much faster than ε-CL when compared under the same conditions using the same lipase. Michaelis-Menten kinetics of the 7-, 12-, 13-, 16-, and 17-membered lactones showed that the 17-membered 16-hexadecanolide had the highest enzymatic polymerizability [72]. The V_{max} and V_{max}/K_m for the macrolides increased with increasing ring size from the 7- to 17-membered ring. On the other hand, a significant change in the K_m value was not observed among these ring sizes, indicating that the affinity of the enzyme with macrolides was not dependent on the ring size. Thus, the high polymerizability is mainly ascribed to the V_{max}, i.e., the reaction process of the lipase-lactone complex to the EM is the key step of the polymerization [73]. Michaelis-Menten kinetics of the poly-

Table 1 Michaelis-Mentene kinetics of the polymerization of lactones with lipase PF [73, 90]

Monomer	$K_{m\ (lactone)}$ mol L^{-1}	$V_{max\ (lactone)}$ mol L^{-1} h^{-1}, ×10^2	$V_{max\ (lactone)}/K_{m\ (lactone)}$ h^{-1}, ×10^2
ε-CL	0.61	0.66	1.1
UDL	0.58	0.78	1.4
DDL	1.1	2.3	2.1
PDL	0.80	6.5	8.1
HDL	0.63	7.2	11

merization of lactones performed by Kobayashi and Uyama were summarized in Table 1 [73, 90].

The kinetics of the bulk polymerization of the 6-, 7-, 12-, 13-, 16-, and 17-membered lactones initiated by the zinc 2-ethylhexanoate/butyl alcohol system was studied and compared to that of the lipase-catalyzed polymerization. The ring strain, which decreases with increasing lactone size, is partially released in the transition state of the elementary reaction of the polyester chain growth, which eventually leads to a faster propagation for more strained monomers during the chemical polymerizations. During the enzymatic polymerizations, the rate-determining step involves the formation of the lactone-lipase complex. The latter reaction is promoted by the hydrophobicity of the lactone monomer, which is higher for the larger lactone rings [94].

2.5
Responsibility of Catalytic Amino Acid Residues of the Enzyme

Lipase-catalyzed ring-opening polymerization of lactones was carried out over a wide temperature range of between 40 and 120 °C. The reaction temperature of higher than 80 °C was greater than that used in the conventional enzymatic reactions. It has been reported that under relatively dry conditions, enzymes, such as lipases, proteases, and esterases, were catalytically active at temperatures around 90–120 °C [3–5]. Turner et al. demonstrated that lipase actually catalyzed the transesterification of octadecanol with palmityl stearate at 130 °C using lipase CA [6].

In order to further verify the involvement of catalytic amino acid residues in enzyme-catalyzed ring-opening polymerization, catalytic amino acids of the catalytic domain of the PHB depolymerase were replaced and evaluated for their polymerization activities. PHB depolymerases have structures that consist of a catalytic domain, a putative linker region, and a substrate binding domain (SBD). Three strictly conserved amino acids, serine, aspartate, and histidine, constitute the catalytic triad at the active center of the catalytic domain. The conserved serine is part of the so-called lipase-box pentapeptide (Gly – X_1 – Ser – X_2 – Gly) that has been found in all known serine hydrolases. Three kinds of site-specific mutants (S139A, ^{139}Ser was substituted to Ala; D214G, ^{214}Asp to Gly; H273D, ^{273}His to Asp) and the wild-type PHB depolymerase from *Alcaligenes faecalis* T1 have been compared with respect to the ring-opening polymerization of β-BL. As a result, BL was polymerized at 80° in bulk by the wild-type enzymes to yield polymers consisting of cyclic and linear structures with a high monomer conversion. In contrast, none of the mutant enzymes showed an obvious polymerization activity. These results demonstrated the involvement of the catalytic amino acid residues of PHB depolymerase in the enzyme-catalyzed polymerization even at 80 °C [95].

2.6
Polymerization of Lactones by PHB Depolymerase

Characteristic polymerizabilities of the lactones were observed by PHB depolymerase from *Alcaligenes faecalis* T1 with respect to the ring size of the lactones. That is, the 4-membered β-PL and 6-membered δ-valerolactone (δ-VL) showed the highest polymerization activity, and δ-VL seemed to be the upper limit for the molecular recognition of the narrow active site cleft of PhaZ*Afa*. On the other hand, ε-CL, UDL, and DDL, which showed excellent polymerization activities by the lipases, were scarcely polymerized by PhaZ*Afa*. These results are summarized in Table 2. The characteristic reactivities of the lactones were explained based on the tertiary structure model of the active site of PhaZ*Afa* [96]. The substrate-binding domain lacking PhaZ*Afa* showed higher polymerizabilities than PhaZ*Afa* for the polymerization of the lactones [97]. The PHB depolymerase from *Pseudomonas*

Table 2 Polymerization of lactones with PhaZ*Afa* and lipase from *Candida cylindracea* (CC)

Entry	Monomer	Enzyme	wt%	temp (°C)	time(h)	conv (%)	M_w	Refs
1	β-PL	PhaZ*Afa*	3	60	48	98	16400	[96]
2	β-PL	CC	3	60	48	87	40000	[22]
3	β-BL	PhaZ*Afa*	5	80	72	96	2000	[96]
4	β-BL	CC	5	80	120	73	3300	[27]
5	δ-VL	PhaZ*Afa*	5	80	48	90	3700	[96]
6	δ-VL	CC	50	60	120	77	2400	[37]
7	ε-CL	PhaZ*Afa*	10	80	48	19	3 mer	[96]
8	ε-CL	CC	10	75	6	92	18500	[7]
9	UDL	PhaZ*Afa*	5	80	48	0	–	[96]
10	UDL	CC	5	60	120	84	19700	[67]

PhaZ*Afa*: PHBDP from *Alcaligene faecalis*, CCL: lipase from *Candida cylindracea* *lemoignei* also catalyzed the polymerization of ε-CL, which resulted in an 18% conversion of the monomer in 48 h at 70 °C to oligomers with an average degree of polymerization (DP) of 4. Under identical conditions, a 30% conversion of the trimethylene carbonate to oligomers with an average DP of 12 was found. Based on these studies, the PHB depolymerase appeared promising for use in ring-opening polymerizations [98].

3
Regioselective Ring-Opening Polymerization of Lactones

Regioselectivity is one of the characteristic properties of enzymatic reactions, and this property can be used for lipase-catalyzed polymerization in the production of environmentally benign and functionally superior polymeric materials. ε-CL showed excellent polymerizability by lipase and is also industrially available as a raw material for the production of biodegradable polymers; therefore, extensive studies have been carried out using ε-CL and lipase. Such examples are the regioselective ring-opening polymerization of ε-CL with various nucleophiles, such as sugars, terminal functionalized polyester synthesis, and branched and block copolymer syntheses.

3.1
End-Functionalized Polyesters

End-functionalized polyesters were synthesized by the ring-opening polymerization of lactones in the presence of functional hydroxy, carboxy, and methylene groups. An alcohol could initiate the ring-opening polymerization of lactones by lipase [90, 99]. The lipase catalysis chemoselectively induced the ring-opening polymerization of 2-methylene-4-oxa-12-dodecanolide yielding a polyester having the reactive exo-methylene group in the main chain. The present polymer could not be obtained using a conventional chemical initiator [100]. The chemospecific ring-opening polymerization of α-methylenemacrolides having various groups, i.e., aromatics, ether, and

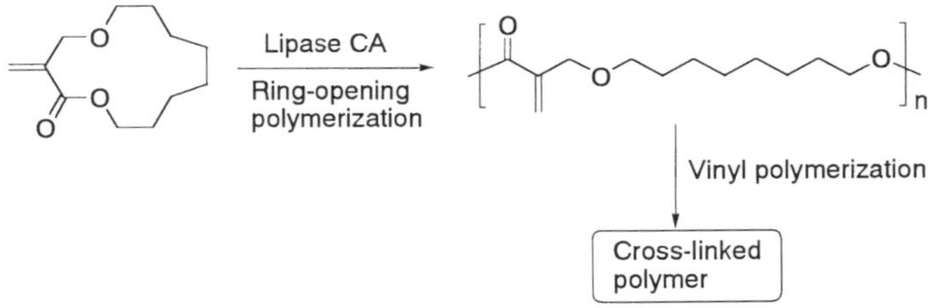

Scheme 10

amine, was enzymatically, anionically, and radically carried out. That is, polymerization with the lipase catalyst successfully yielded polymers only through the ring-opening process, whereas vinyl groups were further polymerized by anionic and radical initiators to produce a cross-linked polymer gel (Scheme 10) [100, 101].

3.2
Polymerization of Naturally Derived Lactones

Polymeric amphiphiles are attractive as pharmaceutical and hygienic products, particularly bearing sugar residues. The conventional selective preparation of sugar-containing polymers requires protection and deprotection of the sugar moieties. However, these methods lack selectivity and often compromise the concept of green chemistry. More straightforward and regioselective modification methods, such as esterification, are now needed. For such purposes, the enzymatic method will be effective.

Amphiphilic PCL macromonomers were prepared by first using the sugar-derived di- to hexahydroxyl group initiators for the regioselective ring-opening polymerization of ε-CL, followed by the selective functionalization of the hydroxyl end group of PCL by lipase (Scheme 11a) [102–104]. The hydrophilic hydroxyethyl cellulose was grafted by the hydrophobic ε-CL

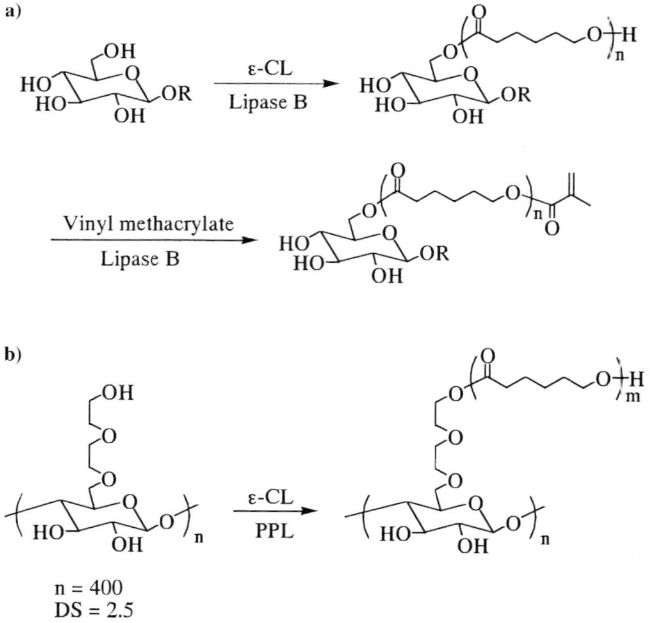

Scheme 11

polyester chain by the ring-opening graft copolymerization of ε-CL in bulk using PPL. The ε-CL could be directly introduced to the cellulose surfaces either by simple addition of a liquid monomer or through the gas phase using *Candida antarctica* lipase B as a new approach to introduce the ε-CL polymer chains into cellulosic materials [105]. The enzymatic polymerization of ε-CL with hydroxyethyl cellulose film produced hydroxyethyl cellulose-graft-PCL having a substitution degree of from 0.1 to 0.32 (Scheme 11b) [106].

Well-defined sophorolipid biosurfactant analogs were also prepared via enzymatic synthesis for the evaluation of the bioactivity and as building blocks for the preparation of glycolipid-based polymers [107]. The direct enzymatic ring-opening polymerization of lactone sophorolipids (SLs) was carried out with lipases. The reaction proceeded with the formation of monoacylated lactone SLs and the subsequent conversion of the intermediates to oligomers and polymers (Scheme 12) [108].

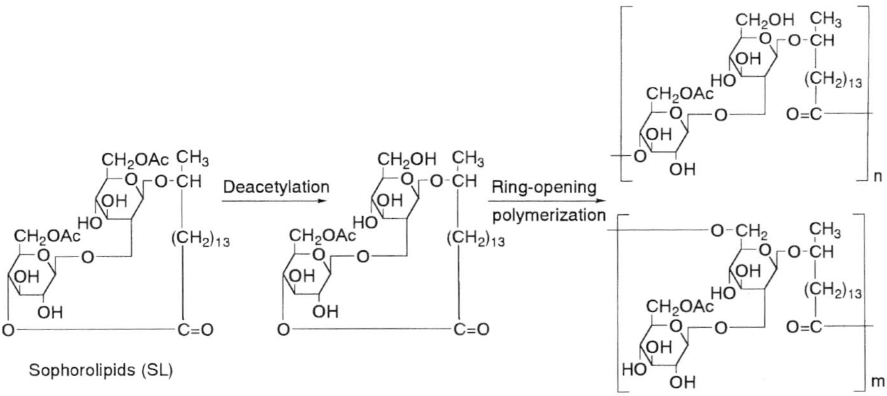

Scheme 12

3.3
Structurally Characteristic Polyesters

A star-shaped polymer was prepared by the lipase-catalyzed ring-opening polymerization of ε-CL and ethyl glucopyranoside as a mutifunctional initiator [109]. Polymers with a treelike structure are known as dendritic. A PCL monosubstituted first-generation dendrimer was prepared by the lipase-catalyzed polymerization of ε-CL and the selective acylation of the hydroxy end group of the PCL chain by lipase were carried out [110]. Hyperbranched aliphatic copolymers were prepared by the selective polymerization of ε-CL to the hydroxy group of 2,2-bis(hydroxymethyl)butyric acid using immobilized lipase B (Scheme 13) [111]. New brush copolymers comprising poly(PDL) with unusual thermal and crystalline properties were synthesized by chemoenzymatic methods. The 2-hydroxyethyl methacrylate-terminated

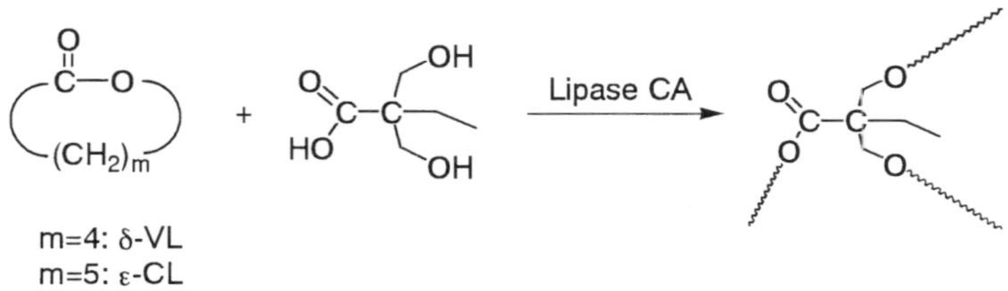

Scheme 13

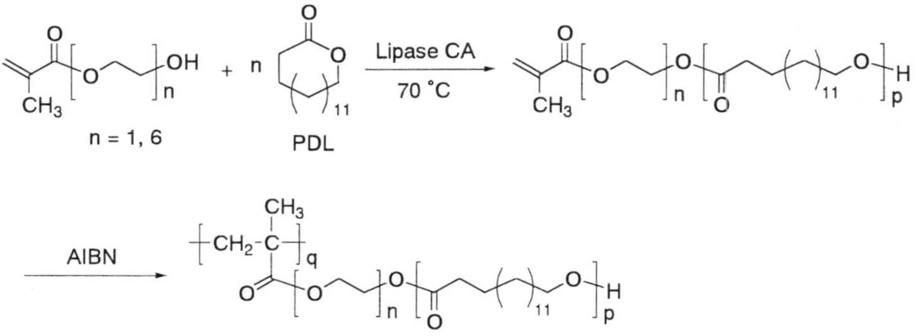

m=4: δ-VL
m=5: ε-CL

Scheme 14

n = 1, 6

PDL

poly(PDL) was synthesized by ring-opening polymerization of PDL to the hydroxy group of 2-hydroxyethyl methacrylate using lipase CA. The methacrylate moiety was polymerized by the radical initiator AIBN to produce novel brush-type copolymers (Scheme 14) [112].

4
Enantioselective Ring-Opening Polymerization of Lactones

Enantioselection for racemic lactones was exhibited by the lipase-catalyzed ring-opening polymerization of lactones to produce optically active polymers and optically active lactones that remained as unreacted material. Usually, lipase can react with racemic lactones by producing the acyl-enzyme intermediates; however, their rate constants differed between the two enantiomers. The enantioselectivity for racemic lactones are varied according to the enzyme origin as well as the reaction conditions, such as monomer conversion, monomer concentrations, solvent, and temperatures. The enantioselectivity toward racemic lactones was not always high but occurred in diluted solution and at a low monomer conversion, because the concentration of the opposite enantiomer increased with the polymerization. There are some strategies to increase the optical purity of the resulting polymer by lipase-catalyzed ring-opening polymerization. A more diluted monomer solution, lower monomer conversion, and enzyme selection will be effective. Also, the copolymerization of racemic lactones with achiral lactones are effective.

The four-membered racemic α-methyl-β-propiolactone (MPL) was enantioselectively polymerized by lipase from *Pseudomonas fluorescens* to produce (S)-enriched poly(MPL) with an M_n of 2900 [24]. Also, the four-membered racemic β-BL was enantioselectively polymerized by thermophilic lipase to produce the (R)-enriched polyesters having a maximum e.e. of 40% and having a relatively low molecular weight of $M_w = 900$ [26]. A highly (S)-enriched substituted PCL was prepared using the seven-membered substituted ε-CL. That is, racemic 4-methyl-caprolactone [(R, S)-4-MeCL] and 4-ethyl-caprolactone [(R, S)-4-EtCL] were enantioselectively polymerized by lipase to produce the corresponding polyesters having > 95% e.e. and having molecular weights of about 5000 (Scheme 15) [113]. Both the (R)- and (S)-3-methyl-4-oxa-6-hexanolides (MOHELs) were polymerized by lipase; however, the apparent initial rate of the S-isomer was seven times faster than that of the R-isomer [88].

Copolymerization of racemic lactones with an achiral lactone is an effective way to prepare a chiral polyester. The enantioselective copolymerization of racemic lactones with achiral lactones, such as ε-CL and DDL, has been established by lipase. As a typical example, β-BL and ε-CL were copolymerized by lipase in isooctane to produce the S-enriched optically active copolyester with 69% e.e. of the β-BL unit, and R – β-BL with 100% e.e. remained unreacted. This indicated that the S-isomer of β-BL preferentially reacted during the copolymerization (Scheme 16a) [114]. Also, during the copolymerization of the racemic β-BL with DDL, the (S)-isomer of β-BL preferentially reacted to give the (S)-enriched optically active copolymer with an enantiomeric excess of the β-BL unit of 69%, which is much higher than that for the homopolymerization of β-BL. δ-CL was also enantioselectively copolymerized

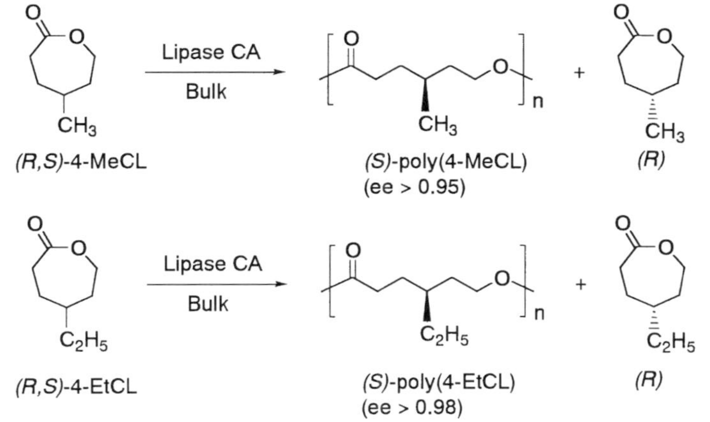

Scheme 15

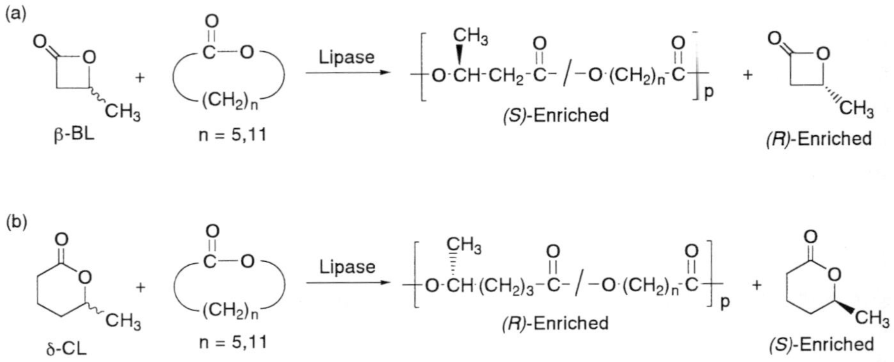

Scheme 16

with DDL by lipase to give the optically active polyester with the highest e.e. value of 76% (Scheme 16b) [114].

The enantioselective polymerization of $(R, S) - \beta$-BL was observed during the early stage of the reaction. The R-enantiomer of β-BL produces the natural-type poly(R-3HB), which is depolymerized by the PHB depolymerase of the environmental degrading microbes, and the S-enantiomer produces the unnatural-type poly(S-3HB), which cannot be depolymerized by the PHB depolymerase. When racemic β-BL was used for the ring-opening polymerization, the R-enantiomer of β-BL was preferentially incorporated into the enzyme to produce the R-3HB-rich polymer. However, the R-enantiomeric purity decreased with increasing S-enantiomer concentration of the monomer mixture [96]. On the other hand, when lipase was used instead of the PHB depolymerase for the ring-opening polymerization of the racemic β-BL, optical purity of the producing PHB was significantly lower because the stereospecificity of lipase for β-BL may be significantly lower than that of the PHB depolymerase.

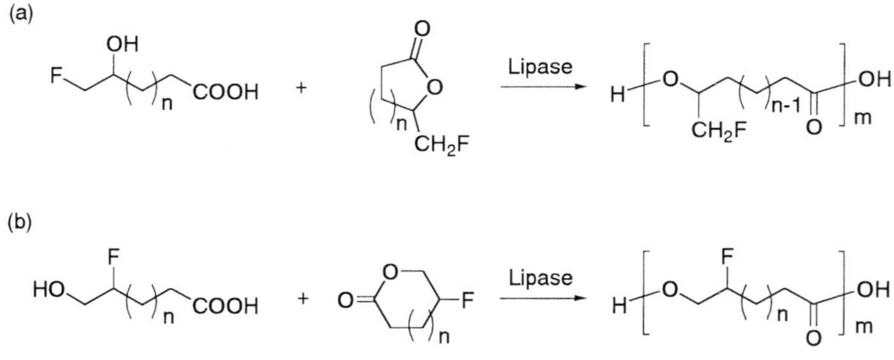

Scheme 17

Fluorinated lactones, 10-fluorodecan-9-olide, 11-fluoroundecan-10-olide, 12-fluorododecan-11-olide, and 14-fluorotetradecan-13-olide, were polymerized by lipase to produce the optically active polyesters with an M_w of 3000 to 8000, while 10-fluorodecan-11-olide gave an optically inactive polymer with an M_w of 11 000 by lipase-catalyzed ring-opening polymerization (Scheme 17) [115].

5
Polymerization of Cyclic Diacid Anhydrides with Diols or Oxiranes

A new type of enzymatic polymerization of the ring-opening addition-condensation polymerization of a cyclic diacid anhydride and a glycol was presented. By using diacid anhydride instead of free diacid, the liberating water is reduced by polycondensation leading to a favorable equilibrium shift for polymerization. Succinic anhydride and glutaric anhydride with diols were polymerized by lipase in an organic solvent to produce the corresponding polyesters (Scheme 18a) [116]. Novel biodegradable polyesters bearing functional groups were obtained by lipase-catalyzed ring-opening copolymerization of oxiranes and succinic anhydride (Scheme 18b,c). Oxiranes, such as glycidyl phenyl ether, benzyl glycidate, glycidyl methyl ether, and styrene oxide, were copolymerized with dicarboxylic anhydrides, such as suc-

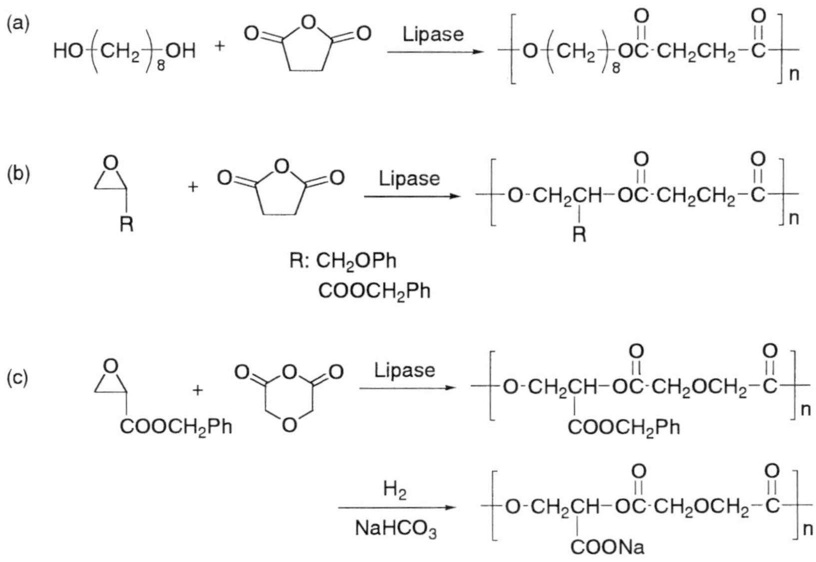

Scheme 18

cinic anhydride, phthalic anhydride, and maleic anhydride, by the action of an enzyme in a stepwise reaction to produce the corresponding polyesters containing ether linkages having a maximum M_w of 13 500 [117–120]. Benzyl glycidate and dicarboxylic anhydrides were polymerized by lipase followed by debenzylation to yield water-soluble ester-type polycarboxylates (Scheme 18c). They showed an excellent biodegradablility and calcium sequestration capacity [120].

6
Polycarbonate

The carbonate linkage in the aliphatic polymer chain may be expected to be enzymatically hydrolyzable and more hydrolytically stable than an ester linkage. The first synthesis of polycarbonate was carried out by the thermal polymerization of trimethylene carbonate (TMC: 1,3-dioxan-2-one) by Carothers et al. in 1932. Recently, it was revealed that aliphatic polycarbonates showed a good biodegradability, biocompatibility, and low toxicity, and extensive studies have been started. The biodegradation of poly(TMC) has been studied in reference to biomedical applications, such as sutures with improved flexibility and drug delivery systems [121–125]. They are prepared by the ring-opening polymerization of cyclic carbonate monomers by anionic, cationic, and coordination catalysts. However, the polymerization of cyclic carbonates often accompanies the decarboxylation reaction by the chemical catalysts resulting in ether linkages. More recently, the enzyme-catalyzed synthesis of polycarbonates has been reported in which no undesirable decarboxylation occurred. There are two routes to the synthesis of polycarbonates, i.e., by ring-opening polymerization of a cyclic carbonate diester and polycondensation of a carbonate and diol [126–131].

During condensation polymerizations, removal of the leaving group is necessary to shift the equilibrium in favor of the polymerization. On the other hand, no leaving group is produced during ring-opening polymerization, which favors a high-molecular-weight polycarbonate. In this section, the syntheses of polycarbonates via the lipase-catalyzed ring-opening polymerization of a cyclic carbonate monomer are reviewed.

6.1
Ring-Opening Polymerization of Trimethylene Carbonate

The lipase-catalyzed ring-opening polymerization of a six-membered cyclic carbonate was first reported by three independent groups in 1997 [132–134]. Six-membered cyclic TMCs with/without a methyl substituent using lipase were polymerized in the presence of lipase at a temperature between 60 and 100° to yield the corresponding polycarbonates (Scheme 19) [64]. No

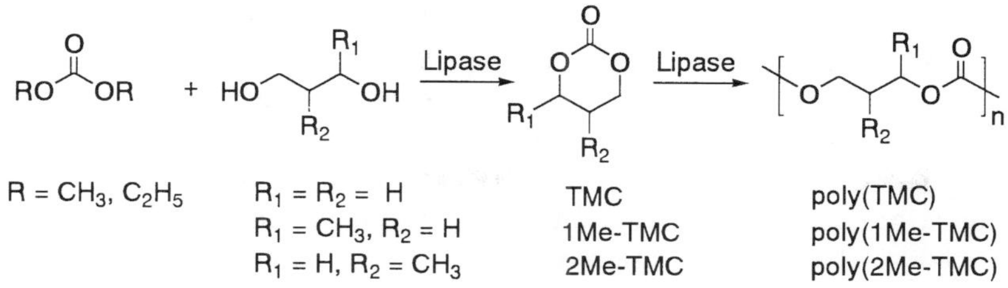

Scheme 19

elimination of carbon dioxide by the enzymatic polymerization of TMC was detected [132].

The mechanism for the proposed lipase-catalyzed ring-opening polymerization of cyclic carbonates is basically the same as that for lactones (Scheme 20) [48, 67, 134]. The initiation step is the formation of an enzyme-activated monomer (EM) by the ring-opening reaction of a cyclic carbonate and a serine residue of the lipase. EM was then nucleophilically attacked by water containing the enzyme to produce diols liberating carbon dioxide due to the decarboxylation of the transiently produced monocarbonate ester. In the propagation step, EAM successively reacts with the diols, forming the high-molecular-weight polycarbonate and regenerating the free lipase.

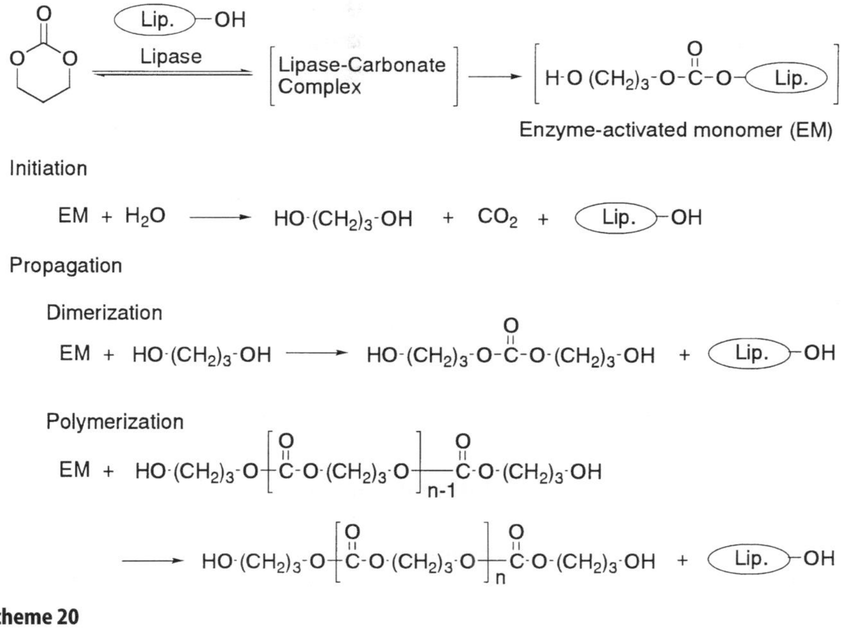

Scheme 20

Copolymerization is the one way to synthesize polymeric materials with desired properties and functions. The cyclic carbonate monomers are successfully copolymerized with various cyclic monomers, such as cyclic carbonates, lactones with/without substituents, lactide, and cyclic phosphates. TMC was copolymerized with lactide by PPL to produce poly(lactide-co-TMC)s having carbonate contents from 0 to 100% and having molecular weights of up to 21 000. The glass transition temperature (T_g) of the copolymer was dependent on the carbonate content, and the T_g values linearly decreased from 35° (polylactide) to −8° [poly(TMC)] [47]. TMC was also copolymerized with medium to large ring-sized lactones. As an example, TMC was copolymerized with PDL in toluene by lipase CA at 70 °C to yield random copolymers [135]. All the poly(PDL-TMC)s were highly crystalline, even those with an equimolar comonomer content and close-to-random distribution. Thermal stability improves with randomization of the comonomer distribution [136].

The cyclic TMC monomer with/without a methyl substituent was synthesized using dimethyl or diethyl carbonate and aliphatic 1,3-diols, such as 1,3-propanediol and 2-methyl-1,3-propanediol, using immobilized lipase CA in an organic solvent (Scheme 19) [137]. TMC is also produced by the enzymatic degradation of poly(TMC) using lipase as the chemical recycling product [138]. That is, the enzymatic degradation of poly(TMC)s having an M_n of 3000 ∼ 48 000 using lipase CA in acetonitrile at 70 °C afforded the corresponding cyclic monomer, TMC, in a yield of up to 80% [137]. Thus the obtained TMC was readily polymerized again by lipase. These results are summarized in Scheme 21 [137, 138].

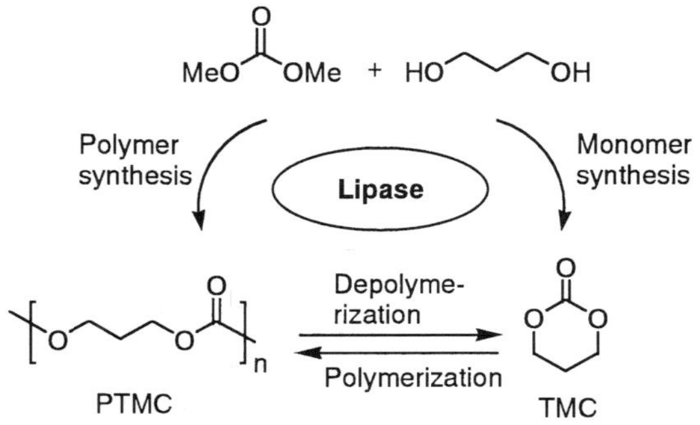

Scheme 21

6.2
Ring-Opening Polymerization of Substituted Trimethylene Carbonate

The new route to aliphatic polycarbonates decorated with pendant carboxyl groups was revealed by the copolymerization of 5-methyl-5-benzyloxycarbonyl-1,3-dioxan-2-one (MBC) with TMC using lipase AK by Al-Azemi et al.

This is the first example of polycarbonates with a carboxyl group as a pendant. The microstructures and thermal properties of poly(MBC-co-TMC) copolymers with different monomer ratios were presented [139–142]. Lipase-catalyzed ring-opening homopolymerization of MBC was also reviewed [143]. The regioselective oligomerization of TMC to glucoside was carried out using lipase. TMC was oligomerized with ethyl glucopyranoside by lipase CA to produce glucopyranoside and oligo(TMC) conjugates [144].

In order to enhance the catalytic efficiency of lipase for the polymerization of cyclic carbonates, lipase was immobilized on porous silica beads and microparticles. Six-membered 5-benzyloxy-1,3-dioxan-2-one (BTMC) was successfully copolymerized with TMC and 5,5-dimethyl-trimethylene carbonate (DTC) by porcine pancrease lipase (PPL) immobilized on silica particles. DTC was also successfully polymerized by the narrow distribution of micron-sized glass beads [79, 145]. The highest molecular weight (M_n = 26 400) of poly(BTMC-co-DTC) was obtained at around a 0.1% concentration of immobilized PPL on silica particles with a size of 75–150 μm. The produced poly(BTMC-co-DTC) was debenzylated to yield a water-soluble polycarbonate having pendant hydroxy groups on the main chain, which have potential biomedical applications [79, 145–147]. DTC was also successfully polymerized by the PPL immobilized on a narrow distribution of micron-sized glass beads. The immobilized PPL showed outstanding recyclability [145].

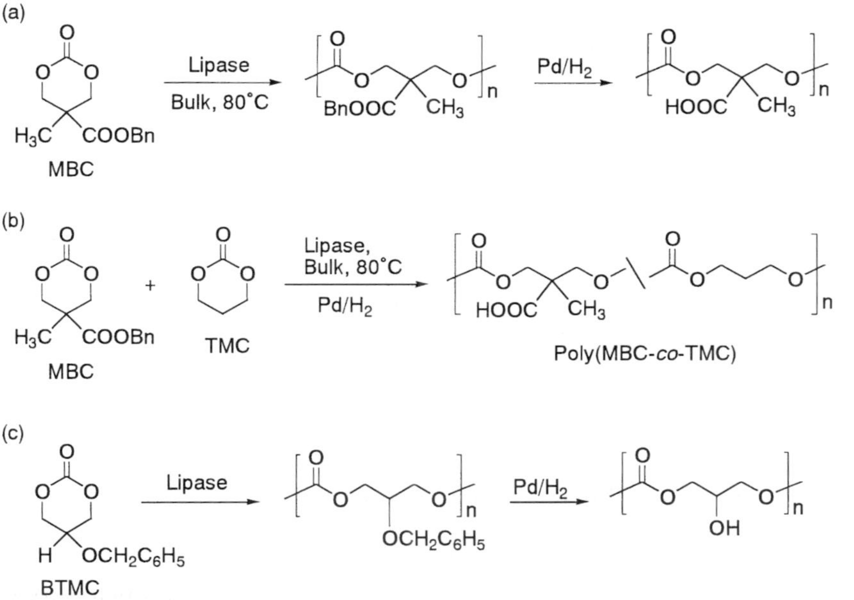

Scheme 22

6.3
Cyclic Carbonate Oligomers for Ring-Opening Polymerization

Cyclic carbonate oligomers act as a novel starting material for the production of polycarbonates using lipase-catalyzed polymerization and also as a low-molecular-weight material for chemical recycling. The cyclic carbonate oligomer often shows better polymerizability than that of conventional cyclic monomers. Novel cyclic dicarbonates were polymerized by lipase to produce the corresponding polycarbonates. The ring-opening polymerization of cyclic dicarbonates cyclobis(hexamethylene carbonate) and cyclobis(diethylene glycol carbonate) and the enzymatic copolymerization with lactones, such as 12-dodecanolide, were carried out using lipase CA (Scheme 23) [148].

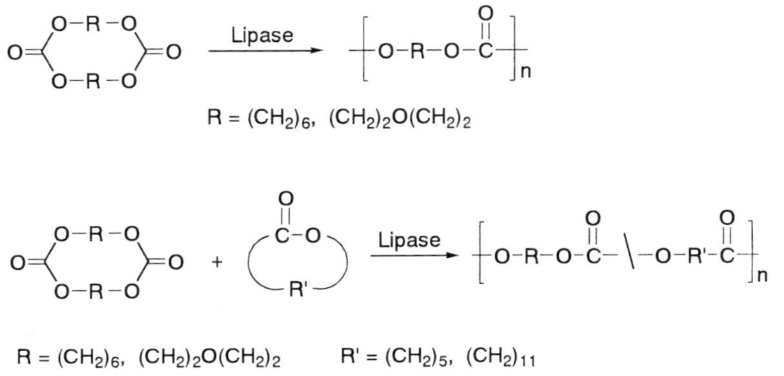

Scheme 23

7
Polythioester and Polyphosphate via Ring-Opening Polymerization

Enzyme-catalyzed ring-opening polymerization should open a new frontier for more types of polyesters, such as structural and elemental variations. Such examples may be the enzyme-catalyzed preparation of thio- and phospho-polyesters.

7.1
Polythioester

The synthetic polythioester was reported 50 years ago, but no commercial production of the polythioester has yet been partially established due to the complex preparation methods [149, 150]. Enzyme-catalyzed polymerization may become one of the environmentally benign methods for industrial applications. Microbial polyesters containing thioester linkages in their backbone were recently reported for the first time by Lütke-Eversloh and Steinbüchel et al. and revealed a novel class of biopolymers [151–153]. The first example was the copolymer of 3-hydroxybutyrate and 3-mercaptopropionate. Aliphatic polyesters containing thioester linkages were enzymatically prepared by the copolymerization of a lactone with mercaptoalkanoic acid as shown in Scheme 24a [154]. ε-CL and mercaptoalkanoic acid were reacted under atmospheric pressure to produce an oligomer, and further polymerization occurred by reducing the pressure by transesterification reaction for the equilibrium shift toward polymerization. The enzymatic copolymerization of ε-CL with 11-mercaptoundecanoic acid (11MU) and 3-mercaptopropionic acid (3MP) using immobilized lipase from *Candida antarctica* at a reduced pressure afforded the corresponding poly(ε-CL)-containing thioester moieties having M_ws of 22 000 and 20 000, respectively. It was also reported that 11MU was homopolymerized by lipase CA (Scheme 24b) [155].

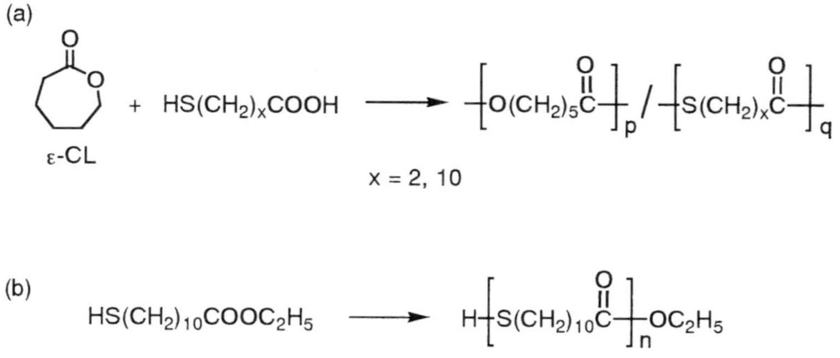

Scheme 24

7.2
Polyphosphate

Polyphosphates are of great interest in a variety of applications including flame retardation, adhesion promotion, plasticizing, and processability either by their solubility in common solvents or by the lowering of the T_g. The enzyme-catalyzed ring-opening polymerization of a cyclic phosphate (ethylene isopropyl phosphate) was achieved in bulk using PPL as shown in Scheme 25 [156]. The ring-opening polymerization of ethylene isobutyl phosphate was also performed using immobilized lipase on porous silica beads at 70 °C to yield the corresponding polyphosphate with M_n values ranging from 1642 to 5783 [157]. The novel enzymatic ring-opening copolymerization of ethyl ethylene phosphate (EEP) and TMC was performed in bulk at 100 °C using PPL or *Candida rugosa* lipase to yield random copolymers having molecular weights ranging from 3200 to 10 200. The degradability of the

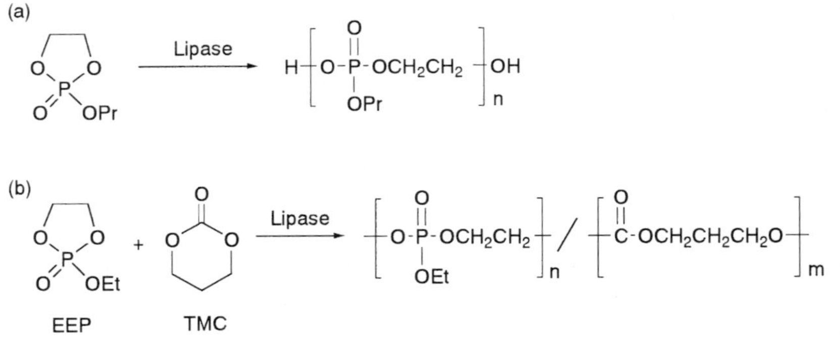

Scheme 25

copolymers was improved by the introduction of an EEP unit to the copolymer chain [158].

8
Degradative Transformation into Cyclic Oligomers as Repetitive Chemical Recycling

The establishment of a sustainable polymer production and recycling system using an enzyme may contribute to the recent environmental and energy problems including carbon dioxide emission. From the standpoint of saving the limited carbon resources and energy consumption for polymer production and polymer processing, even naturally derived and biodegradable plastics should be recycled as much as possible. Among the methods for polymer recycling, the enzymatic method may have the greatest potential for establishing a sustainable polymer recycling system. Polymers containing enzymatically hydrolyzable moieties, such as esters and carbonate esters, are degraded into oligomers by an enzyme, and under water-limited and diluted conditions cyclic oligomers may be produced that are repolymerizable by the enzyme. In this section, three kinds of commercially available biodegradable polymers, bio-based polymers, microbial polyesters, and chemically synthesized polyesters, are reviewed with respect to chemical recycling.

8.1
Lipase-Catalyzed Transformation of Biodegradable Polymers into Cyclic Oligomers

The lipase CA-catalyzed degradation of PCL with a molecular weight of around 40 000 readily took place in toluene at 60 °C to give oligomers with a molecular weight of less than 500. The produced oligomer was polymerized in bulk by the same lipase. The degradation-polymerization could be controlled by the presence or absence of the solvent [50, 89]. There are two routes for the chemical recycling of PCL. One is the enzymatic conversion of PCL into 6-hydroxyhexanoate (CL) oligomers. The other is the selective ring-closing depolymerization of PCL into dicaprolactone (DCL) (Scheme 26) [63]. The DCL was readily polymerized by lipase CA to produce PCL. The DCL also copolymerized with conventional cyclic monomers, such as PDL and TMC, to produce the corresponding copolyester and poly(ester-carbonate).

Poly(butylene adipate) (PBA) is a typical biodegradable synthetic plastic prepared by the polycondensation of 1,4-butanediol and adipic acid. PBA is readily transformed into the repolymerizable cyclic oligomers mainly consisting of BA dimer by lipase CA (Scheme 27). Commercially available PBA with an M_w of 22 000 was degraded into a cyclic BA oligomer with an M_w of 600 by lipase CA within 1 h. This cyclic BA oligomer was readily repolymerized by lipase that produced the corresponding polyesters with an M_w of 52 000. The molecular weight of the regenerated polymer was higher than that of the parent polymer. This is ascribed to the polymerizability difference between the ring-opening polymerization of the cyclic BA oligomer and polycondensation of the diol with a diacid. Similar results were obtained for poly(butylene succinate) and their copolymers [77].

Poly(lactic acid) may be one of the most promising green plastics. Poly(lactic acid) can be repeatedly chemically recycled with lipase via repolymerizable cyclic oligomers having a molecular weight of a few hundred. Poly(L-lactic acid) (PLLA) having an M_w of 120 000 was transformed into cyclic oligomers by lipase CA at 100 °C. Similar results were obtained by the lipase-catalyzed

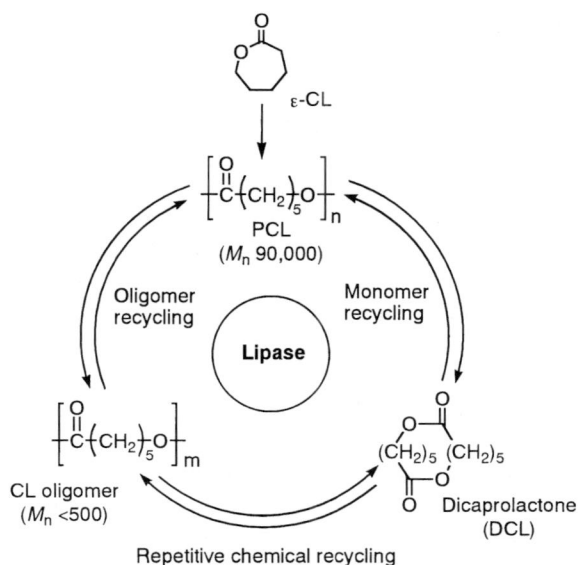

Scheme 26

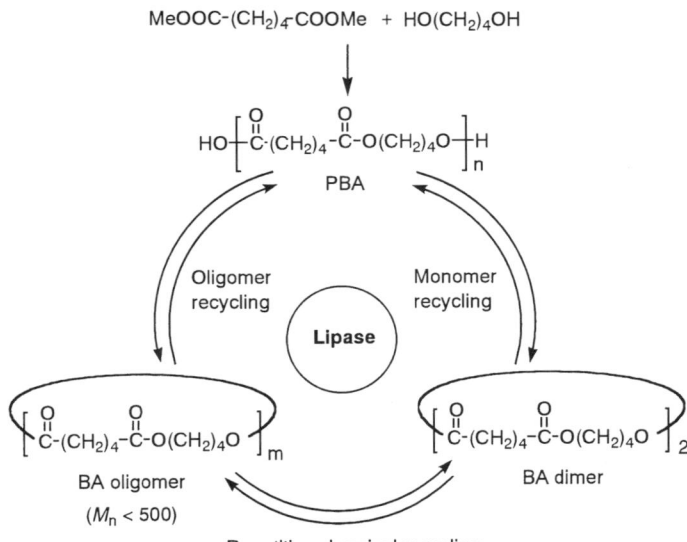

Scheme 27

degradation of poly(DL-lactic acid) (PDLLA). That is, PDLLA having an M_w of 84 000 was quantitatively transformed into cyclic oligomers with the peak-top oligomerzation degree of the octamer (8-mer) by lipase RM at 60 °C. No linear-type fraction was detected in the MALDI-TOF mass spectrum [159].

A variety of microbial polyesters is known. As the current interest in poly(R-3-hydroxyalkanoate) (PHA) is largely based on its biodegradability, to consider the large-scale recycling of this polymer is to ignore one of its most important properties. The thermal treatments will almost certainly produce free crotonic acid from the hydroxybutyrate and double bonds at the end of the PHA chain. Also, hydrolytic degradation using acid or base catalysts produces oligomers mainly having crotonate end groups by the elimination of water molecules at the terminal hydroxy groups of the polymer chain. On the other hand, PHA can be readily transformed into the corresponding cyclic oligomers by lipase without the elimination of water. The produced cyclic oligomers are readily repolymerized by lipase or conventional chemical catalysts to produce the parent PHA. Similar results were obtained for the unnatural-type poly(R, S-3-hydroxybutanoate) [160].

8.2
Continuous Degradation of Polyesters Using the Enzyme-Packed Column

The above-mentioned batch process required a large quantity of enzymes relative to the polymer for rapid reaction, and a more efficient procedure will be needed. In order to save enzymes, a continuous flow system was developed

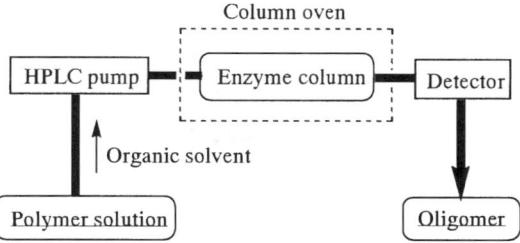

Fig. 1 Schematic diagramm of the continuous degradation system

for the degradation of the polyester (Fig. 1). The polymer solution was continuously passed through the immobilized lipase-packed column using an HPLC pump. The combined eluents contained cyclic oligomers exclusively. As a typical example, PCL in toluene solution was degraded into the corresponding cyclic oligomers by continuous passage through the immobilized lipase packed column at 40 °C. The obtained cyclic oligomer mainly consisting of a cyclic dimer, dicaprolactone, could be repolymerized using lipase for 6 h at 70 °C under a slightly reduced pressure to yield a high-molecular-weight PCL having an M_n of greater than 90 000 [160, 161].

The continuous flow method required a large amount of organic solvent, such as toluene, to dissolve the polymers. Therefore, supercritical carbon dioxide was partially used in place of the organic solvent. RS-PHB, PCL, and PBA are all quantitatively degraded into the corresponding cyclic oligomers at 40 °C. The composition of the supercritical carbon dioxide and organic solvent, such as toluene, was responsible for the degradation of the polymer.

8.3
Poly(ester/Carbonate Urethane)

The novel enzymatically degradable polyurethane that consisted of a diurethane moiety as the hard segment and a carbonate linkage as the biodegradable unit was prepared by the polycondensation of biodegradable urethanediol and dimethyl carbonate using lipase CA (Scheme 28). The produced poly(carbonate urethane) with an M_w of 20 000 was readily transformed by lipase CA into the corresponding repolymerizable cyclic oligomers mainly consisting of a monomer and dimer. The cyclic oligomers were more readily repolymerized by lipase to produce a higher-molecular-weight poly(carbonate urethane) with an M_w of greater than 40 000 when compared with that of the parent poly(carbonate urethane) [162]. In a similar way, a series of enzymatically recyclable poly(ester-urethane)s was chemoenzymatically prepared by the two routes (Scheme 29). That is, poly(ester-urethane) was prepared by the ring-opening polymerization of a cyclic ester-urethane monomer prepared by the reaction of diurethanediol and the dicarboxy-

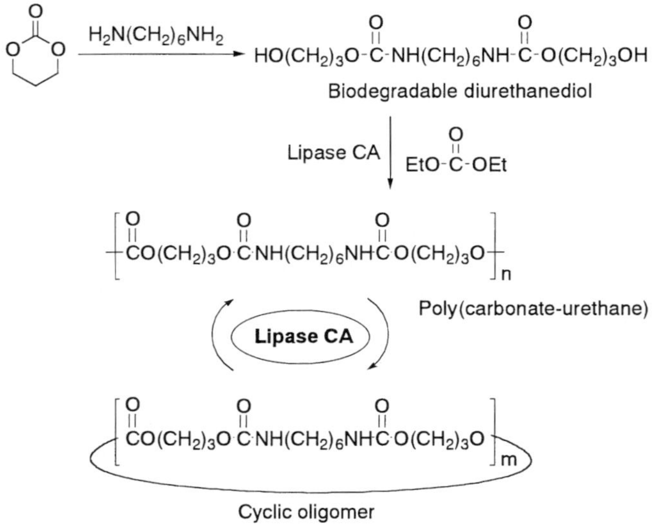

Scheme 28

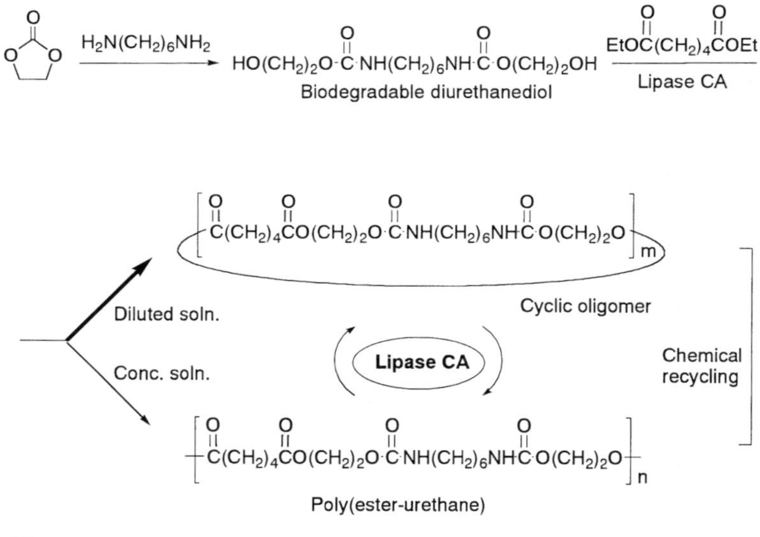

Scheme 29

late ester using lipase, and by the direct polycondensation of diurethanediol and dicarboxylate ester. A significantly higher-molecular-weight poly(ester-urethane) with an M_w of 103 000 was produced by the ring-opening polymerization of the cyclic ester-urethane monomer when compared with the polycondensation of the dicarboxylate ester with diurethanediol [78]. The poly(ester-urethane) was readily transformed by lipase into the corresponding cyclic oligomers, which were readily repolymerized by ring-opening polymerization using lipase, as the chemical recycling [78].

9
Green Solvent for Lipase-Catalyzed Polymerization and Degradation

As a green solvent, water, supercritical carbon dioxide (scCO$_2$), and an ionic liquid were tested and evaluated with respect to the polymer synthesis and degradation by lipase. Though water is the most abundant solvent in nature, its application to polymer synthesis is relatively limited because of monomer and polymer solubilities and an unfavorable equilibrium shift for polymerization. The lipase-catalyzed polymerization of lactones has been realized even in aqueous media [163]. This is due to the hydrophobic nature of the catalytic domain of lipase. For example, macrolides such as UDL and DDL were polymerized by lipase in water to produce the corresponding polyesters [10]. On the other hand, scCO$_2$ may have potential as a solvent for polymer synthesis and recycling. Moreover, the availability of CO$_2$ as a byproduct of many industrial processes, its possible recycling, and easily accessible critical parameters account for its steadily increasing use [164, 165]. Similar results were obtained, such that in scCO$_2$, ε-CL was polymerized by lipase CA to produce the PCL with the highest molecular weight (M_w) of 74 000. The enzyme and scCO$_2$ were repeatedly used for the polymerization [166]. Takamoto et al. reported that the lipase-catalyzed degradation of PCL in the presence of acetone produced an oligomer with a molecular weight of less than 500 in scCO$_2$, which can be polymerized by the same catalyst [167]. Similar results by Matsumura et al. were obtained such that PCL was transformed in scCO$_2$ in the presence of a small amount of water and lipase to produce repolymerizable oligomers having an M_n of about 500 [168]. The produced CL oligomer was again polymerized with lipase CA to yield a PCL having an M_n of greater than 80 000 [168].

An ionic liquid also has vast possibilities because the liquid properties as a solvent can be altered by designing the molecular structure. There are a few reports of enzymatic ring-opening polymerization in ionic liquids. The lipase-catalyzed polymerization of ε-CL was carried out in ionic solvents, such as the 1-butyl-3-methyl-imidazolium salts [169]. In order to establish the compatibility of enzymes with ionic liquids, the lipase-catalyzed transesterification of 2-hydroxymethyl-1,4-benzodioxane was studied [170].

10
Concluding Remarks

It has been revealed that lipase can act as a powerful catalyst for the ring-opening polymerization of cyclic monomers and oligomers containing oxo-, thio-, and phosphoesters in addition to carbonate esters. Regioselective and enantioselective polymerization is realized by lipase-catalyzed ring-opening polymerization. Lipase-catalyzed ring-opening polymerization may not only contribute to establishing a green and sustainable polymer industry but also enable the direct production of green polymers that cannot be produced by industrially feasible methods. Although enzymatic synthesis via ring-opening polymerization has been shown to have many prominent characteristics, some problems must be resolved with respect to large-scale industrial applications. Such problems may include the versatile application of enzyme-catalyzed polymerization, increasing the molecular weight of polymers, development of the biodegradable cross-linkage that allows chemical recycling, and establishment of a sustainable biochemical complex. Extensive studies are now needed with respect to development of the production systems containing bioreactors and polymerization strategies, more active and selective biocatalysts, and feasibility studies for industrial production.

References

1. Zaks A, Klibanov AM (1984) Science 224:1249
2. Klibanov AM (1986) CHEMITECH 354
3. Volkin DB, Staubli A, Langer R, Klibanov AM (1991) Biotech Bioeng 37:843
4. Parvaresh F, Roberts H, Thomas D, Legoy M-D (1992) Biotech Bioeng 39:467
5. Turner NA, Duchateau DB, Vulfson EN (1995) Biotech Lett 17:371
6. Turner NA, Vulfson EN (2000) Enzyme Microb Technol 27:108
7. Uyama H, Kobayashi S (1993) Chem Lett 1149
8. Uyama H, Takeya K, Kobayashi S (1993) Proc Jpn Acad 69 Ser B:203
9. Knani D, Gutman AL, Kohn DH (1993) J Polym Sci Part A Polym Chem 31:1221
10. Namekawa S, Uyama H, Kobayashi S (1998) Polym J 30:269
11. Dong H, Wang H, Cao S, Shen J (1998) Biotechnol Lett 20:905
12. Cordova A, Iversen T, Hult K, Martinelle M (1998) Polymer 39:6519
13. Kobayashi S, Shoda S, Uyama H (1995) Adv Polym Sci 121:1
14. Kobayashi S (1999) J Polym Sci Part A Polym Chem 37:3041
15. Kobayashi S, Uyama H, Kimura S (2001) Chem Rev 101:3793
16. Matsumura S (2002) Macromol Biosci 2:105
17. Kobayashi S, Uyama H (2002) Curr Organ Chem 6:209
18. Uyama H (2001) Kobunshi Ronbunshu 58:382
19. Kobayashi S, Uyama H (2003) ACS Symp Ser 840:128
20. Brandstadt K, Saam J, Sharma A (2003) ACS Symp Ser 840:141
21. Kumar A, Kalra B, Gross R (2003) ACS Symp Ser 840:172
22. Matsumura S, Beppu H, Tsukada K, Toshima K (1996) Biotechnol Lett 18:1041
23. Namekawa S, Uyama H, Kobayashi S (1996) Polym J 28:730
24. Svirkin YY, Xu J, Gross RA, Kaplan DL, Swift G (1996) Macromolecules 29:4591
25. Nobes GAR, Kazlauskas RJ, Marchessault RH (1996) Macromolecules 29:4829
26. Xie W, Li J, Chen D, Wang PG (1997) Macromolecules 30:6997
27. Matsumura S, Suzuki Y, Tsukada K, Toshima K, Doi Y, Kasuya K (1998) Macromolecules 31:6444

28. Osanai Y, Toshima K, Matsumura S (2000) Chem Lett 576
29. Osanai Y, Toshima K, Matsumura S (2001) Macromol Biosci 1:171
30. Matsumura S, Beppu H, Toshima K (1998) ACS Symp Ser 684:74
31. Matsumura S, Beppu H, Nakamura K, Osanai S, Toshima K (1996) Chem Lett 795
32. Matsumura S, Beppu H, Toshima K (1999) Chem Lett 249
33. Abe Y, Matsumura S, Imai K (1986) J Jpn Oil Chem Soc 35:937
34. Matsumura S, Yoshikawa S (1990) ACS Symp Ser 433:124
35. Panova A, Taktak S, Randriamahefa S, Cammas-Morion S, Guerin P, Kaplan DL (2003) Biomacromolecules 4:19
36. Nobes GAR, Kazlauskas RJ, Marchessault RH (1996) Macromolecules 29:4829
37. Kobayashi S, Takeya K, Suda S, Uyama H (1998) Macromol Chem Phys 199:1729
38. Kullmer K, Kikuchi H, Uyama H, Kobayashi S (1998) Macromol Rapid Commun 19:127
39. Nishida H, Yamashita M, Nagashima M, Endo T, Tokiwa Y (2000) J Polym Sci Part A Polym Chem 38:1560
40. Feng Y, Knufermann J, Klee D, Hocker H (1999) Macromol Chem Phys 200:1506
41. Feng Y, Klee D, Keul H, Hocker H (2000) Macromol Chem Phys 201:2670
42. Feng Y, Klee D, Hocker H (2001) Macromol Biosci 1:66
43. Feng Y, Knuefermann J, Klee D, Hoecker H (1999) Macromol Rapid Commun 20:88
44. Feng Y, Klee D, Hoecker H (2004) Macromol Biosci 4:587
45. Matsumura S, Mabuchi K, Toshima K (1997) Macromol Rapid Commun 18:477
46. Matsumura S, Mabuchi K, Toshima K (1998) Macromol Symp 130:285
47. Matsumura S, Tsukada K, Toshima K (1999) Int J Biol Macromol 25:161
48. MacDonald RT, Pulapura SK, Svirkin YY, Gross RA, Kaplan DL, Akkara J, Swift G, Wolk S (1995) Macromolecules 28:73
49. Henderson LA, Svirkin YY, Gross RA, Kaplan DL, Swift G (1996) Macromolecules 29:7759
50. Matsumura S, Ebata H, Toshima K (2000) Macromol Rapid Commun 21:860
51. Ebata H, Toshima K, Matsumura S (2001) Chem Lett 798
52. Kumar A, Kalra B, Dekhterman A, Gross RA (2000) Macromolecules 33:6303
53. Mei Y, Kumar A, Gross R (2002) Macromolecules 35:5444
54. Cordova A, Iversen T, Hult K (1999) Polymer 40:6709
55. Serata N, Yanagi C, Kunugi S (2002) Biocatal Biotransform 20:111
56. Panova AA, Kaplan DL (2001) Polym Mater Sci Eng 84:538
57. Dong H, Cao S, Li Z, Han S, You D, Shen J (1999) J Polym Sci Part A Polym Chem 37:1265
58. Divakar S (2004) J Macromol Sci Pure Appl Chem A41:537
59. Qi G, Sun J, Zhou Q (2003) Gaoxiao Huaxue Gongcheng Xuebao 17:655
60. Kumar A, Gross RA (2000) Biomacromolecules 1:133
61. Uyama H, Suda S, Kikuchi H, Kobayashi S (1997) Chem Lett 1109
62. Sivalingam G, Madras G (2004) Biomacromolecules 5:603
63. Ebata H, Toshima K, Matsumura S (2000) Biomacromolecules 1:511
64. Deng F, Gross RA (1999) Int J Biol Macromol 25:153
65. Kikuchi H, Uyama H, Kobayashi S (2002) Polym J 34:835
66. Uyama H, Takeya K, Hoshi N, Kobayashi S (1995) Macromolecules 28:7046
67. Uyama H, Takeya K, Kobayashi S (1995) Bull Chem Soc Jpn 68:56
68. Uyama H, Kikuchi H, Takeya K, Kobayashi S (1996) Acta Polym 47:357
69. Kobayashi S, Uyama H, Namekawa S, Hayakawa H (1998) Macromolecules 31:5655
70. Bisht KS, Henderson LA, Gross RA, Kaplan DL, Swift G (1997) Macromolecules 30:2705

71. Focarete ML, Scandola M, Kumar A, Gross R (2001) J Polym Sci Part B Polym Phys 39:1721
72. Namekawa S, Uyama H, Kobayashi S (1998) Proc Jpn Acad 74 Ser B:65
73. Kobayashi S, Uyama H (1999) Macromol Symp 144:237
74. Gazzano M, Malta V, Focarete ML, Scandola M, Gross RA (2003) J Polym Sci Part B Polym Phys 41:1009
75. Taden A, Antonietti M, Landfester K (2003) Macromol Rapid Commun 24:512
76. Müller S, Uyama H, Kobayashi S (1999) Chem Lett 1317
77. Okajima S, Kondo R, Toshima K, Matsumura S (2003) Biomacromolecules 4:1514
78. Soeda Y, Toshima K, Matsumura S (2005) Macromol Biosci 5 5:277
79. He F, Wang Y, Feng J, Zhuo R, Wang X (2003) Polymer 44:3215
80. He F, Li S, Vert M, Zhuo R (2003) Polymer 44:5145
81. Kumar A, Gross RA, Wang Y, Hillmyer MA (2002) Macromolecules 35:7606
82. Namekawa S, Uyama H, Kobayashi S (2001) Macromol Chem Phys 202:801
83. Namekawa S, Uyama H, Kobayashi S (2000) Biomacromolecules 1:335
84. Uyama H, Kikuchi H, Takeya K, Hoshi N, Kobayashi S (1996) Chem Lett 107
85. Noda S, Kamiya N, Goto M, Nakashio F (1997) Biotechnol Lett 19:307
86. Maruyama T, Noda S, Kamiya N, Goto M (2003) J Chem Eng Jpn 36:307
87. Uyama H, Kuwabara M, Tsujimoto T, Kobayashi S (2002) Polym J 34:970
88. Kobayashi S, Uyama H, Namekawa S (1998) Polym Degrad Stab 59:195
89. Kobayashi S, Uyama H, Takamoto T (2000) Biomacromolecules 1:3
90. Namekawa S, Suda S, Uyama H, Kobayashi S (1999) Int J Biol Macromol 25:145
91. Matsumoto M, Odachi D, Kondo K (1999) Biochem Eng J 4:73
92. Mei Y, Kumar A, Gross R (2003) Macromolecules 36:5530
93. Panova AA, Kaplan DL (2003) Biotechnol Bioeng 84:103
94. Duda A, Kowalski A, Penczek S, Uyama H, Kobayashi S (2002) Macromolecules 35:4266
95. Suzuki Y, Taguchi S, Saito T, Toshima K, Matsumura S, Doi Y (2001) Biomacromolecules 2:541
96. Suzuki Y, Taguchi S, Hisano T, Toshima K, Matsumura S, Doi Y (2003) Biomacromolecules 4:537
97. Suzuki Y, Ohura T, Kasuya K, Toshima K, Doi Y, Matsumura S (2000) Chem Lett 318
98. Kumar A, Gross RA, Jendrossek D (2000) J Org Chem 65:7800
99. Uyama H, Suda S, Kobayashi S (1998) Acta Polym 49:700
100. Uyama H, Kobayashi S, Morita M, Habaue S, Okamoto Y (2001) Macromolecules 34:6554
101. Habaue S, Asai M, Morita M, Okammoto Y, Uyama H, Kobayashi S (2003) Polymer 44:5195
102. Cordova A (2001) Biomacromolecules 2:1347
103. Li J, Xie W, Cheng HN, Nickol RG, Wang PG (1999) Macromolecules 32:2789
104. Kumar R, Gross RA (2002) J Am Chem Soc 124:1850
105. Gustavsson MT, Persson PV, Iversen T, Hult K, Martinelle M (2004) Biomacromolecules 5:106
106. Li J, Xie W, Cheng HN, Nickol RG, Wang PG (1999) Macromolecules 32:2789
107. Bisht K, Gross RA (2001) ACS Symp Ser 786:222
108. Hu Y, Ju LK (2003) Biotechnol Prog 19:303
109. Deng F, Gross RA (2001) ACS Symp Ser 786:195
110. Cordova A, Hult A, Hult K, Ihre H, Iversen T, Malmstrom E (1998) J Am Chem Soc 120:13521
111. Skaria S, Smet M, Frey H (2002) Macromol Rapid Commun 23:292

112. Kalra B, Kumar A, Gross RA, Baiardo M, Scandola M (2004) Macromolecules 37:1243
113. Al-Azemi TF, Kondaveti L, Bisht KS (2002) Macromolecules 35:3380
114. Kikuchi H, Uyama H, Kobayashi S (2000) Macromolecules 33:8971
115. Runge M, O'Hagan D, Haufe G (2000) J Polym Sci Part A Polym Chem 38:2004
116. Kobayashi S, Uyama H (1993) Makromol Rapid Commun 14:841
117. Soeda Y, Okamoto T, Toshima K, Matsumura S (2002) Macromol Biosci 2:429
118. Soeda Y, Toshima K, Matsumura S (2001) Chem Lett 1:76
119. Matsumura S, Okamoto T, Tsukada K, Toshima K (1998) Macromol Rapid Commun 19:295
120. Matsumura S, Okamoto T, Tsukada K, Mizutani N, Toshima K (1999) Macromol Symp 144:219
121. Suyama T, Tokiwa Y (1997) Enzyme Microb Technol 20:122
122. Suyama T, Hosoya H, Tokiwa Y (1998) FEMS Microbiol Lett 161:255
123. Albertsson AC, Eklund M (1994) J Polym Sci Part A Polym Chem 32:265
124. Buchholz B (1993) J Mater Sci Mater Med 4:381
125. Wang H, Dong JH, Qiu KY, Gu ZW (1998) J Polym Sci Part A Poly Chem 36:1301
126. Abramowicz DA, Keese CR (1989) Biotechnol Bioeng 33:149
127. Rodney RL, Stagno JL, Beckman EJ, Russell AJ (1999) Biotechnol Bioeng 62:259
128. Rodney RL, Allinson BT, Beckman EJ, Russell AJ (1999) Biotechnol Bioeng 65:485
129. Matsumura S, Harai S, Toshima K (1999) Proc Jpn Acad Ser B 75:117
130. Matsumura S, Harai S, Toshima K (2000) Macromol Chem Phys 201:1632
131. Gross RA, Kalra B, Kumar A (2001) Appl Microbiol Biotechnol 55:655
132. Matsumura S, Tsukada K, Toshima K (1997) Macromolecules 30:3122
133. Kobayashi S, Kikuchi H, Uyama H (1997) Macromol Rapid Commun 18:575
134. Bisht KS, Svirkin YY, Henderson LA, Gross RA, Kaplan DA, Swift G (1997) Macromolecules 30:7735
135. Kumar A, Garg K, Gross RA (2001) Macromolecules 34:3527
136. Focarete ML, Gazzano M, Scandola M, Kumar A, Gross RA (2002) Macromolecules 35:8066
137. Tasaki H, Toshima K, Matsumura S (2003) Macromol Biosci 3:436
138. Matsumura S, Harai S, Toshima K (2001) Macromol Rapid Commun 22:215
139. Al-Azemi TF, Bisht KS (1999) Macromolecules 32:6536
140. Al-Azemi TF, Harmon JP, Bisht KS, Kirpal S (2000) Biomacromolecules 1:493
141. Al-Azemi TF, Bisht KS (2002) J Polym Sci Part A Polym Chem 40:1267
142. Al-Azemi TF, Bisht KS (2001) Polym Mater Sci Eng 84:1045
143. Bisht KS, Al-Azemi TF (2003) ACS Symp Ser 840:156
144. Bisht KS, Deng F, Gross RA, Kaplan DL, Swift G (1998) J Am Chem Soc 120:1363
145. He F, Wang Y, Zhuo R (2003) Chinese J Polym Sci 21:5
146. Feng J, He F, Zhuo R (2002) Macromolecules 35:7175
147. Feng J, Zhuo R, He F, Wang X (2003) Macromol Symp 195:237
148. Namekawa S, Uyama H, Kobayashi S, Kricheldorf HR (2000) Macromol Chem Phys 201:261
149. Marvel CS, Kotch A (1951) J Am Chem Soc 73:1100
150. Lütke-Eversloh T, Steinbüchel A (2002) In: Matsumura S, Steinbüchel A (eds) Biopolymers, vol 9. Wiley, Weinheim, p 63
151. Lütke-Eversloh T, Bergander K, Luftmann H, Steinbüchel A (2001) Microbiology 147:11
152. Lütke-Eversloh T, Bergander K, Luftmann H, Steinbüchel A (2001) Biomacromolecules 2:1061

153. Lütke-Eversloh T, Kawada J, Marchessault RH, Steinbüchel A (2002) Biomacromolecules 3:159
154. Iwata S, Toshima K, Matsumura S (2003) Macromol Rapid Commun 24:467
155. Kato M, Toshima K, Matsumura S (2005) Biomacromolecules 6:2275
156. Wen J, Zhuo RX (1998) Macromol Rapid Commun 19:641
157. He F, Zhuo RX, Liu L, Jin D, Feng J, Wang XL (2001) React Function Polym 47:153
158. Feng J, Zhuo R, He F (2003) Sci China Ser B Chem 46:160
159. Takahashi Y, Okajima S, Toshima K, Matsumura S (2004) Macromol Biosci 4:346
160. Osanai Y, Toshima K, Matsumura S (2004) Macromol Biosci 4:936
161. Osanai Y, Toshima K, Matsumura S (2003) Green Chem 5:567
162. Soeda Y, Toshima K, Matsumura S (2004) Macromol Biosci 4:721
163. Kobayashi S, Uyama H, Suda S, Namekawa S (1997) Chem Lett 105
164. Chaudhary AK, Beckman EJ, Russell AJ (1995) J Am Chem Soc 117:3728
165. Hile DD, Pishko MV (1999) Macromol Rapid Commun 20:511
166. Loeker FC, Duxbury CJ, Kumar R, Gao W, Gross RA, Howdle SM (2004) Macromolecules 37:2450
167. Takamoto T, Uyama H, Kobayashi S (2001) Macromol Biosci 1:215
168. Matsumura S, Ebata H, Kondo R, Toshima K (2001) Macromol Rapid Commun 22:1325
169. Uyama H, Takamoto T, Kobayashi S (2002) Polym J 34:94
170. Nara S, Harjani JR, Salunkhe MM (2002) Tetrahedron Lett 43:2979

Enzymatic Synthesis of Polyesters via Polycondensation

Hiroshi Uyama[1] · Shiro Kobayashi[2,3] (✉)

[1]Department of Materials Chemistry, Graduate School of Engineering, Osaka University, 565-0871 Suita, Japan

[2]Department of Materials Chemistry, Graduate School of Engineering, Kyoto University, 615-8510 Kyoto, Japan

[3]*Present address:*
R & D Center for Bio-based Materials, Kyoto Institute of Technology, Matsugasaki, Sakyo-ku, 606-8585 Kyoto, Japan
kobayash@kit.ac.jp

1	Introduction	134
2	Lipase-Catalyzed Polycondensation of Dicarboxylic Acids or Their Derivatives with Glycols	135
2.1	Dicarboxylic Acids	135
2.2	Dicarboxylic Acid Alkyl Esters	137
2.3	Dicarboxylic-Acid-Activated Esters	138
2.4	Cyclic and Polymeric Anhydrides	142
3	Lipase-Catalyzed Polycondensation of Oxyacid Derivatives	143
4	Enzymatic Synthesis of Functional Polyesters by Polycondensation	145
4.1	Chiral Polyesters	145
4.2	Sugar-Containing Polyesters	146
4.3	Reactive Polyesters	148
4.4	Other Functional Polymers	151
5	In Vitro PHA Polymerase-Catalyzed Polymerization to PHA	152
6	Enzymatic Synthesis of Polycarbonates	154
7	Concluding Remarks	154
	References	155

Abstract In vitro synthesis of polyesters by polycondensation using isolated enzymes as catalyst via nonbiosynthetic pathways is reviewed. In most cases, lipase was used as catalyst and the polyesters were obtained from oxyacids, their esters, and dicarboxylic acid derivatives/glycols. Enzymatic polymerization proceeded under mild reaction conditions in comparison with chemical processes. By utilizing characteristic properties of lipases, regio- and enantioselective polymerizations proceeded to give functional polymers, most of which are difficult to synthesize by conventional methodologies.

Keywords Enzyme · Lipase · Polyester · Polycondensation · Biodegradable polymer

Abbreviations

BOD	biochemical oxygen demand
DP	degree of polymerization
ee	enantiomeric excess
Lipase A	*Aspergillus niger* lipase
Lipase CA	*Candida antarctica* lipase
Lipase CR	*Candida rugosa* lipase
Lipase MM	*Mucor miehei* lipase
Lipase PC	*Pseudomonas cepacia* lipase
Lipase PF	*Pseudomonas fluorescens* lipase
PEG	poly(ethylene glycol)
PHA	poly(hydroxyalkanoate)
PHB	poly(3-hydroxybutyrate)
PPL	porcine pancreas lipase

1
Introduction

Lipase is an enzyme that catalyzes the hydrolysis of fatty acid esters normally in an aqueous environment in living systems. However, some isolated lipases are stable in organic solvents and can act as catalyst for reverse reactions, esterifications, and transesterifications (Scheme 1) in organic media [1–5]. So far, chiral drugs, liquid crystals, acylated sugar-based surfactants, and functional triglycerides have been synthesized through lipase catalysis [6–10].

Aliphatic polyesters are among the most used biodegradable polymers in environmental and medical fields. Many studies concerning syntheses of aliphatic polyesters by fermentation and chemical processes have been performed for developing new boidegradable materials [11–13]. Recently, another approach to synthesizing biodegradable polyesters has been developed

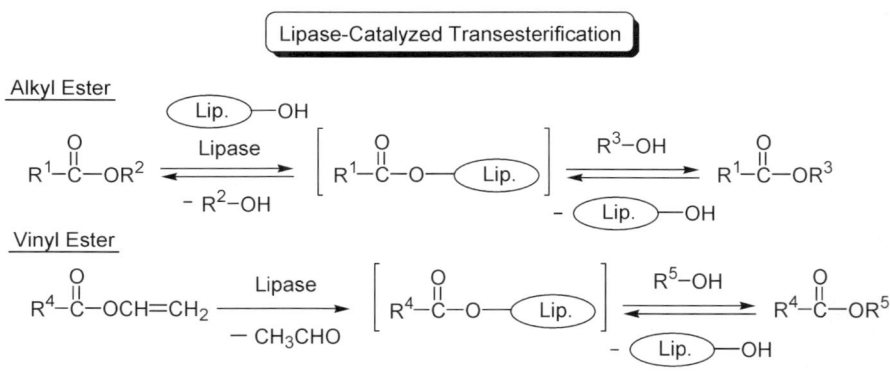

Scheme 1

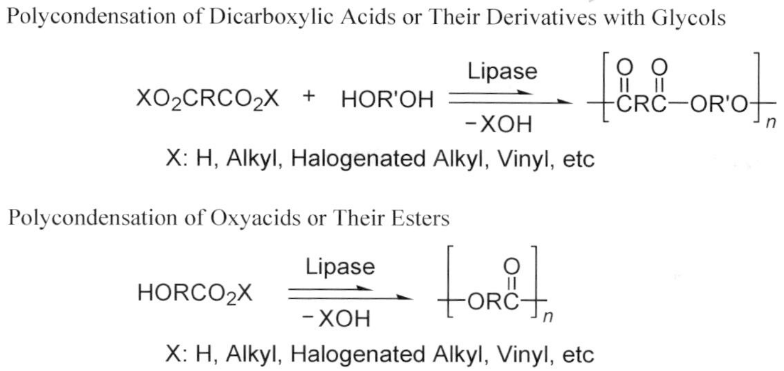

Polycondensation of Dicarboxylic Acids or Their Derivatives with Glycols

$$XO_2CRCO_2X \ + \ HOR'OH \ \xrightleftharpoons[-XOH]{Lipase} \ {+\!\!\!\!\!\!\!-}CRC-OR'O{-\!\!\!\!\!\!\!+}_n \ \ (\text{with } C=O,C=O)$$

X: H, Alkyl, Halogenated Alkyl, Vinyl, etc

Polycondensation of Oxyacids or Their Esters

$$HORCO_2X \ \xrightleftharpoons[-XOH]{Lipase} \ {+\!\!\!\!\!\!\!-}ORC{-\!\!\!\!\!\!\!+}_n$$

X: H, Alkyl, Halogenated Alkyl, Vinyl, etc

Scheme 2

by lipase-catalyzed polymerization [14–28]. By using lipase catalysis, functional aliphatic polyesters have been synthesized by various polymerization modes (Scheme 2). This review deals with lipase-catalyzed synthesis of polyesters by polycondensation.

2
Lipase-Catalyzed Polycondensation of Dicarboxylic Acids or Their Derivatives with Glycols

It is generally accepted that an enzymatic reaction is virtually reversible, and hence the equilibrium can be controlled by selecting the appropriate reaction conditions. Based on this concept, many hydrolases, which are enzymes that catalyze a bond-cleavage reaction by hydrolysis, have been employed as catalyst for the reverse reaction of hydrolysis, leading to polymer production by a bond-forming reaction. Lipase catalysis of esterification and transesterification has produced useful polyesters mainly under anhydrous conditions. First, enzymatic polymerization of dicarboxylic acids or their esters with glycols is described. So far, various dicarboxylic acid derivatives, dicarboxylic acids, their activated and nonactivated esters, and cyclic and polymeric anhydrides have been polymerized with glycols through lipase catalysis to give polyesters [17, 18, 22]

2.1
Dicarboxylic Acids

Many dicarboxylic acids are commercially available; however, their enzymatic reactivity is relatively low. *Mucor miehei* lipase (lipase MM) in immobilized form induced the polycondensation of adipic acid and 1,4-butanediol in diisopropyl ether [29]. A horizontal two-chamber reactor was employed to facilitate the use of the molecular sieves. A low dispersity polyester with a degree of polymerization (DP) of 20 was obtained by two-stage polymerization.

High-molecular-weight polyesters were enzymatically obtained by the polymerization of sebacic acid and 1,4-butanediol under vacuum [30, 31]. In lipase MM-catalyzed polymerization in hydrophobic solvents of high boiling points such as diphenyl ether and veratrole using a programmed vacuum profile, the molecular weight reached higher than 4×10^4. The increase in the molecular weight of aromatic polyesters was also observed by polymerization under vacuum [32]. In the polymerization of isophthalic acid and 1,6-hexanediol using an immobilized lipase derived from *Candida antarctica* (lipase CA) at 70 °C, a polyester with a molecular weight of 5.5×10^4 was formed, whereas lipase MM produced only the corresponding oligomer.

The effects of substrates and solvents on the formation, molecular weight, and end-group structure of a polymer in polycondensation using lipase CA as catalyst were systematically investigated [33]. Diphenyl ether was found to be the preferred solvent that gave the polyester of the highest molecular weight. Concerning the effect of the monomer structure, the longer-chain-length diacids (sebacic and adipic acid) and diols (1,8-octanediol and 1,6-hexanediol) gave a higher enzymatic reactivity than the shorter-chain-length diacids (succinic and glutaric acid) and 1,4-butanediol.

In 1998 it was reported that aliphatic polyesters were synthesized by enzymatic polymerization of dicarboxylic acids and glycols in a solvent-free system by the present authors and by Roberts' group [34, 35]. Lipase CA efficiently catalyzed polymerization under mild reaction conditions, despite the heterogeneous mixture of monomers and catalyst. The methylene chain length of the monomers greatly affected the polymer yield and molecular weight. A polymer with a molecular weight greater than 1×10^4 was obtained by reaction under reduced pressure. A small amount of adjuvant was effective for polymer production when both monomers were solid at the reaction temperature [36].

In the polymerization of adipic acid and 1,6-hexanediol, loss of enzymatic activity was small during polymerization, whereas less than half of the activity remained when glycols were used with a methylene chain length of less than 4 [37]. A scale-up experiment produced a polyester from adipic acid and 1,6-hexanediol in a yield of more than 200 kg. This solvent-free system claimed a large potential as an environmentally friendly synthetic process of polymeric materials owing to mild reaction conditions and to the fact that no organic solvents or toxic catalysts were used.

The enzymatic synthesis of a telechelic polyester having a hydroxy group, a prepolymer of polyurethane, at both ends was reported [38]. The lipase CA-catalyzed polymerization of adipic acid with an excess of 1,4-butanediol was performed under reduced pressure. The content of water formed by dehydration polymerization greatly affected the enzyme activity and polymerization rates.

The polymerization of adipic acid and 1,8-octanediol in bulk was investigated using lipase CA immobilized on different resins as well as lipase CA free of immobilized resin [39]. The immobilized lipase induced polymerization more efficiently to produce a polyester of higher molecular weight. Under a wide range of reaction conditions, the molecular weight index of the resulting polymer without fractionation was less than 1.5, suggesting that lipase CA catalyzes chain growth with chain-length selectivity.

The effects of the feed ratio on the lipase CA-catalyzed polymerization of adipic acid and 1,6-hexanediol were examined using NMR and MALDI-TOF mass spectroscopies [40]. In case of the stoichiometric substrates, ^{1}H NMR analysis showed that the hydroxyl-terminated product was preferentially formed at the early stage of polymerization. As the reaction proceeded, the carboxyl-terminated product was mainly formed. Even in the use of an excess of dicarboxylic acid monomer, the hydroxy-terminated polymer was formed largely at the early reaction stage, which is a specific polymerization behavior due to the unique enzyme catalysis.

A dehydration reaction is generally realized in nonaqueous media. Since water as a product of dehydration is in equilibrium with the starting materials, water as solvent disfavors dehydration to proceed in an aqueous medium due to "the law of mass action". Nevertheless, the present authors have found that lipase catalysis provides a dehydration polymerization of a dicarboxylic acid and glycol in water [41, 42]. Dehydration in an aqueous medium is a new aspect in organic chemistry. Lipases CA and MM as well as lipases from *Candida rugosa*, *Pseudomonas cepacia*, and *Pseudomonas fluorescens* (lipases CR, PC, and PF, respectively) were active for the polymerization of sebacic acid and 1,8-octanediol. In the polymerization of α,ω-dicarboxylic acid and glycol, the polymerization behavior greatly depended on the methylene chain length of monomers. A polymer was obtained in good yields from 1,10-decanediol, whereas no polymer formation was observed in using 1,6-hexanediol, suggesting that the combination of monomers with appropriate hydrophobicity is favored for polymer formation.

2.2
Dicarboxylic Acid Alkyl Esters

Alkyl esters often show low reactivity for lipase-catalyzed transesterifications with alcohols. Therefore, it is difficult to obtain high-molecular-weight polyesters by lipase-catalyzed polycondensation of dialkyl esters with glycols. Lipase CA- or MM-catalyzed polycondensation of dimethyl succinate and 1,6-hexanediol in toluene quickly reached equilibrium between the starting materials and the polymer [43]. Adsorption of methanol by molecular sieves or elimination of methanol by nitrogen bubbling shifted the thermodynamic equilibrium. Polyesters in the molecular weight range of several thousand were prepared from α,ω-alkylene dicarboxylic acid dialkyl esters, and, whatever the monomer structure, cyclic oligomers were formed [44]. The yield of the cyclics depended on the monomer structure, initial monomer concentration, and reaction temperature. Ring-chain equilibrium was observed, and the molar distribution of the cyclic species obeyed the Jacobson–Stockmayer equation. Oligomer formation was observed by the reaction of dimethyl adipate and neopentyl glycol by lipase CA catalyst [45].

In the polymerization of dimethyl terephthalate and diethylene glycol catalyzed by lipase CA in toluene, unique macrocyclic oligomers were formed [46]. Their formation was ascribed to the presence of a driving force leading to a π-stacking phenomenon. Under the selected conditions, the exclusive formation of the cyclic dimer was observed. In the lipase-catalyzed polymerization of dimethyl isophthalate and 1,6-hexanediol in toluene with nitrogen bubbling, a mixture of linear and cyclic polymers was formed [47]. Protease was also effective as catalyst for aromatic polyester synthesis; *Bacillus licheniformis* protease catalyzed the oligomerization of esters of terephthalic acid with 1,4-butanediol [48].

The molecular weight greatly improved by polymerization under vacuum to remove the formed alcohols, leading to a shift in equilibrium toward the product polymer [49]; a polyester with a molecular weight of 2×10^4 was obtained by the lipase MM-catalyzed polymerization of sebacic acid and 1,4-butanediol in diphenyl ether or veratrole under reduced pressure.

Recently, room-temperature ionic liquids have received much attention as green designer solvents. The present authors first demonstrated that ionic liquids acted as a good medium for lipase-catalyzed production of polyesters [50]. The polycondensation of diethyl adipate and 1,4-butanediol using lipase CA as catalyst efficiently proceeded in 1-butyl-3-methylimidazolinium tetrafluoroborate or hexafluorophosphate under reduced pressure. The polymerization of diethyl sebacate and 1,4-butanediol in 1-butyl-3-methylimidazolinium hexafluorophosphate took place even at room temperature in the presence of lipase PC [51].

2.3
Dicarboxylic-Acid-Activated Esters

Activated esters of halogenated alcohols such as 2-chloroethanol, 2,2,2-trifluoroethanol, and 2,2,2-trichloroethanol have often been used as substrate for enzymatic synthesis of esters, owing to the increase in the electrophilicity (reactivity) of the acyl carbonyl and to the need to avoid significant alcoholysis of the products by decreasing the nucleophilicity of the leaving alcohols [1].

The enzymatic synthesis of biodegradable polyesters from activated diesters was achieved under mild reaction conditions. The polymerization of bis(2,2,2-trichloroethyl) glutarate and 1,4-butanediol proceeded in the presence of porcine pancreas lipase (PPL) at room temperature in diethyl ether to produce polyesters with a molecular weight of 8.2×10^3 [52]. The polycondensation of various bis(2,2,2-trichloroethyl) alkanediaoates and glycols took place by PPL catalyst in anhydrous solvents of low polarity [53]. In the reaction of bis(2-chloroethyl) succinate and 1,4-butanediol catalyzed by lipase PF, only oligomeric products were formed [54]. This may be due to the low enzymatic reactivity of the succinate substrate.

The lipase-catalyzed synthesis of polyesters was achieved in a supercritical fluid. The polymerization of bis(2,2,2-trichloroethyl) adipate and 1,4-butanediol using PPL catalyst proceeded in a supercritical fluoroform solvent to give a polymer with a molecular weight of several thousand [55]. By changing the pressure, the low-dispersity polymer fractions were separated.

A vacuum was applied to shift the equilibrium forward by removal of the activated alcohol formed, leading to the production of high-molecular-weight polyesters [49]. The polycondensation of bis(2,2,2-trifluoroethyl) sebacate and aliphatic diols took place using lipases CR, MM, PC, and PPL as catalysts in diphenyl ether. Under the appropriate reaction conditions, a polymer with a molecular weight higher than 4×10^4 was obtained and lipase MM showed the highest catalytic activity [31, 56]. In the PPL-catalyzed polymerization of bis(2,2,2-trifluoroethyl) glutarate with 1,4-butanediol in 1,2-dimethoxybenzene, the periodical vacuum method increased the molecular weight to nearly 4×10^4 [57].

An irreversible procedure for lipase-catalyzed acylation using vinyl esters as acylating agent has been developed, where a leaving group of vinyl alcohol tautomerizes to acetaldehyde (Scheme 1) [1]. In these cases, reaction with the vinyl esters proceeds much faster to produce the desired compounds in higher yields in comparison with alkyl esters.

Divinyl esters, as was reported first by the present authors, are efficient monomers for polyester production under mild reaction conditions [58]. In the lipase PF-catalyzed polymerization of divinyl adipate and 1,4-butanediol in diisopropyl ether at 45 °C, a polyester with a molecular weight of 6.7×10^3 was formed, whereas adipic acid and diethyl adipate did not yield polymeric materials under similar reaction conditions (Scheme 3). The polymerization also proceeded in bulk using lipase CA as catalyst [59].

Lipase-catalyzed polymerization of divinyl ester and glycol is proposed to proceed as follows (Scheme 4). First, the hydroxy group of the serine residue nucleophilically attacks the acyl-carbon of the divinyl ester monomer to produce an acyl-enzyme intermediate ("enzyme-activated monomer", EM) involving elimination of acetaldehyde. The reaction of EM with the glycol produces a 1 : 1 adduct of both monomers. In the propagation stage, the nucleophilic attack of the terminal hydroxy group takes place on the acyl-enzyme intermediate formed from the vinyl ester group of the monomer and 1 : 1 adduct, and then the propagation steps proceed similarly.

Lipases CA, MM, PC, and PF showed high catalytic activity toward the polymerization of divinyl adipate or divinyl sebacate with α,ω-glycols with different chain lengths [60]. A combination of divinyl adipate, 1,4-butanediol,

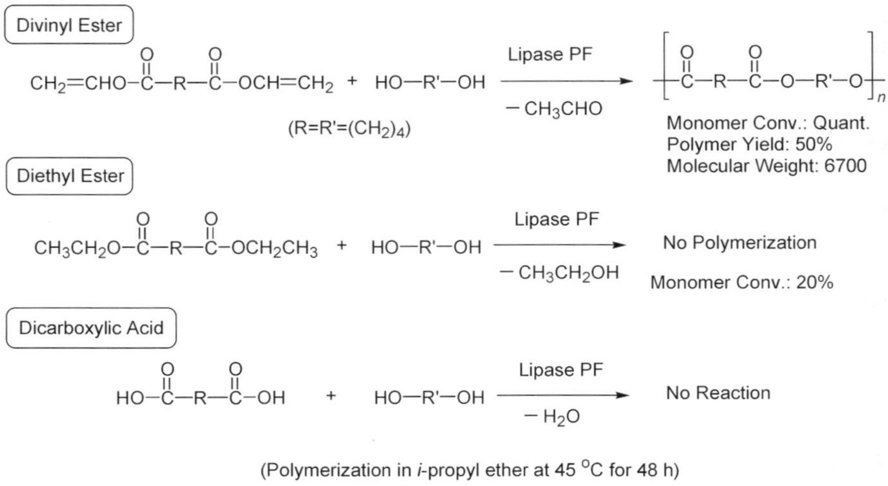

Scheme 3

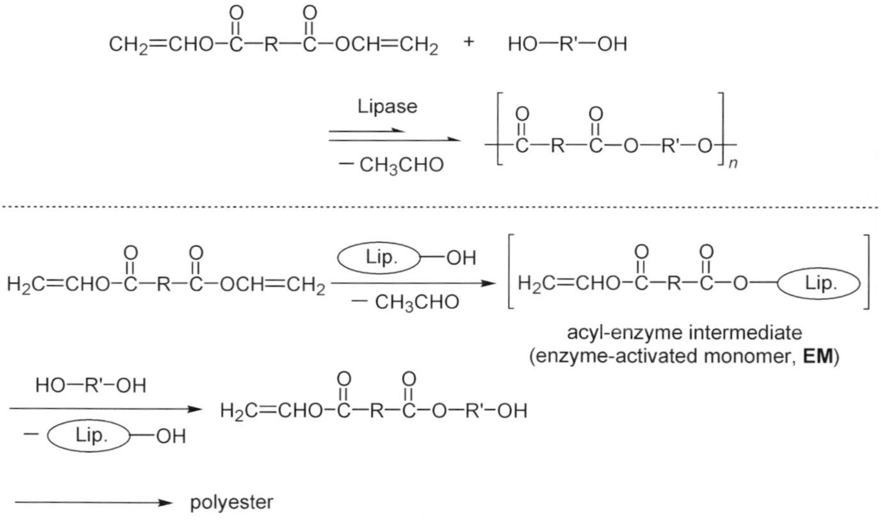

Divinyl Ester: CH₂=CHO-C(=O)-R-C(=O)-OCH=CH₂ + HO-R'-OH →(Lipase PF, -CH₃CHO) [-C(=O)-R-C(=O)-O-R'-O-]ₙ
(R=R'=(CH₂)₄)
Monomer Conv.: Quant.
Polymer Yield: 50%
Molecular Weight: 6700

Diethyl Ester: CH₃CH₂O-C(=O)-R-C(=O)-OCH₂CH₃ + HO-R'-OH →(Lipase PF, -CH₃CH₂OH) No Polymerization
Monomer Conv.: 20%

Dicarboxylic Acid: HO-C(=O)-R-C(=O)-OH + HO-R'-OH →(Lipase PF, -H₂O) No Reaction

(Polymerization in *i*-propyl ether at 45 °C for 48 h)

Scheme 4

CH₂=CHO-C(=O)-R-C(=O)-OCH=CH₂ + HO-R'-OH →(Lipase, -CH₃CHO) [-C(=O)-R-C(=O)-O-R'-O-]ₙ

H₂C=CHO-C(=O)-R-C(=O)-OCH=CH₂ + Lip.-OH →(-CH₃CHO) [H₂C=CHO-C(=O)-R-C(=O)-O-Lip.]
acyl-enzyme intermediate
(enzyme-activated monomer, **EM**)

HO-R'-OH / -Lip.-OH → H₂C=CHO-C(=O)-R-C(=O)-O-R'-OH

→ polyester

and lipase PC yielded a polymer with a molecular weight of 2.1×10^4. The yield of the polymer from divinyl sebacate was higher than that from divinyl adipate, whereas the opposite tendency was observed in the polymer molecular weight. Polyester formation was observed in various organic solvents, and of these, diisopropyl ether gave the best results.

During the lipase-catalyzed polymerization of divinyl esters and glycols, there was a competition between the enzymatic transesterification and hydrolysis of the vinyl end group, resulting in the limitation of polymer growth [61]. A mathematical model showing the kinetics of the polymerization predicts the product composition (terminal structure) [62]. A comparison of the experimental data and model predictions suggests that the molecular weight and terminal group functionality of polyesters can be controlled by selection of biocatalysts. Reaction calorimetry was used to monitor the kinetics of the polymerization [63]. The reaction rate increased as a function of monomer concentration. As the polymerization proceeded, the rate constant for the polyester synthesis was significantly reduced. A batch-stirred reactor was developed to minimize temperature and mass-transfer effects [64]. Using this reactor, poly(1,4-butylene adipate) with a molecular weight of 2.3×10^4 was synthesized in only 1 h at 60 °C.

Aromatic polyesters were efficiently synthesized from aromatic diacid divinyl esters. Lipase CA induced the polymerization of divinyl esters of isophthalic acid, terephthalic acid, and p-phenylene diacetic acid with glycols to give polyesters containing aromatic moiety in the main chain [65]. The highest molecular weight (7.2×10^3) was attained from a combination of divinyl isophthalate and 1,10-decanediol. Enzymatic polymerization of divinyl esters and aromatic diols also yielded the aromatic polyesters [66].

A combinatorial approach to the biocatalytic production of polyesters was demonstrated [67]. A library of polyesters was synthesized in 96 deep-well plates from a combination of divinyl esters and glycols with lipases of different origin. In this screening, lipase CA was confirmed to be the most active biocatalyst for polyester production. As acyl acceptor, 2,2,2-trifluoroethyl esters and vinyl esters were examined, and the former produced a polymer of higher molecular weight. Various monomers such as carbohydrates, nucleic acids, and a natural steroid diol were used as acyl acceptor.

Supercritical carbon dioxide (scCO$_2$) was employed as solvent for the polycondensation of divinyl adipate and 1,4-butanediol [68]. Quantitative consumption of both monomers was achieved to yield a polyester with a molecular weight of 3.9×10^3, indicating that scCO$_2$ was a good medium for enzymatic polycondensation.

Fluorinated polyesters were synthesized by the enzymatic polymerization of divinyl adipate with fluorinated diols [69]. Lipase CA was effective in producing the fluorinated polyesters. The highest molecular weight (5.2×10^3) was achieved by polymerization using 3,3,4,4,5,5,6,6-octafluorooctan-1,8-diol in bulk.

Bis(2,3-butanedione monoxime) alkanedioates were employed as monomers for polyester production [70]. Lipase MM-catalyzed polymerization of a glutarate derivative with 1,6-hexanediol in diisopropyl ether produced a polymer with a molecular weight of 7.0×10^3 in 80% yield.

Lipase-catalyzed copolymerization of divinyl esters, glycols, and lactones produced ester copolymers with a molecular weight greater than 1×10^4 (Scheme 5) [71]. Lipases CA and PC showed high catalytic activity for this copolymerization. ^{13}C NMR analysis showed that the resulting prod-

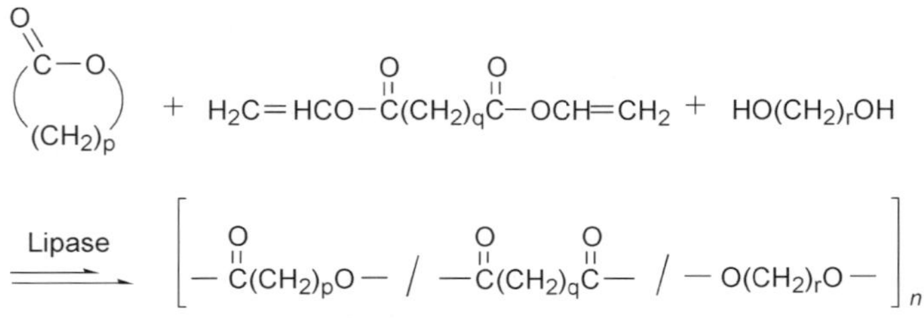

Scheme 5

uct was not a mixture of homopolymers, but a copolymer derived from the monomers, indicating that two different modes of polymerization, polycondensation and ring-opening polymerization, simultaneously take place through enzyme catalysis in one-pot to produce ester copolymers. Furthermore, this result strongly suggests the frequent occurrence of transesterification between the resulting polyesters during polymerization [72, 73].

2.4
Cyclic and Polymeric Anhydrides

Acid anhydride derivatives are also good acylating reagents through lipase catalysis. A new type of enzymatic polymerization involving lipase-catalyzed ring-opening poly(addition-condensation) of cyclic anhydride with glycols was demonstrated (Scheme 6) [74]. The polymerization of succinic anhydride with 1,8-octanediol using lipase PF as catalyst proceeded at room temperature to produce a polyester. Glutaric and diglycolic anhydrides were polymerized with α,ω-alkylene glycols in the presence of lipase CA in toluene to give polyesters [75]. Under appropriate reaction conditions, the molecular weight reached 1×10^4.

Polyester synthesis was carried out by insertion-dehydration of glycols into polyanhydrides using lipase CA as catalyst (Scheme 6) [76]. The insertion of 1,8-octanediol into poly(azelaic anhydride) took place at $30 \sim 60\,°C$ to give the corresponding polyester with a molecular weight of several thou-

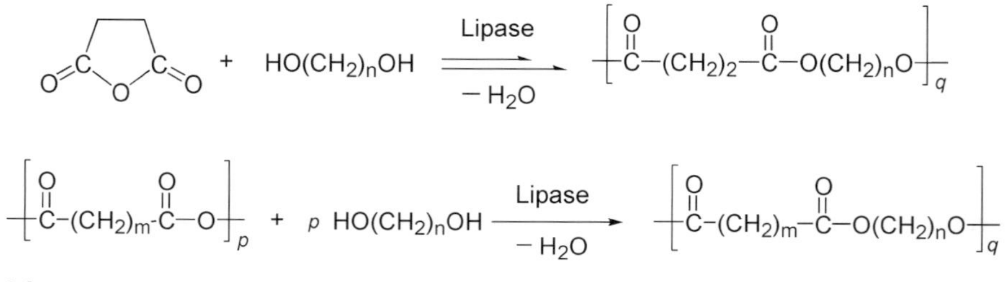

Scheme 6

sand. Effects of the reaction parameters on the polymer yield and molecular weight were systematically investigated [75]. The dehydration reaction also proceeded in water and supercritical carbon dioxide. The reaction behaviors depended on the monomer structure and reaction media.

Lipase-catalyzed synthesis of polyesters from cyclic anhydrides and oxiranes was reported [77, 78]. The polymerization took place by PPL catalyts, and the molecular weight reached 1×10^4 under the selected reaction conditions. During polymerization, the enzymatically formed acid group from the anhydride may open the oxirane ring to give a glycol, which is then reacted with the anhydride or acid by lipase catalysis, yielding the polyesters.

3
Lipase-Catalyzed Polycondensation of Oxyacid Derivatives

In 1985, a lipase-catalyzed polymerization of an oxyacid monomer was first reported. 10-hydroxydecanoic acid was polymerized in benzene using poly(ethylene glycol) (PEG)-modified lipase soluble in the medium [79]. The DP value of the product was more than 5. PEG-modified esterase from hog liver and lipase from *Aspergillus niger* (lipase A) induced the oligomerization of glycolic acid [80].

The polymerization of ricinoleic acid proceeded using lipase CR or lipase from *Chromobacterium viscosum* as catalyst at 35 °C in water, hydrocarbons, or benzene to give a polymer with a molecular weight of around 1×10^3 [81]. These lipases also induced the polymerization of 12-hydroxyoctadecanoic acid, 16-hydroxyhexadecanoic acid, and 12-hydroxydodecanoic acid. Oligoester from ricinoleic acid (estolide) possessing industrial applications in various fields was synthesized using lipase CR immobilized on ceramics [82]. The coloring of estolide improved by selecting mild reaction conditions (40 °C). In the lipase CA-catalyzed oligomerization of cholic acid, a hydroxy group at 3-position was regioselectively acylated to give an oligoester with a molecular weight of 920 (Scheme 7) [83]. The polymerization of lactic acid was performed by using lipase MM as catalyst. Only a low-molecular-weight polymer was formed even under a variety of reaction conditions. The add-

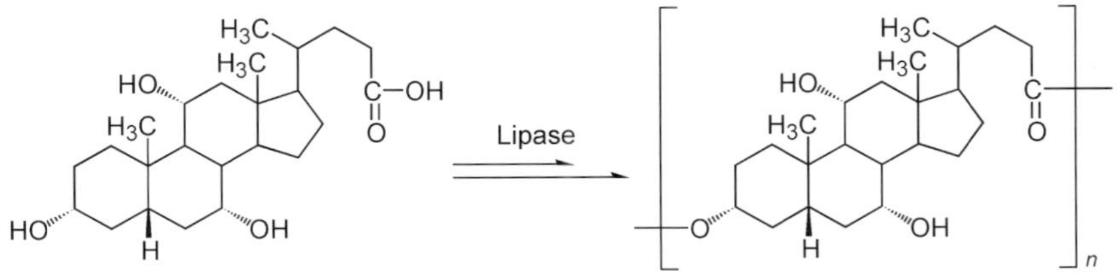

Scheme 7

ition of a small amount of dicarboxylic acid or cyclic anhydrides enhanced the molecular weight [84].

Polyesters of relatively high molecular weight were enzymatically produced from 10-hydroxydecanoic acid [85] and 11-hydroxyundecanoic acid [86] using a large amount of lipase CR as catalyst (10-weight fold for the monomer). In the case of 11-hydroxyundecanoic acid, the corresponding polymer with a molecular weight of 2.2×10^4 was obtained in the presence of activated molecular sieves. PPL-catalyzed polymerization of 12-hydroxydodecanoic acid at 75 °C for 56 h produced a polyester with a molecular weight of 2.9×10^3 having hydroxyl and carboxyl end groups [87].

Lipase CA also polymerized hydrophobic oxyacids efficiently [88]. The DP value was beyond 100 in the polymerization of l6-hydroxyhexadecanoic acid, 12-hydroxydodecanoic acid, or 10-hydroxydecanoic acid under vacuum at high temperature (90 °C) for 24 h, whereas a polyester with a lower molecular weight was formed from 6-hydroxyhexanoic acid under similar reaction conditions. This difference may be due to the lipase-substrate interaction. Similar behaviors were reported in the lipase-catalyzed ring-opening polymerization of lactones in different ring sizes [24, 25].

Lipase-catalyzed polymerization of oxyacid esters was reported. PPL catalyzed the polymerization of methyl 6-hydroxyhexanoate [89]. A polymer with a DP of up to 100 was synthesized by polymerization in hexane at 69 °C for more than 50 d. The PPL-catalyzed polymerization of methyl 5-hydroxypentanoate for 60 d produced a polymer with a DP of 29. Solvent effects were systematically investigated; hydrophobic solvents such as hydrocarbons and diisopropyl ether were suitable for the enzymatic production of high-molecular-weight polymer. Various hydroxyesters, ethyl esters of 3- and 4-hydroxybutyric acids, 5- and 6-hydroxyhexanoic acids, 5-hydroxydodecanoic acid, and 15-hydroxypentadecanoic acid were polymerized by *Pseudomonas* sp. lipase at 45 °C to give the corresponding polyesters with molecular weights of several thousand [90].

Ester-thioester copolymers were enzymatically synthesized (Scheme 8) [91]. The lipase CA-catalyzed copolymerization of ε-caprolactone with 11-

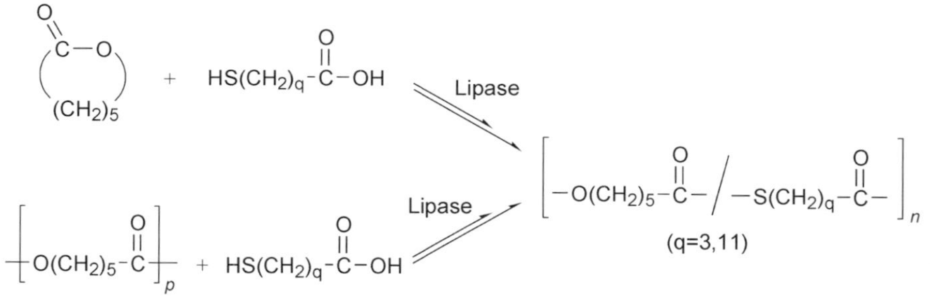

Scheme 8

mercaptoundecanoic acid or 3-mercaptopropionic acid under reduced pressure produced a polymer with a molecular weight greater than 2×10^4. The thioester unit of the resulting polymer was lower than the feed ratio. The transesterification between poly(ε-caprolactone) and 11-mercaptoundecanoic acid or 3-mercaptopropionic acid also took place by lipase CA catalyst.

4
Enzymatic Synthesis of Functional Polyesters by Polycondensation

Functional polyesters were synthesized through the specific catalysis of lipase, and their properties and functions were evaluated. Enantio- and regioselective polycondensations produced chiral and sugar-containing polyesters, respectively [20, 23]. Using lipase catalyst reactive polyesters were conveniently obtained, some of which were crosslinked to biodegradable coatings. Recently, polyester-based biomaterials have been developed by lipase-catalyzed polymerizations.

4.1
Chiral Polyesters

PPL catalyzed an enantioselective polymerization of bis(2,2,2-trichloroethyl) *trans*-3,4-epoxyadipate with 1,4-butanediol in diethyl ether to give a highly optically active polyester (Scheme 9) [92]. The molar ratio of the diester to the diol was adjusted to 2 : 1 to produce the (-) polymer with an enantiomeric purity of > 96%. The polymerization of racemic bis(2-chloroethyl) 2,5-dibromoadipate with an excess of 1,6-hexanediol using lipase A as catalyst produced optically active trimer and pentamer [93]. The polycondensation of 1,4-cyclohexanedimethanol with fumarate esters using PPL as catalyst afforded moderate diastereoselectivity for the *cis/trans* monocondensate and markedly increased diastereoselectivity for the dicondensate product [94].

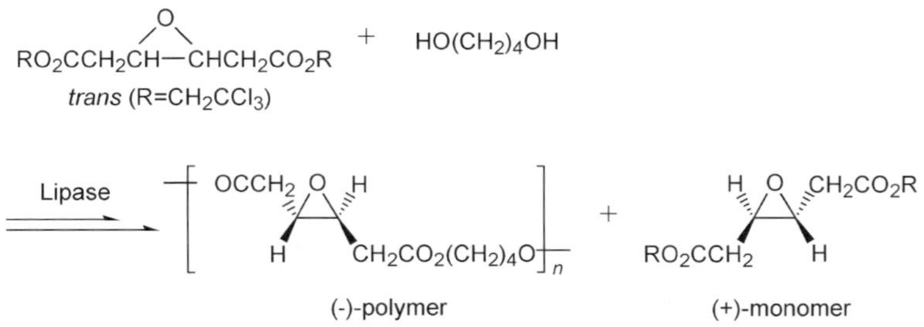

Scheme 9

An optically active oligoester was enantioselectively prepared from racemic 10-hydroxyundecanoic acid using lipase CR as catalyst. The resulting oligomer was enriched in the (S) enantiomer to a level of 60% enantiomeric excess (ee), and the residual monomer was recovered with a 33% ee favoring the antipode [95]. Lactic acid was converted to the corresponding oligomer with a DP of up to 9 by lipase CA catalyst [96]. HPLC analysis showed that the (R)-enantiomer possessed higher enzymatic reactivity. Optically active oligomers (DP < 6) were also synthesized from racemic ε-substituted-ε-hydroxy esters using PPL as catalyst [97]. The enantioselectivity increased as a function of bulkiness of the monomer substituent. The enzymatic copolymerization of the racemic hydroxyacid esters with methyl 6-hydroxyhexanoate produced optically active polyesters with molecular weights greater than 1×10^3. Enzymatic enantioselective oligomerization of a symmetrical hydroxy diester, dimethyl β-hydroxyglutarate, produced a chiral oligomer (dimer or trimer) with 30 ~ 37% ee [98].

4.2
Sugar-Containing Polyesters

For regioselective acylation of sugars, proteases were often used as catalyst [1]. Polyesters having a sugar moiety in the main chain were synthesized via protease catalysis. In the polycondensation of sucrose with bis(2,2,2-trifluoroethyl) adipate catalyzed by an alkaline protease from *Bacillus* sp. showing an esterase activity, the regioselective acylation of sucrose at the 6 and 1′-positions was claimed to yield a sucrose-containing polyester (Scheme 10) [99]. The reaction proceeded slowly; the molecular weight was greater than 1×10^4 after 7 d [100].

Two-step synthesis of sugar-containing polyesters by lipase CA catalyst was reported (Scheme 11) [101]. Lipase CA catalyzed the condensation of sucrose

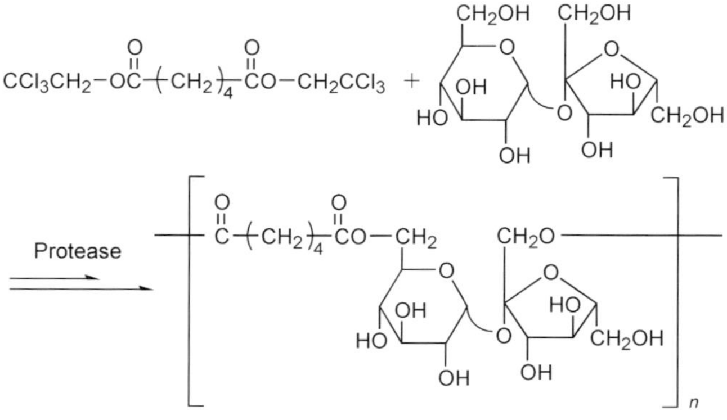

Scheme 10

with an excess of divinyl adipate to produce sucrose 6,6'-O-divinyl adipate, which was reacted with α,ω-alkylene diols by the same catalyst, yielding polyesters containing a sucrose unit in the main chain. This method conveniently yields sugar-containing polyesters with relatively high molecular weights. Similarly, a trehalose-containing polyester was obtained from trehalose 6,6'-O-divinyl adipate through the catalysis of lipase CA.

The present authors first demonstrated that lipase CA produced reduced sugar-containing polyesters regioselectively from divinyl sebacate and sorbitol, in which sorbitol was exclusively acylated at the 1 and 6-positions (Scheme 12) [102]. Mannitol and *meso*-erythritol were also regioselectively polymerized with divinyl sebacate. The enzymatic formation of a high-molecular-weight sorbitol-containing polyester was confirmed by the combinatorial approach [67].

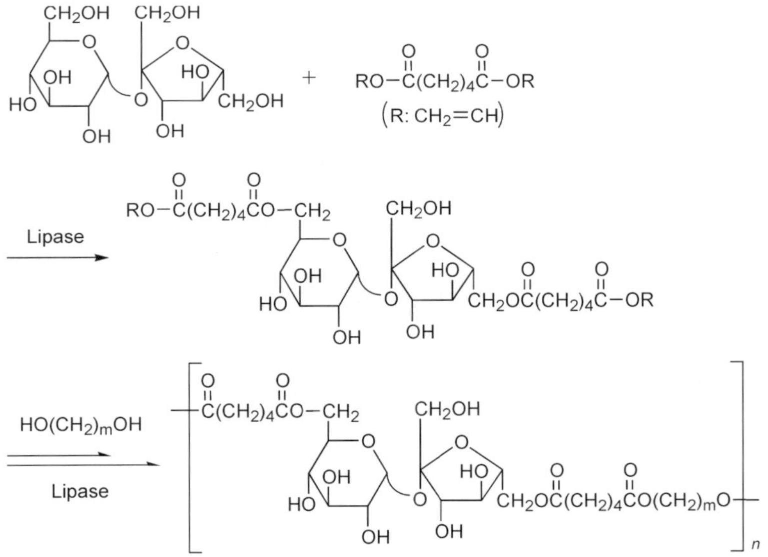

Scheme 11

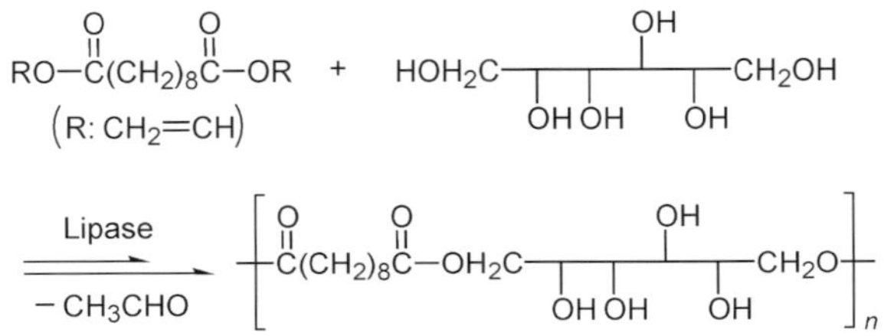

Scheme 12

The lipase CA-catalyzed polycondensation of adipic acid and sorbitol also took place in bulk [103]. In polymerization at 90 °C, the molecular weight reached 1×10^4; however, the regioselectivity decreased (85%), probably due to the high temperature and/or the bulk condition. These data suggest that the divinyl ester is a suitable monomer for regioselective synthesis of sugar-containing polymers. The copolymerization of adipic acid, sorbitol, and 1,8-octanediol enhanced the molecular weight of the sugar-containing polyesters. The melting and crystallization temperatures as well as the melting enthalpy decreased with increasing sorbitol content [104]. This is attributed to the polyol units along the polyester chain that disrupt crystallinity. The biocompatibility of the sugar-containing polyester from adipic acid, sorbitol, and 1,8-octanediol was examined using a mouse fibroblast 3T3 cell line in vitro [105].

4.3
Reactive Polyesters

Polyols such as glycerol were also regioselectively polymerized with divinyl esters by lipase catalyst to produce polyesters having a secondary hydroxy group in the main chain [106]. NMR analysis of a polymer obtained from divinyl sebacate and glycerol using lipase CA as catalyst at 60 °C in bulk showed that 1,3-diglyceride was a main unit and that a small amount of the branching unit (triglyceride) was contained [107]. The regioselectivity of the acylation between primary and secondary hydroxy groups was 74 : 26. In polymerization at 45 °C, the regioselectivity was perfectly controlled to give a linear polymer consisting exclusively of 1,3-glyceride units (Scheme 13).

The polymerization of divinyl sebacate with 1,2,4-butanetriol or 1,2,6-hexanetriol at 60 °C produced a polymer containing a branched unit. In polymerization at lower temperatures, the regioselectivity was perfectly controlled to give a linear polymer consisting of exclusively α,ω-disubstituted unit [108]. The lipase origin and feed ratio of monomers greatly affected the microstructure of the polymer. The lipase MM-catalyzed polymerization of divinyl

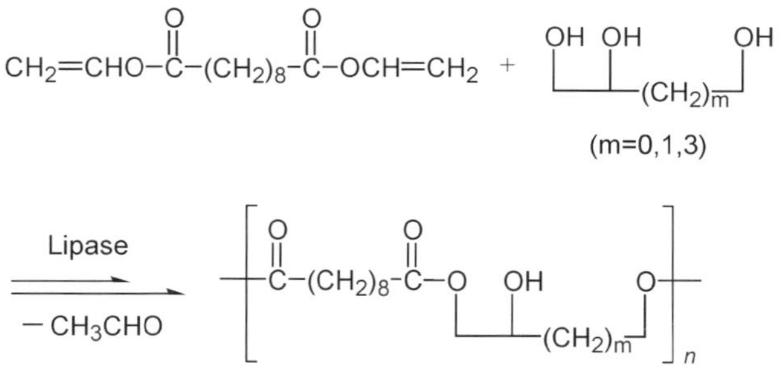

Scheme 13

sebacate and glycerol produced a linear polymer consisting of 1,2- and 1,3-disubstituted units, whereas the 1,3-disubstituted and trisubstituted units were observed in a polymer obtained using lipase PC as catalyst. Interestingly, the highly branched polyester with branching factor > 0.7 was formed by the lipase CA-catalyzed reaction of poly(azelaic anhydride) and triols such as glycerol [18].

New crosslinkable polyesters were synthesized by lipase CA-catalyzed polymerization of divinyl sebacate and glycerol in the presence of unsaturated higher fatty acids derived from renewable plant oils (Scheme 14) [109, 110]. The polymerization under reduced pressure improved the polymer yield and molecular weight. The curing of the polymer obtained using linoleic or linolenic acid proceeded by cobalt naphthenate catalyst or thermal treatment to give a crosslinked transparent film. Biodegradability of the obtained film was evaluated by biochemical oxygen demand (BOD) measurement in an activated sludge. The degradation took place gradually, and the biodegradability reached 45% after 42 d, indicating the good biodegradability of this crosslinked film.

Epoxide-containing polyesters were enzymatically synthesized via two routes using unsaturated fatty acids as starting substrate (Scheme 15) [111]. Lipase catalysis was used for both polycondensation and epoxidation of unsaturated fatty acid groups. One route was synthesis of aliphatic polyesters containing an unsaturated group in the side chain from divinyl sebacate, glycerol, and unsaturated fatty acids, followed by an epoxidation of unsaturated fatty acid moieties in the side chain of the resulting polymer. In another route, epoxidized fatty acids were prepared from unsaturated fatty acids and hydrogen peroxide in the presence of lipase catalyst, and subsequently the epoxidized fatty acids were polymerized with divinyl sebacate

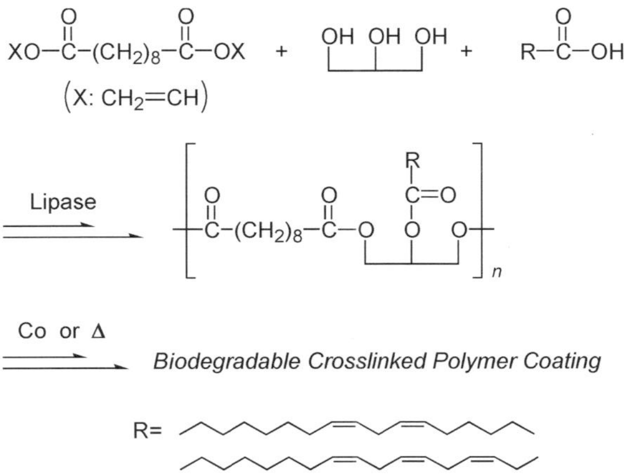

Scheme 14

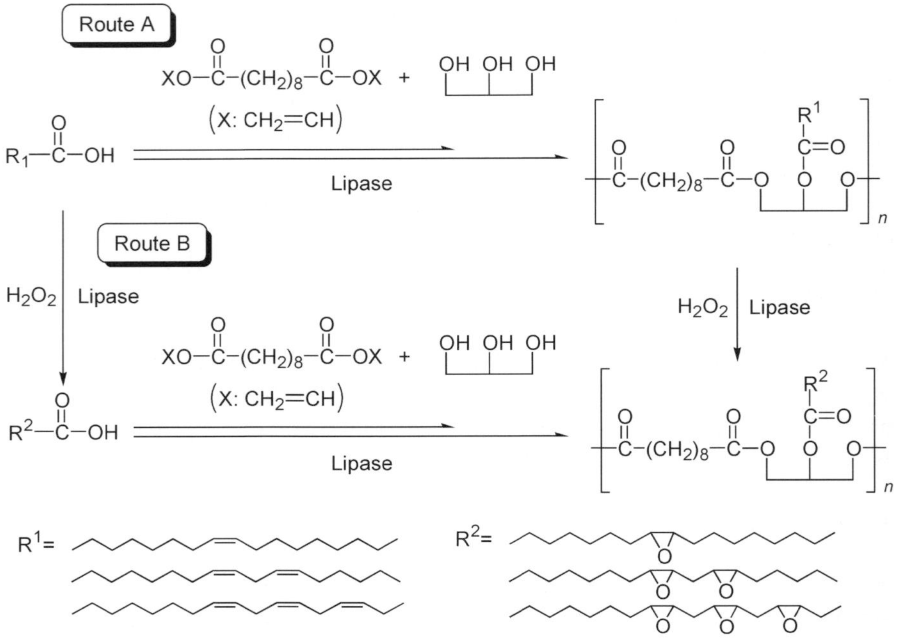

Scheme 15

and glycerol. The polymer structure was confirmed by NMR and IR, and for both routes, a high epoxidized ratio was achieved. Curing of the resulting polymers proceeded thermally, yielding transparent polymeric films with a high gloss surface. Pencil-scratch hardness of the present film improved, compared with that of the cured film obtained from a polyester having an unsaturated fatty acid in the side chain. The obtained film showed good biodegradability, evaluated by BOD measurement in an activated sludge.

Long-chain unsaturated α,ω-dicarboxylic acid methyl esters and their epoxidized derivatives were polymerized with 1,3-propanediol or 1,4-butanediol in the presence of lipase CA catalyst to produce reactive polyesters [112]. The molecular weight of the polymer from 1,4-butanediol was higher than that from 1,3-propanediol. All the resulting polymers had a melting point ranging from 23 to 75 °C.

Unsaturated ester oligomers were synthesized by lipase-catalyzed polymerization of diesters of fumaric acid and 1,4-butanediol [113]. Isomerization of the double bond did not occur to give all-*trans* oligomers showing crystallinity, whereas industrial unsaturated polyesters having a mixture of *cis* and *trans* double bonds are amorphous [114]. The enzymatic polymerization of bis(2-chloroethyl) fumarate with xylylene glycol produced an unsaturated oligoester containing aromaticity in its backbone [115].

A configurationally pure poly(1,6-hexanediyl maleate) was enzymatically synthesized by the lipase CA-catalyzed polymerization of dimethyl maleate and 1,6-hexanediol in toluene. No melting point was observed in the linear polymer of the cis structure, whereas the cyclic oligomer (byproduct) was semicrystalline [116]. In the enzymatic copolymerization of dimethyl maleate, dimethyl fumarate, and 1,6-hexanediol in toluene, the content of the cyclization was found to mainly depend on the configuration and concentration of the monomers [117].

Chemoenzymatic synthesis of alkyds (oil-based polyester resins) was demonstrated [118]. PPL-catalyzed transesterification of triglycerides with an excess of 1,4-cyclohexanedimethanol mainly produced 2-monoglycerides, followed by thermal polymerization with phthalic anhydride to give alkyd resins with a molecular weight of several thousand. The reaction of the enzymatically obtained alcoholysis product with toluene diisocyanate produced alkyd-urethane.

Structural control of polymer terminals has been extensively studied since terminal-functionalized polymers, typically macromonomers and telechelics, are often used as prepolymers for the synthesis of functional polymers. The enzymatic polymerization of 12-hydroxydodecanoic acid in the presence of 11-methacryloylaminoundecanoic acid conveniently produced a methacrylamide-type polyester macromonomer [119, 120]. Lipases CA and CC were active for the macromonomer synthesis.

4.4
Other Functional Polymers

Poly(ethylene glycol) (PEG)-containing polyesters were synthesized via lipase catalysis. The lipase CA-catalyzed polymerization of dimethyl 5-hydroxyisophthalate, α,ω-alkylene glycol, and PEG having a hydroxy group at both ends gave PEG-containing polyesters [121]. The chemoenzymatic synthesis of amphiphilic polyesters was examined by the lipase CA-catalyzed polymerization of dimethyl 5-hydroxyisophthalate and PEG, followed by coupling with alkyl bromide (Scheme 16) [122]. The supramolecular organization of the resulting polymeric nanomicelles in an aqueous medium was confirmed by NMR and light scattering. In vivo studies by encapsulating anti-inflammatory agents in the polymeric nanomicelles and by applying topically resulted in a significant reduction in inflammation. The reduction ratio in inflammation using the polymeric nanomicelles was about 60%.

A novel chemoenzymatic route to polyester polyurethanes was developed without employing highly toxic isocyanate intermediates [123]. First, diurethane diols were prepared from cyclic carbonates and primary diamines, which were subsequently polymerized with dicarboxylic acids and glycols using lipase CA as catalyst, yielding polyurethanes under mild reaction conditions (Scheme 17).

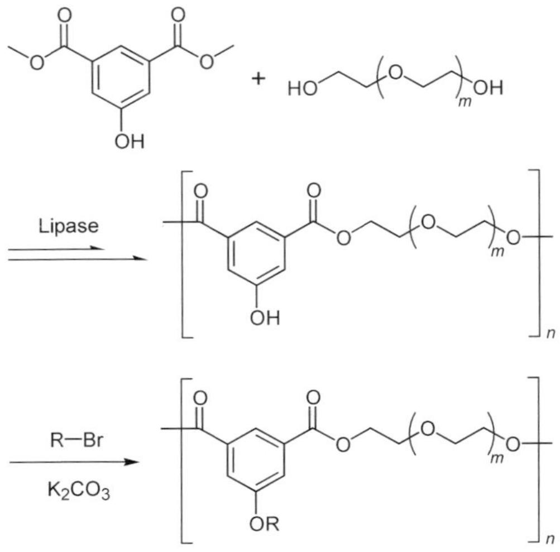

Scheme 16

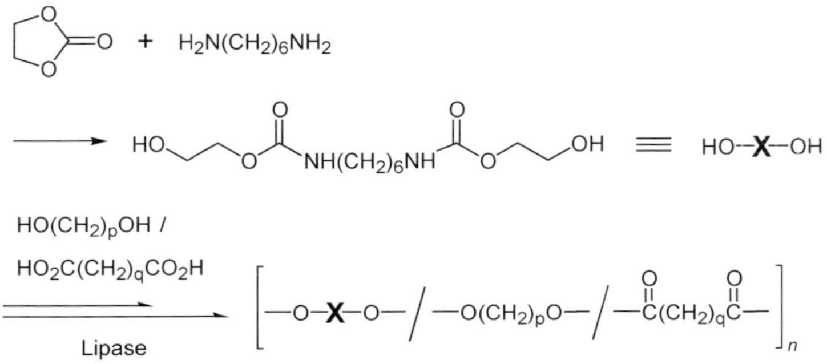

Scheme 17

A silicon oligomer was synthesized by the polycondensation of diethoxydimethylsilane using lipid-coated lipase from *Rhizopus delemar* as catalyst in isooctane containing a small amount of water [124]. Polymerization is proposed to be initiated at the OH head group of the coating lipid.

5
In Vitro PHA Polymerase-Catalyzed Polymerization to PHA

Poly[(R)-3-hydroxybutylate] (PHB) is accumulated within the cells of a wide variety of bacteria as an intracellular energy and carbon storage material [125, 126]. Poly(hydroxyalkanoate) (PHA)s are commercially produced as biodegradable plastics. In the biosynthetic pathways of PHA, the last step is the chain growth polymerization of hydroxyalkanoate CoA esters catalyzed by PHA polymerase (synthase), yielding PHA with a well-defined structure (Scheme 18). The active site of PHA polymerase is cysteine residue; however, the mechanism of this enzymatic polymerization has not been elucidated clearly.

PHA polymerase from *Alcaligenes eutrophus* polymerized the CoA monomers of (R)-hydroxyalkanoate in vitro to give high-molecular-weight homopolymers and copolymers [127, 128]. The molecular weight of the polymers was found to be proportional to the molar ratio of monomer to enzyme, and the obtained polymer had a narrow molecular weight distribution, suggesting that polymerization proceeds in a living fashion.

Random copolymers were obtained from a mixture of the two CoA esters in the presence of polymerase, whereas block copolymers were synthesized when the two monomers were reacted sequentially with the enzyme. In the polymerization of racemic 3-hydroxybutyryl CoA, only the (R)-monomer was polymerized. Furthermore, the presence of the (S)-monomer did not reduce the polymerization rate of the (R)-isomer. These data indicate that the (S)-monomer does not act as competitive inhibitor for the polymerase.

In the PHA production catalyzed by purified *Ralstonia eutropha* PHA synthase from recombinant cells, the substrate selectivity was examined [129]. A kinetic study showed that 3-hydroxybutyryl CoA is the optimal substrate for the enzyme. 3-hydroxyvaleryl CoA exhibits a slower catalytic rate and lower binding ability due to its lower reactivity of 3-hydroxyvaleryl CoA, as compared to that of 3-hydroxybutyryl CoA. The change of hydroxyl group from the β to the γ position causes dramatic decreases in the binding ability of 4-hydroxybutyryl CoA.

The polymerization behavior of 3-hydroxybutyryl CoA by purified recombinant PHA synthase from *Chromatium vinosum* was different from that of *Alcaligenes eutrophus* [130]. This enzyme lost its activity during polymerization, and the yield and molecular weight were lower than those of *Alcaligenes eutrophus*. The molecular weight did not depend on the feed ratio of the monomer and enzyme.

A novel system for surface-initiated enzymatic polymerization to a PHA film on solid surfaces was demonstrated [131]. PHA synthase from *Ralstonia eutropha* H16 was expressed as a poly-histidine fusion in *Escherichia coli* and immobilized onto several solid substrates through Ni^{2+}-nitrilotriacetic acid. The surface-initiated polymerization of 3-(R)-hydroxybutyryl CoA by immobilized PHA synthase on silicon produced a polymer film with a uniform thickness on the surface.

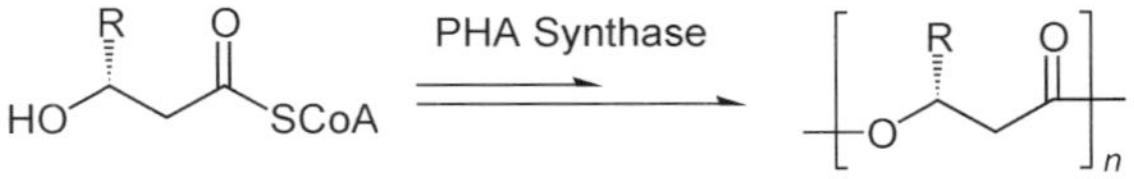

Scheme 18

In combination of this polymerase with purified propionyl CoA transferase of *Clostridium propionicum*, a two-enzyme in vitro PHB biosynthesis system was established that allowed PHB synthesis from (R)-hydroxybutyric acid as substrate [132]. In this way, the PHB synthesis was independent of the consumption of the expensive CoA, and hence PHA could be readily produced in a semipreparative scale.

6
Enzymatic Synthesis of Polycarbonates

The enzymatic synthesis of polycarbonate was first reported in 1989 [133]. Lipase CR catalyzed the polycondensation of carbonic acid diphenyl ester with bisphenol-A in an aqueous acetone to give an oligocarbonate with a molecular weight of 900.

Diethyl carbonate was polymerized with 1,3-propanediol or 1,4-butanediol by lipase CA catalyst in a successive two-step polymerization (Scheme 19) [134, 135]. The bulk reaction of an excess of diethyl carbonate and glycols under ambient pressure gave oligomeric products, followed by polymerization under vacuum to give aliphatic polycarbonates with molecular weights greater than 4×10^4.

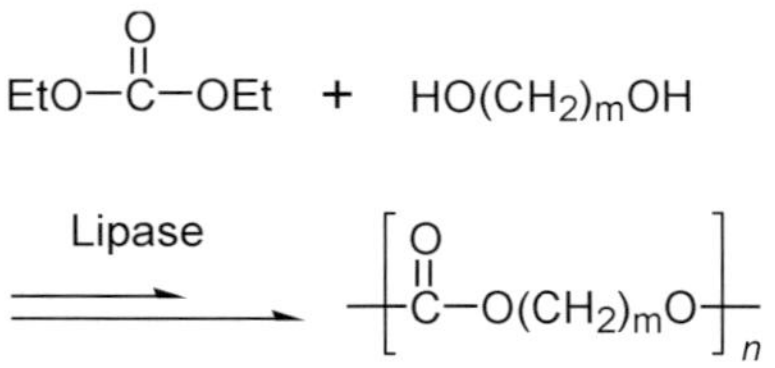

$$\text{EtO}-\overset{\overset{\displaystyle O}{\|}}{C}-\text{OEt} + \text{HO(CH}_2)_m\text{OH}$$

$$\xrightarrow{\text{Lipase}} \left[\overset{\overset{\displaystyle O}{\|}}{C}-\text{O(CH}_2)_m\text{O}\right]_n$$

Scheme 19

Activated dicarbonate, 1,3-propanediol divinyl dicarbonate, was used as a new monomer for the enzymatic synthesis of polycarbonates [136]. Lipase CA-catalyzed polymerization with α,ω-alkylene glycols produced polycarbonates with molecular weights of up to 8.5×10^3. Aromatic polycarbonates with a DP greater than 20 were enzymatically obtained from activated dicarbonate and xylylene glycols in bulk [66].

7
Concluding Remarks

This article provides an overview of in vitro syntheses of polyesters using an isolated enzyme as catalyst by polycondensations. In nonbiosynthetic pathways to polyesters, the following advantages should be claimed in comparison with fermentation and chemical processes: (i) structural variation of monomers and polymers, (ii) nontoxic catalyst and mild reaction conditions, and (iii) enantio- and regioselective polymerizations to produce functional polymers, which are very difficult to obtain by conventional methods. These features will expand the synthesis of biodegradable useful polymers with precisely controlled structures.

References

1. Wong C-H, Whitesides GM (1994) Enzymes in synthetic organic chemistry. Pergamon, Oxford
2. Whitesides GM, Wong C-H (1985) Angew Chem Int Ed 24:617
3. Jones JB (1986) Tetrahedron 42:3351
4. Klibanov AM (1990) Acc Chem Res 23:114
5. Santaniello E, Ferraboschi P, Grisenti P, Manzocchi A (1992) Chem Rev 92:1071
6. Anderson EM, Larsson KM, Kirk O (1998) Biocatal Biotrans 16:181
7. Sugai T (1999) Curr Org Chem 3:373
8. Seoane G (2000) Curr Org Chem 4:283
9. Roberts SM (2000) J Chem Soc Perkin Trans 1:611
10. Biermann U, Friedt W, Lang S, Lühs W, Machmüller G, Metzger JO, Rüsch gen. Klaas M, Schäfer HJ, Schneider MP (2000) Angew Chem Int Ed 39:2206
11. Gross RA, Kalra B (2002) Science 202:803
12. Albertsson AC, Varma IK (2002) Adv Polym Sci 157:1
13. Mecking S (2004) Angew Chem Int Ed 43:1078
14. Kobayashi S, Shoda S, Uyama H (1995) Adv Polym Sci 121:1
15. Ritter H (1993) Trends Polym Sci 1:171
16. Kobayashi S (1999) J Polym Sci A Polym Chem 37:3041
17. Kobayashi S, Uyama H, Kimura S (2001) Chem Rev 101:3793
18. Kobayashi S, Uyama H, Ohmae M (2001) Bull Chem Soc Jpn 74:613
19. Gross RA, Kumar A, Kalra B (2001) Chem Rev 101:2097
20. Uyama H, Kobayashi S (2002) J Mol Catal B Enz 19–20:117
21. Matsumura S (2002) Macromol Biosci 2:105
22. Kobayashi S, Uyama H (2002) Curr Org Chem 6:209
23. Kobayashi S, Uyama H (2003) ACS Symp Ser 840:128
24. Kobayashi S, Uyama H (1999) Macromol Symp 144:237
25. Namekawa S, Uyama H, Kobayashi S (1999) Int J Biol Macromol 25:145
26. Kobayashi S, Shoda S, Uyama H (1996) In: Salamone JC (ed) The polymeric materials encyclopedia. CRC Press, Boca Raton, FL, p 2102
27. Kobayashi S, Shoda S, Uyama H (1997) In: Kobayashi S (ed) Catalysis in precision polymerization. Wiley, Sussex, p 417
28. Kobayashi S, Uyama H (2001) Enzymatic polymerization. In: Kroschwitz JI (ed) Encyclopedia of polymer science and technology, 3rd edn. Wiley, New York
29. Binns F, Roberts SM, Taylor A, Williams CF (1993) J Chem Soc Perkin Trans I p 899
30. Linko Y-Y, Wang Z-L, Seppälä J (1995) J Biotechnol 40:133
31. Wang ZL, Hiltunen K, Orava P, Seppälä J, Linko YY (1996) J Macromol Sci Pure Appl Chem A33:599
32. Wu X, Linko YY, Seppälä J (1998) Ann NY Acad Sci 864:399–404

33. Mahapatro A, Kalra B, Kumar A, Gross RA (2003) Biomacromolecules 4:544
34. Uyama H, Inada K, Kobayashi S (1998) Chem Lett 1285
35. Binns F, Harffey P, Roberts SM, Taylor A (1998) J Polym Sci A Polym Chem 36:2069
36. Uyama H, Inada K, Kobayashi S (2000) Polym J 32:440
37. Binns F, Harffey P, Roberts SM, Taylor A (1999) J Chem Soc Perkin Trans I 2671
38. Tajiri Y, Masumoto M, Shiraishi K, Nishimura K, Kobayashi S (2004) Polym Prepr Jpn 53:5437
39. Mahapatro A, Kumar A, Kalra B, Gross RA (2004) Macromolecules 37:35
40. Nakaoki T, Danno M, Kurokawa K (2003) Polym J 35:791
41. Kobayashi S, Uyama H, Suda S, Namekawa S (1997) Chem Lett 105
42. Suda S, Uyama H, Kobayashi S (1999) Proc Jpn Acad 75B:201
43. Mezoul G, Lalot T, Brigodiot M, Maréchal E (1995) J Polym Sci A Polym Chem 33:2691
44. Berkane C, Mezoul G, Lalot T, Brigodiot M, Maréchal E (1997) Macromolecules 30:7729
45. Gryglewicz S (2001) J Mol Catal B Enz 15:9
46. Lavalette A, Lalot T, Brigodiot M, Maréchal E (2002) Biomacromolecules 3:226
47. Mezoul G, Lalot T, Brigodiot M, Maréchal E (1996) Polym Bull 36:541
48. Park HG, Chang HN, Dordick JS (1994) Biocatalysis 11:263
49. Linko Y-Y, Seppälä J (1996) CHEMTECH 26(8):25
50. Uyama H, Takamoto T, Kobayashi S (2002) Polym J 34:94
51. Nara SJ, Harjani JR, Salunkhe MM, Mane AT, Wadgaonkar PP (2003) Tetrahedron Lett 44:1371
52. Wallace JS, Morrow CJ (1989) J Polym Sci A Polym Chem 27:3271
53. Jarvie AW, Samra BK, Wiggett AJ (1996) J Chem Res (S) 129
54. Linko Y-Y, Wang Z-L, Seppälä J (1994) Biocatalysis 8:269
55. Chaudhary AK, Beckman EJ, Russell AJ (1995) J Am Chem Soc 117:3728
56. Linko Y-Y, Wang Z-L, Seppälä J (1995) Enzyme Microb Technol 17:506
57. Brazwell EM, Filos DY, Morrow CJ (1995) J Polym Sci A Polym Chem 33:89
58. Uyama H, Kobayashi S (1994) Chem Lett 1687
59. Chaundhary AK, Lopez J, Beckman EJ, Russell AJ (1997) Biotechnol Prog 13:318
60. Uyama H, Yaguchi S, Kobayashi S (1999) J Polym Sci A Polym Chem 37:2737
61. Chaundhary AK, Beckman EJ, Russell AJ (1997) Biotechnol Bioeng 55:227
62. Chaundhary AK, Beckman EJ, Russell AJ (1998) AIChE J 44:753
63. Chaundhary AK, Beckman EJ, Russell AJ (1998) Biotechnol Bioeng 59:428
64. Kline BJ, Lele SS, Lenart PJ, Beckman EJ, Russell AJ (2000) Biotechnol Bioeng 67:424
65. Uyama H, Yaguchi S, Kobayashi S (1999) Polym J 31:380
66. Rodney RL, Allinson BT, Beckman EJ, Russell AJ (1999) Biotechnol Bioeng 65:485
67. Kim DY, Dordick JS (2001) Biotechnol Bioeng 76:200
68. Takamoto T, Uyama H, Kobayashi S (2001) e-Polymers no 4
69. Mesiano AJ, Beckman EJ, Russell AJ (2000) Biotechnol Prog 16:64
70. Athawale VD, Gaonkar SR (1994) Biotechnol Lett 16:149
71. Namekawa S, Uyama H, Kobayashi S (2000) Biomacromolecules 1:335
72. Namekawa S, Uyama H, Kobayashi S (2001) Macromol Chem Phys 202:801
73. Takamoto T, Kerep P, Uyama H, Kobayashi S (2001) Macromol Biosci 1:223
74. Kobayashi S, Uyama H (1993) Macromol Chem Rapid Commun 14:841
75. Uyama H, Wada S, Fukui T, Kobayashi S (2003) Biochem Eng J 16:145
76. Uyama H, Wada S, Kobayashi S (1999) Chem Lett 893
77. Matsumura S, Okamoto T, Tsukada K, Toshima K (1998) Macromol Rapid Commun 19:295

78. Matsumura S, Okamoto T, Tsukada K, Mizutani N, Toshima K (1999) Macromol Symp 144:219
79. Ajima A, Yoshimoto T, Takahashi K, Tamaura Y, Saito Y, Inada Y (1985) Biotechnol Lett 7:303
80. Ohya Y, Sugitou T, Ouchi T (1995) J Macromol Sci Pure Appl Chem A32:179
81. Matsumura S, Takahashi J (1986) Macromol Chem Rapid Commun 7:369
82. Yoshida Y, Kawase M, Yamaguchi C, Yamane T (1995) Yukagaku 44:328
83. Noll O, Ritter H (1996) Macromol Rapid Commun 17:553
84. Kiran KR, Divakar S (2003) World J Microbiol Biotechnol 19:859
85. O'Hagan D, Zaidi NA (1993) J Chem Soc Perkin Trans 1:2389
86. O'Hagan D, Zaidi NA (1994) Polymer 35:3576
87. Shuai X, Jedlinski Z, Kowalczuk M, Rydz J, Tan H (1999) Eur Polym J 35:721
88. Mahapatro A, Kumar A, Gross RA (2004) Biomacromolecules 5:62
89. Knani D, Gutman AL, Kohn DH (1993) J Polym Sci A Polym Chem 31:1221
90. Dong H, Wang HD, Cao SG, Shen JC (1998) Biotechnol Lett 20:905
91. Iwata S, Toshima K, Matsumura S (2003) Macromol Rapid Commun 24:467
92. Wallace JS, Morrow CJ (1989) J Polym Sci A Polym Chem 27:2553
93. Margolin AL, Crenne JY, Klibanov AM (1987) Tetrahedron Lett 28:1607
94. Geresh S, Elbaz E, Glaser R (1993) Tetrahedron 49:4939
95. O'Hagan D, Parker AH (1998) Polym Bull 41:519
96. Szakács-Schmidt A, Albert L, Kelemen-Horváth I (1999) Biomed Chromatogr 13:252
97. Knani D, Kohn DH (1993) J Polym Sci A Polym Chem 31:2887
98. Gutman AL, Bravdo T (1989) J Org Chem 54:5645
99. Patil DR, Rethwisch DG, Dordick JS (1991) Biotechnol Bioeng 37:639
100. Dordick JS (1992) Ann NY Acad Sci 672:352
101. Park OJ, Kim DY, Dordick JS (2000) J Polym Chem A Polym Chem Ed 70:208
102. Uyama H, Klegraf E, Wada S, Kobayashi S (2000) Chem Lett 800
103. Kumar A, Kulshrestha AS, Gao W, Gross RA (2003) Macromolecules 36:8219
104. Fu H, Kulshrestha AS, Gao W, Gross RA, Baiardo M, Scandola M (2003) Macromolecules 36:9804
105. Mei Y, Kumar A, Gao W, Gross RA, Kennedy SB, Washburn NR, Amis EJ, Elliott JT (2004) Biomaterials 25:4195
106. Kline BJ, Beckman EJ, Russell AJ (1998) J Am Chem Soc 120:9475
107. Uyama H, Inada K, Kobayashi S (1999) Macromol Rapid Commun 20:171
108. Uyama H, Inada K, Kobayashi S (2001) Macromol Biosci 1:40
109. Tsujimoto T, Uyama H, Kobayashi S (2001) Biomacromolecules 2:29
110. Tsujimoto T, Uyama H, Kobayashi S (2002) Macromol Biosci 2:329
111. Uyama H, Kuwabara M, Tsujimoto T, Kobayashi S (2003) Biomacromolecules 4:211
112. Warwel S, Demes C, Steinke G (2001) J Polym Sci A Polym Chem 39:1601
113. Geresh S, Gilboa Y (1990) Biotechnol Bioeng 36:270
114. Geresh S, Gilboa Y, Abrahami S, Abrahami A (1993) Polym Engi Sci 33:311
115. Geresh S, Gilboa Y (1991) Biotechnol Bioeng 37:883
116. Mezoul G, Lalot T, Brigodiot M, Maréchal E (1995) Macromol Rapid Commun 16:613
117. Mezoul G, Lalot T, Brigodiot M, Maréchal E (1996) Macromol Chem Phys 197:3581
118. Kumar GS, Ghogare A, Mukesh D (1997) J Appl Polym Sci 63:35
119. Pavel K, Ritter H (1991) Macromol Chem 192:1941
120. Noll O, Ritter H (1997) Macromol Rapid Commun 18:53
121. Kumar R, Tyagi R, Parmar VS, Samuelson LA, Kumar J, Watterson AC (2003) Mol Diversity 6:287

122. Kumar R, Chen MH, Parmar VS, Samuelson LA, Kumar J, Nicolosi R, Yoganathan S, Watterson AC (2004) J Am Chem Soc 126:10640
123. Nishino H, Mori T, Okahata Y (2002) Chem Commun 2684
124. McCabe RW, Taylor A (2002) Chem Commun 934
125. Steinbüchel A (2001) Macromol Biosci 1:1
126. Taguchi S, Doi Y (2004) Macromol Biosci 4:145
127. Lenz RW, Farcet C, Dijkstra PJ, Goodwin S, Zhang S (1999) Int J Biol Macromol 25:55
128. Su L, Lenz RW, Takagi Y, Zhang S, Goodwin S, Zhong L, Martin DP (2000) Macromolecules 33:229
129. Zhang S, Yasuo T, Lenz RW, Goodwin S (2000) Biomacromolecules 1:244
130. Jossek R, Reichelt R, Steinbüchel A (1998) Appl Microbiol Biotechnol 49:258
131. Kim YR, Paik HJ, Ober CK, Coates GW, Batt CA (2004) Biomacromolecules 5:889
132. Jossek R, Steinbüchel A (1998) FEMS Microbiol Lett 168:319
133. Abramowicz DA, Keese CR (1989) Biotechnol Bioeng 33:149
134. Matsumura S, Harai S, Toshima K (1999) Proc Jpn Acad 75B:117
135. Matsumura S, Harai S, Toshima K (2000) Macromol Chem Phys 201:1632
136. Rodney RL, Stagno JL, Beckman EJ, Russell AJ (1999) Biotechnol Bioeng 62:259

Enzymatic Polymerization to Polysaccharides

Shiro Kobayashi[1,2] (✉) · Masashi Ohmae[2]

[1]*Present address:*
R & D Center for Bio-based Materials, Kyoto Institute of Technology, Matsugasaki, Sakyo-ku, 606-8585 Kyoto, Japan
kobayash@kit.ac.jp

[2]Department of Materials Chemistry, Graduate School of Engineering, Kyoto University, 615-8510 Kyoto, Japan

1	Introduction	160
2	Characteristics of Enzymatic Polymerization	161
3	Synthesis of Polysaccharides Via Enzymatic Polymerization	164
3.1	Polysaccharides and Glycosylation	164
3.2	Cellulose	167
3.2.1	Brief Historical Background	167
3.2.2	Synthesis Via Enzymatic Polymerization	168
3.2.3	Reaction Mechanism	171
3.2.4	High-Order Molecular Structures	174
3.2.5	Extension to Cellulose-Related Saccharides	177
3.3	Amylose	182
3.4	Xylan	183
3.5	Chitin	183
3.5.1	Synthesis Via Enzymatic Polymerization	183
3.5.2	Mechanistic Aspects	185
3.5.3	High-Order Molecular Structures	190
3.5.4	Chitin Oligomer Derivatives	192
3.6	Glycosaminoglycans	193
3.6.1	Hyaluronan and Its Derivatives	194
3.6.2	Chondroitin and Its Derivatives	200
3.6.3	Synthesis of Glycosaminoglycans by Other Methods	202
3.7	Unnatural Polysaccharides	203
4	Concluding Remarks	205
	References	207

Abstract Polysaccharides are formed by repeated glycosylations between a sugar donor and a sugar acceptor. This chapter reviews the in vitro synthesis of polysaccharides by "enzymatic polymerization." As catalyst, a hydrolase enzyme is used where the monomer was designed based on the concept of a "transition-state analog substrate" (TSAS), sugar fluoride monomers for the polycondensation, and sugar oxazoline monomers for the ring-opening polyaddition. Enzymatic polymerization enabled the first in vitro synthesis of natural polysaccharides such as cellulose, xylan, chitin, hyaluronan, and chondroitin, and also of unnatural polysaccharides such as a cellulose-xylan hybrid, a regularly methylated cellulose, and derivatives of hyaluronan and chondroitin. The polymerization mechanism is discussed here. Mutant enzymes were effective for polysaccharide synthesis, where only polycondensation of monomers was induced by suppressing the hydrolysis activity. The polymerization principle was extended to a stepwise synthesis of oligosaccharides. By using characteristics of the polymerization to produce a polysaccharide by a single step, in situ observations of the polymerization reaction by TEM, POM, SEM, and SANS revealed the formation of various high-order molecular structures from the polysaccharide molecules.

Keywords Cellulose · Chitin · Enzymatic polymerization · Glycosaminoglycan · Polysaccharide

Abbreviations
Ala	alanine
BTH	bovine testicular hyaluronidase
CBD	cellulose binding domain
Ch	chondroitin
Chi-oxa	chitobiose oxazoline monomer
DMF	dimethylformamide
DMSO	dimethyl sulfoxide
DP	degree of polymerization
ECM	extracellular matrix
EG	endoglucanase
GAG	glycosaminoglycan
GalNAc	N-acetyl-D-galactosamine
GlcA	D-glucuronic acid
GlcNAc	N-acetyl-D-glucosamine
Glu	glutamic acid
HA	hyaluronic acid (hyaluronan)
MALDI-TOF	matrix-assisted laser desorption ionization time-of-flight
OTH	ovine testicular hyaluronidase
POM	polarization optical microscopy
SANS	small-angle neutron scattering
SEM	scanning electron microscopy
TEM	transmission electron microscopy
TSAS	transition-state analogue substrate
UDP	uridine 5′-diphospho-
α-CF	α-cellobiosyl fluoride
β-CF	β-cellobiosyl fluoride
α-LF	α-lactosyl fluoride
β-LF	β-lactosyl fluoride
α-MF	α-maltosyl fluoride
β-XF	β-xylobiosyl fluoride

1
Introduction

The advance of polymer science has been due to the development of new polymers produced by the polymerization of new monomers. All polymer-

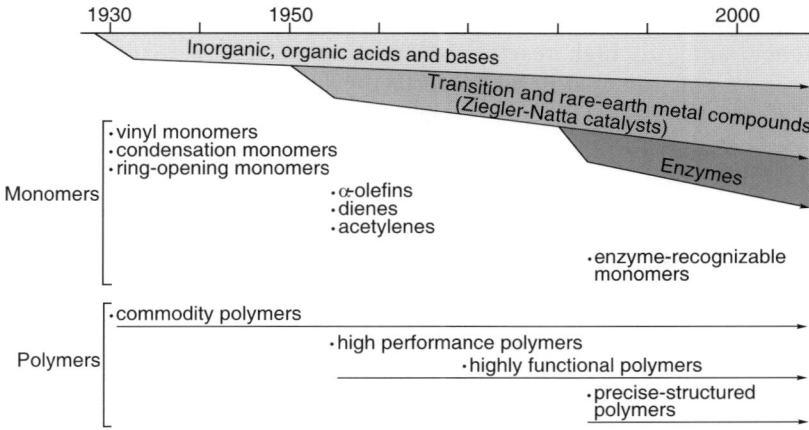

Fig. 1 A rough sketch in historical development of polymerization catalysts along with monomers and product polymers by respective catalysts

izations need a catalyst or an initiator. There have been two major streams for polymerization catalysts. The first utilizes classical catalysts such as acids (Brønsted acids, Lewis acids, and various cations), bases (Lewis bases and various anions), and radical generating species. The second involves the use of transition metals and rare-earth metal compounds as catalysts (Fig. 1). The former stream started with the beginning of polymer chemistry in the 1920s, and the latter was initiated by the discovery of Ziegler–Natta catalysts in the early 1950s followed by metathesis catalysts and rare-earth metal catalysts. These two major streams are still developing for "precision polymerization" (in a wider sense, "precision polymer synthesis"), whose ultimate goal is to allow the formation of defect-free polymers with a precisely controlled, desired structure at the molecular level [1].

A more specific catalysis was developed using enzymes as catalysts for polymerizations ("enzymatic polymerization") starting in the mid-1980s, which can now be regarded as a third stream of polymerization catalyst. Normally, monomers of enzymatic polymerization are to be recognized and activated by the enzyme for the polymerization to occur, and thus polymers with precisely controlled structures are expected. This articles concentrates on in vitro polysaccharide synthesis via enzymatic polymerization.

2
Characteristics of Enzymatic Polymerization

All the substances including biomacromolecules in vivo are produced by enzymatic catalysis via biosynthetic pathways. Enzymatic catalysis has the following advantageous characteristics in general: (i) a high catalytic activity (high turnover number); (ii) reactions under mild conditions with respect to temperature, pressure, solvent, neutral pH, etc., bringing about energetic efficiency; and (iii) high reaction selectivities of regio-, enantio-, chemo-, and stereoregulation, giving rise to perfectly structure-controlled products. If these in vivo characteristics can be realized for in vitro polymer synthesis, we may expect the following outcomes: (i) precise control of polymer structures, (ii) creation of polymers with a new structure, (iii) a clean process (without byproducts), (iv) a low loading process (energy saving), and (v) product polymers would be biodegradable in many cases. These are often difficult to realize by conventional methods.

In fact, it has been proven over the last two decades that enzymatic polymerization is a powerful method for polymer synthesis [2–14]. Thus, the definition of enzymatic polymerization is a "chemical polymer synthesis in vitro (in test tubes) via nonbiosynthetic (nonmetabolic) pathways catalyzed by an isolated enzyme" [2, 4, 8–10, 12–14].

All enzymes are classified into six categories (Table 1) [2, 8–10, 13]. Of these, hydrolase enzymes are most often used for enzymatic polymerization, because they are readily isolable due to their stability and widely available as commercial products.

Very fundamental and important characteristics in enzymatic reactions are described here as to the following two aspects. First, a "key and lock" theory proposed by Fischer in 1894 [15] pointed out the relationship between substrate and enzyme. This theory implies that via biosynthetic pathways in vivo a substrate and an enzyme correspond very strictly in a 1 : 1 fashion like a key and a lock. This phenomenon is nowadays understood in the following way; they recognize each other and form an enzyme-substrate complex, shown as (A) in Fig. 2. The formation of the complex activates

Table 1 Classification of enzymes, their examples, and typical polymers produced in vitro by enzymatic catalysis

Enzymes	Example enzymes	Typical polymers
Oxidoreductases	Peroxidase, laccase, tyrosinase, glucose oxidase	Polyphenols, polyanilines, vinyl polymers
Transferases	Glycosyltransferase, acyltransferase	Polysaccharides, cyclic oligosaccharides, polyesters
Hydrolases	Glycosidase (cellulase, amylase, chitinase, hyaluronidase), lipase, peptidase, protease	Polysaccharides, polyesters, polyamides, polycarbonates, poly(amino acid)s
Lyases	Decarboxylase, aldolase, dehydratase	
Isomerases	Racemase, epimerase	
Ligases	Acyl CoA synthetase	

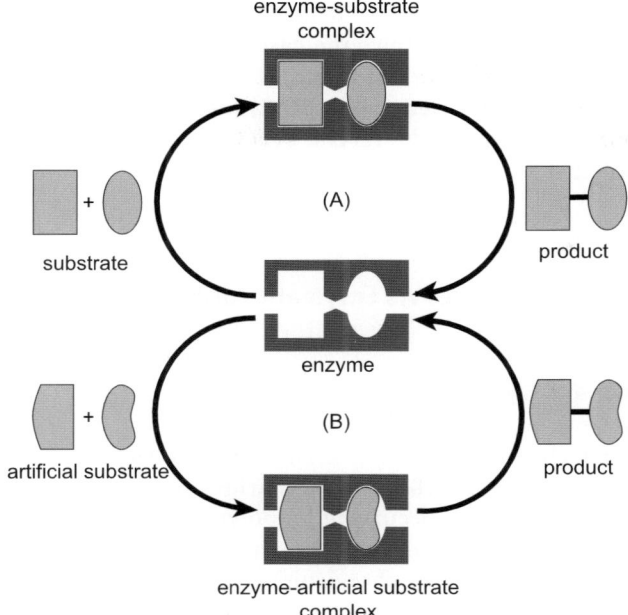

Fig. 2 An enzyme-substrate relationship for enzymatic reactions. **A** "Key and Lock" theory valid for all in vivo reactions. **B** An enzyme forming a complex with an artificial substrate to induce an in vitro enzymatic reaction

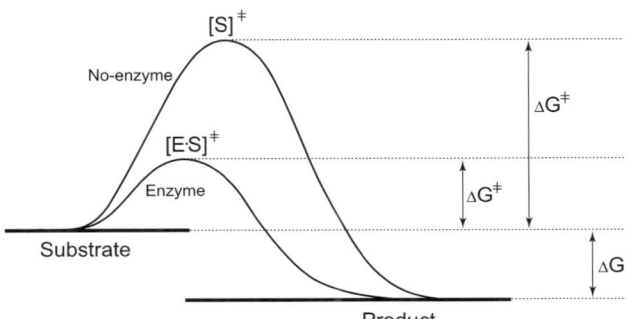

Fig. 3 All the enzymatic reactions proceed under mild reaction conditions due to the formation of the enzyme-substrate complex, which stabilizes the transition state of the reaction by lowering the activation energy in comparison with the no-enzyme case

the substrate to lead to a product with perfect structure control. The complex formation is due to the supramolecular interactions mainly of hydrogen bonding.

Second, Pauling described why enzymatic reactions proceed under such mild reaction conditions [16, 17]. The formation of the enzyme-substrate complex stabilizes the *transition state* and significantly lowers the activation energy compared with the no-enzyme case (Fig. 3). This suggests that once such a complex is formed, an enzymatic reaction is reversible; a hydrolysis enzyme catalyzing a bond cleavage in vivo is able to catalyze a bond formation of monomers (formally the reverse reaction of hydrolysis) to eventually produce a polymer.

3
Synthesis of Polysaccharides Via Enzymatic Polymerization

3.1
Polysaccharides and Glycosylation

In carbohydrate chemistry, a polysaccharide is a molecule having more than ten monosaccharide units [18]. Polysaccharides belong to one of three important classes of naturally occurring biopolymers, the other two being nucleic acids (DNA and RNA) and proteins. Cellulose, amylose, xylan, chitin, hyaluronic acid (hyaluronan), chondroitin, and chondroitin sulfate can be cited as typical natural polysaccharides (Scheme 1).

The former four are called homopolysaccharides since the fundamental sugar repeating unit is of the same structure; D-glucose is linked via a $\beta(1 \rightarrow 4)$ glycosidic bond in cellulose and $\alpha(1 \rightarrow 4)$ in amylose, xylose and N-acetyl-D-glucosamine being via $\beta(1 \rightarrow 4)$ in xylan and chitin, respectively. Polysaccharides having a $\beta(1 \rightarrow 4)$ structure are called "structural polysaccharides," which provide plants and animals with physical strength. Amylose with an $\alpha(1 \rightarrow 4)$ structure is an "energy-storage polysaccharide." The latter three are denoted as heteropolysaccharides since the repeating sugar unit is constituted from different sugar molecules—hyaluronan from a D-glucuronic acid $\beta(1 \rightarrow 3)$ N-acetyl-D-glucosamine disaccharide unit linked via a $\beta(1 \rightarrow 4)$ glycosidic bond and chondroitin from a D-glucuronic acid $\beta(1 \rightarrow 3)$ N-acetyl-D-galactosamine disaccharide unit linked via a $\beta(1 \rightarrow 4)$ glycosidic bond.

Of the three above-mentioned major classes of natural biopolymers, chemical synthesis of nucleic acids (in particular DNA) and proteins had already been achieved by the Merrifield solid-phase synthesis method; a computer-controlled automated synthesizer became commercially available more than 20 years ago. However, the chemical synthesis of polysaccharides like cellulose was far more difficult than that of the former two classes. Polysaccharide synthesis is the repetition of glycosylation, the most fundamental reaction in carbohydrate chemistry (Scheme 2).

The main reasons for the difficulty during glycosylation are (i) controlling the stereochemistry of the anomeric C1 carbon and (ii) regioselectivity control of many hydroxyl groups with similar reactivity.

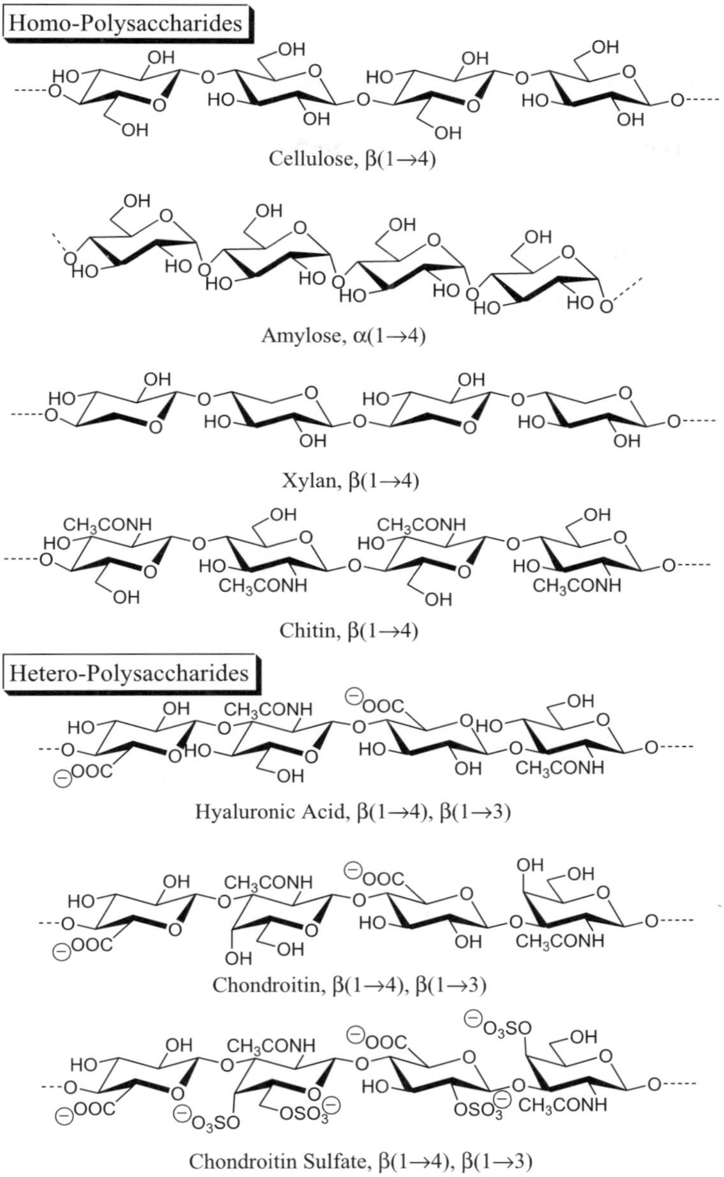

Enzymatic Polymerization to Polysaccharides

Homo-Polysaccharides

Cellulose, β(1→4)

Amylose, α(1→4)

Xylan, β(1→4)

Chitin, β(1→4)

Hetero-Polysaccharides

Hyaluronic Acid, β(1→4), β(1→3)

Chondroitin, β(1→4), β(1→3)

Chondroitin Sulfate, β(1→4), β(1→3)

Scheme 1

To overcome difficulty (i), a leaving group X of the glycosyl donor has been extensively studied for more than 100 years, which is a continued central issue in the area of carbohydrate chemistry. The activation of the C1 carbon is achieved by combination of X with a catalyst. The following glycosylations are well known; they are, chronologically, (1) the Koenigs–Knorr method (1901) using bromine or chlorine at the C1 carbon of glycosyl donors with $AgClO_4$ or $Hg(CN)_2$ as promoter [19]; (2) the Helferich method (1993) and the modified method (1981); (3) the imidate method (1978); (4) the modified imidate method (1980) using trichloroacetyl imidate with $(CH_3)_3SiOSO_2CF_3$ as promoter [20]; and (5) the glycosyl fluoride method (1981) with $AgClO_4 - SnCl_2$. In terms of effectiveness in controlling the stereochemistry, the modified imidate method appears to be most often used one in recent times.

To solve difficulty (ii), the hydroxyl group on which the reaction should not take place is normally to be protected by appropriate groups (R) as shown in Scheme 2. After the reaction, the protected groups are removed (deprotected). Therefore, the protection-deprotection procedure is always involved in the glycosylation reaction.

For reference, a number of isomers can be conceivable; cello-trimer is composed of three D-glucose units connecting through $\beta(1 \rightarrow 4)$ glycosidic linkage, whereas 120 kinds of trisaccharide isomers can be theor-

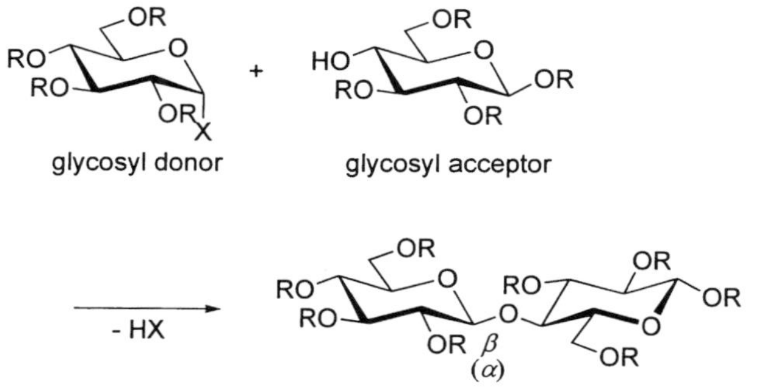

Scheme 2

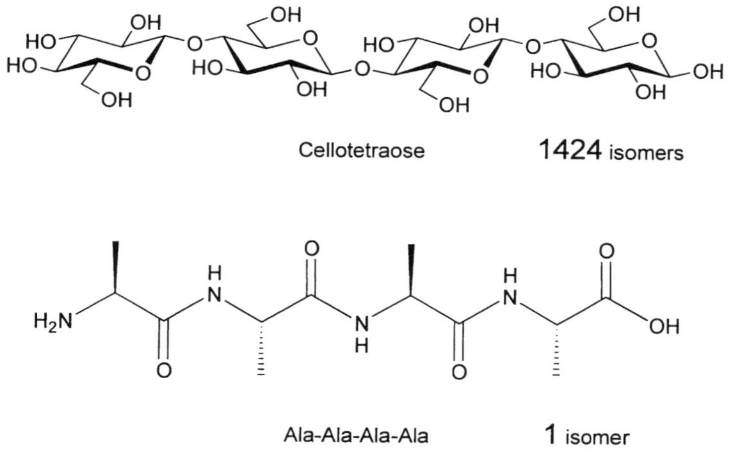

Cellotetraose   **1424** isomers

Ala-Ala-Ala-Ala   **1** isomer

Scheme 3

etically formed from three D-glucose molecules, 1424 isomers from four D-glucose molecules including cello-tetramer, and 17 872 isomers from five D-glucose molecules including cello-pentamer [21]. In contrast, four L-alanine molecules can afford only one tetrapeptide, L-alanyl-L-alanyl-L-alanyl-L-alanine (Scheme 3). This simple example shows the complexity of carbohydrate chemistry.

3.2
Cellulose

3.2.1
Brief Historical Background

Cellulose is the most abundant organic substance on the earth; some hundred billion tons of cellulose are annually photosynthesized with carbon dioxide fixation on land and in the sea. It is one of the major components in higher plant cell walls (60–80%), along with hemicellulose (10–20%) and lignin (10–30%). Cellulose is one of the oldest raw materials in such industries as paper, fibers, and lumber, and it continues to attract broad and new interest, even in advanced materials, due to its biocompatibility, chirality, structure-forming capability, and environmentally benign property [22–25].

Cellulose is a linear polysaccharide with a $\beta(1 \rightarrow 4)$ glycosidic bond structure of a dehydrated D-glucose repeating unit. It is a representative of polysaccharides that had been an important target molecule from the 1920s for Standinger, Mark, and other prominent polymer scientists due to its structure, molecular weight, and derivatization reactions [14, 22]. Its chemical synthesis had been a challenging problem in polymer chemistry since the first attempt in 1941 [26]. Despite considerable efforts at synthesis, all attempts came to naught for half a century. Following are typical examples.

The polycondensation of 2,3,6-tri-O-phenylcarbamoyl-D-glucose in $CHCl_3$/DMSO in the presence of P_2O_5 as a dehydrating agent was carried out to yield a polysaccharide with DP claimed as ca. 60; however, regio- and stereoregulations of the reaction were not accomplished (Scheme 4) [27]. Synthesis of polysaccharides with a well-defined structure from the derivatives of monosaccharides or disaccharides containing a glycosidic OH group, halogen or acetate group, typical orthoesters, and thio-analogs by polycondensation failed to form the desired products [28].

$R = CONHC_6H_5$

Scheme 4

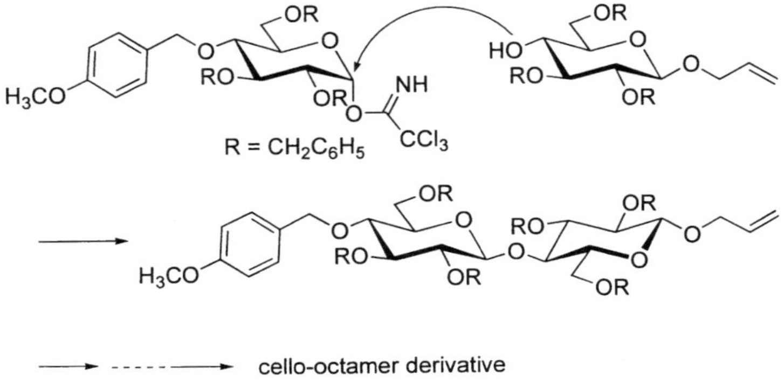

Scheme 5

Cationic ring-opening polymerization of a bicyclic acetal monomer was a possible method to synthesize cellulose [29]. A polysaccharide of DP ~ 20 was obtained; however, a regioselective ring-opening did not take place. The major component was a tetrahydrofuran-ring unit and not a six-membered glucopyranose unit.

In the carbohydrate chemistry approach, repeated glycosylation is able to produce oligo- or polysaccharides. From tribenzylated glucoside four kinds of protecting groups with different reactivities were to be used, and glycosylation was achieved by the imidate method to form a $\beta(1 \rightarrow 4)$ bond (Scheme 5) [30]. Seven times glycosylations gave cello-octamer derivatives, and further deprotection yielded the cello-octamer.

3.2.2
Synthesis Via Enzymatic Polymerization

None of these conventional methods was effective enough for cellulose synthesis. In addition to the present-day knowledge of chemistry at the time, enzymes were introduced as catalysts for cellulose synthesis [14]. Then, taking the hypothesis of an artificial substrate as shown in Fig. 2B into consideration, β-cellobiosyl fluoride (β-CF) was designed as substrate monomer for catalysis by cellulase, an extracellular hydrolysis enzyme. It was found that the reaction proceeded smoothly to produce synthetic cellulose via a one-step polycondensation that liberated the HF molecule (Scheme 6). In the reaction β-CF acted as both donor and acceptor [31].

A typical polymerization example is shown. The monomer β-CF and cellulase (*Trichoderma viride*, 5 wt % for β-CF) in a 5 : 1 mixture of acetonitrile and 0.05 M acetate buffer (pH 5) solution were stirred at 30 °C. The initially homogeneous solution became heterogeneous with a white precipitation of the product. After 12 h, the resulting suspension was heated at 100 °C to inactivate the enzyme and poured into an excess amount of methanol/water

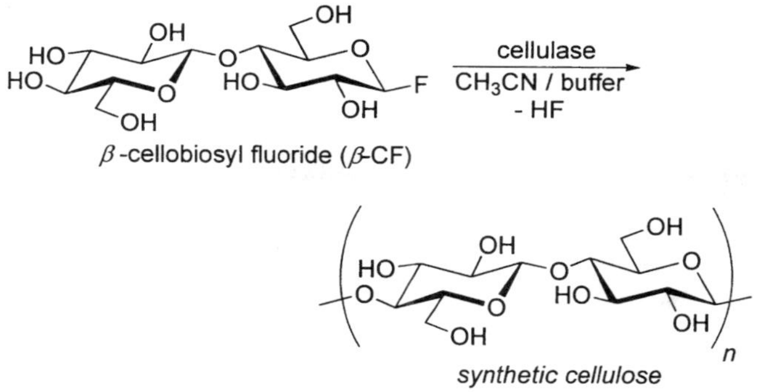

Scheme 6

to separate the insoluble part by filtration. This part was further purified by suspending it in water and isolating it by filtration, giving rise to the water-insoluble product of the white powdery materials in 54% yields. The water-insoluble part was established by various spectroscopic measurements as the first clear-cut example of "synthetic cellulase" having a $\beta(1 \rightarrow 4)$ structure. It decomposed at ca. 260 °C without showing a melting point, which is similar to natural cellulose. Its DP value was around 22, and hence it was a polysaccharide of cellulose.

Some polymerization results are given in Table 2. Polymerization of β-CF catalyzed by *Trichoderma viride* in CH_3CN/buffer (5 : 1) gave the best results in terms of synthetic cellulose yield. Of the organic solvents screened, acetonitrile was most effective for promotion of the reaction. DMF and DMSO were probably too polar to maintain the enzyme activity. Buffer was necessary to neutralize the HF molecule in the reaction mixture. An organic cosolvent was needed to prevent the hydrolysis of β-CF.

The design of the β-CF monomer involves four important features. (1) The cellobiose (disaccharide) structure of β-CF was chosen so that the monomer could be recognized and activated by the enzyme because the smallest unit was speculated to be a cellobiose repeating unit from the cellulose structure and not a glucose unit. (2) Fluorine has an atomic size (covalent radius, 0.64 Å) close to that of oxygen (0.66 Å) constituting the glycosidic bond of cellulose. (3) The fluoride anion is one of the best leaving groups. (4) Of the glycosyl halides only glycosyl fluorides are stable as the unprotected form, which is required for most enzymatic reactions carried out in the presence of water. After all, combination of cellulase and β-CF readily leads to a transition state that is close in structure to that involved in the hydrolysis of cellulose by cellulase ("transition-state analog substrate", TSAS). This cellulose synthesis reaction opened a new door to enzymatic polymerization for the synthesis of not only polysaccharides but also other various types of polymers.

Table 2 Enzymatic polymerization of β-CF catalyzed by cellulase from different origin in various solvents [a]

Entry	Solvent[b]	Cellulase	Product yield/%[c]
1	CH_3CN / buffer (5 : 1)	T. viride	64 (54)
2	C_2H_5CN / buffer (5 : 1)	T. viride	10 (3)
3	$(CH_3)_2CO$ / buffer (5 : 1)	T. viride	13 (7)
4	CH_3OH / buffer (5 : 1)	T. viride	17 (6)
5	C_2H_5OH / buffer (5 : 1)	T. viride	20 (7)
6	CH_3NO_2 / buffer (5 : 1)	T. viride	15 (1)
7	DMF / buffer (5 : 1)	T. viride	~0 (0)
8	DMSO / buffer (5 : 1)	T. viride	~0 (0)
9	buffer	T. viride	0 (0)
10	CH_3CN / buffer (2 : 1)	T. viride	16 (3)
11	CH_3CN / buffer (7.5 : 1)	T. viride	77 (31)
12	CH_3CN / buffer (10 : 1)	T. viride	67 (14)
13	CH_3CN / buffer (5 : 1)	A. niger	30 (8)
14	CH_3CN / buffer (5 : 1)	P. tulipiferae	43 (25)
15	CH_3CN / buffer (5 : 1)	β-glucosidase	0 (0)

[a] Polymerization of β-CF at 30 °C for 12 h : [β-CF] = 2.5×10^{-2} mol/L, cellulase 5 wt % for β-CF
[b] Acetate buffer (0.05 M, pH 5)
[c] 5 : 1 methanol/water insoluble part. Synthetic cellulose yield is given in parentheses

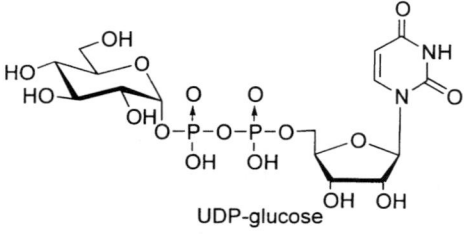

Scheme 7

In contrast to the reaction in Scheme 6, nature uses UDP-α-glucose as monomer for the catalysis of cellulose synthase (Scheme 7). The stereochemistry of the C1 carbon inverts from α to β during a reaction.

3.2.3
Reaction Mechanism

Enzymatic reactions are the most vividly studied phenomena in organic chemistry and biochemistry. Such studies concern the mechanistic and enzyme-structural aspects. Since the 1990s many results of 3D structure analysis of enzymes have become available. For example, Fig. 4 illustrates the structure of the catalytic core domain of the family 5 endoglucanase, which suggests that the catalytic acid/base and enzymatic nucleophile are due to carboxylic acid residues Glu 139 and Glu 228, respectively [32]. In the catalytic site, cellotriose is included for reference.

In the hydrolase-catalyzed hydrolysis of polysaccharides, two carboxylic acid residues in the active site are generally involved. In the cellulase catalyzed hydrolysis of cellulose (Fig. 5A) one residue pulls (protonation on) the glycosidic oxygen atom and the other pushes the C1 carbon atom in the general acid-base mode to facilitate the cleavage of the glycosidic bond (stage a). After the cleavage, a highly reactive intermediate (or transition state) of glycosyl carboxylate is formed. This intermediate may have another structure of an oxocarbenium ion (stage b). Then, a water molecule attacks the C1 carbon to end up with the formation of the hydrolysis product (stage c).

On the other hand, polymerization mechanism is speculated as follows (Fig. 5B). The β-CF monomer is readily recognized at the donor site and activated via a general acid-base mode (a push-pull mechanism) to cleave the C – F bond by kicking the fluoride anion off (stage a) and to form a highly reactive glycosyl carboxylate intermediate (or transition state). This intermediate may have an oxocarbenium ion structure (stage b). The 4-hydroxyl group of the β-CF monomer or of the growing chain end attacks the C1 car-

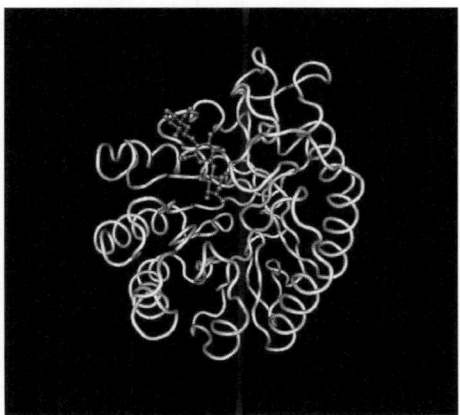

Fig. 4 A crystalline structure of cellulase (core domain of *Bacillus agaradherans*) determined by X-ray analysis. Cellotriose is placed in the active site

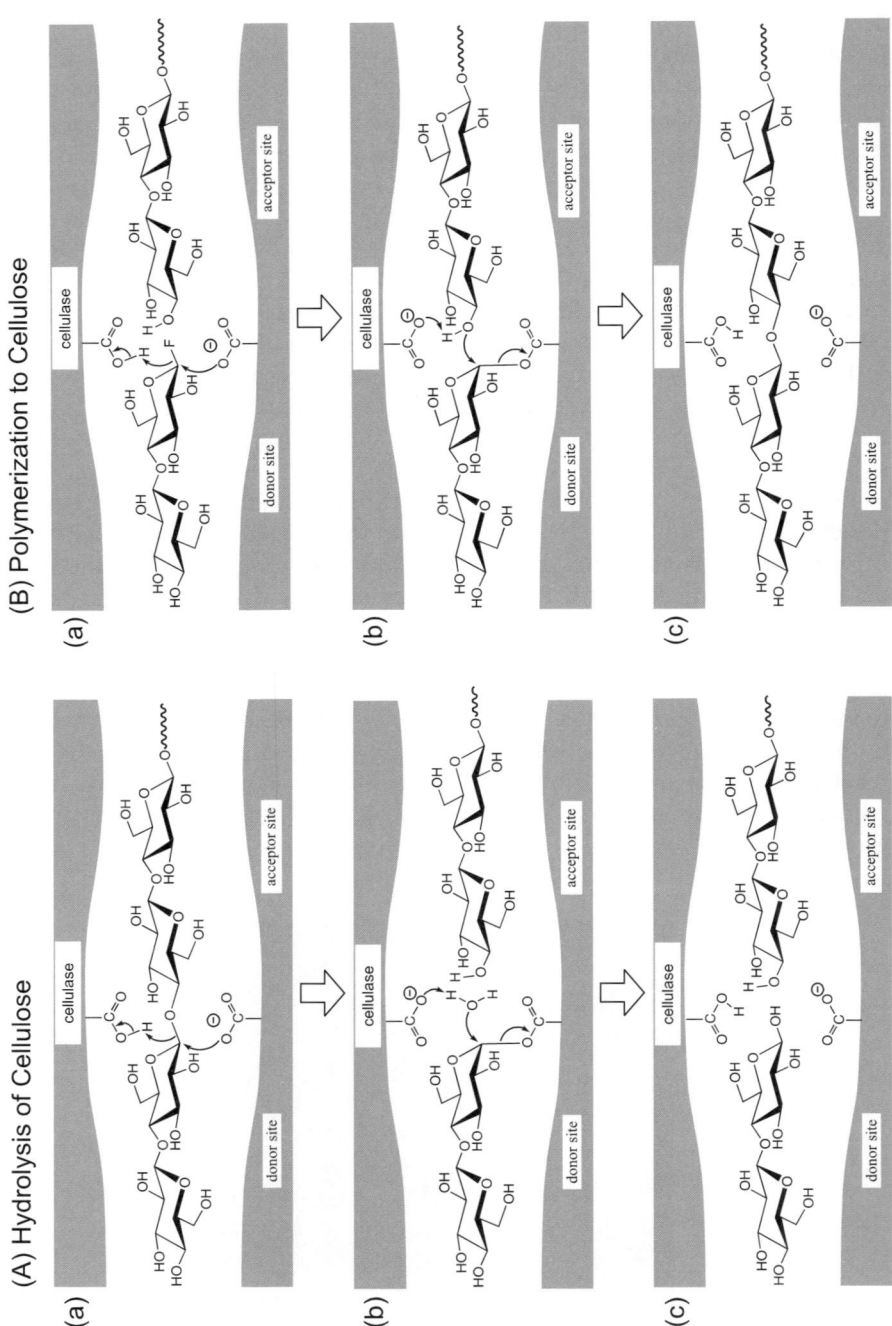

Fig. 5 Possible reaction mechanisms of cellulase catalysis. **A** Hydrolysis of cellulose. **B** Polymaerization of β-CF to synthetic cellulose

bon of this intermediate from the exo-direction because the endo-direction is blocked by the carboxylate group. Thus, $\beta(1 \rightarrow 4)$-glycosidic bond formation is achieved to give a one-monomer (cellobiose) unit-elongated product (stage c), which shifts to the right for the next monomer β-CF coming in at the donor site. Repetition of stages a–c produces synthetic cellulose. Therefore, the polymerization mechanism belongs to a "monomer-activated mechanism".

During glycosylation, the configuration of C1 carbon inverts from β to α from stage a to stage b and inverts again from α to β from stage b to stage c. This double-inversion mechanism brings about the retention of stereochemistry of C1 from the monomer to the product cellulose. This similarity in catalysis function is obvious, which involves an intermediate (stage b) common to both reactions; the only difference is the substrate, which reacts nucleophilically on the C1 carbon at stage b: water in hydrolysis and the 4-hydroxyl group of a glucose unit in polymerization.

The monomer structure of β-CF is essential so that β-CF can be recognized and activated by forming an enzyme-substrate complex, and then a fluoride anion is cleaved off to lead to intermediate b. That is, for the recognition of the monomer by cellulase, its structure must be close to that of cellulose or of stage a, which β-CF satisfies. It is highly probable that the transition state comes very early, which lies rather close to stage a going to stage b. Thus, β-CF can be taken as a TSAS monomer, which is a new concept for monomer design in enzymatic polymerization [8–10, 12–14]. The reason for using the term TSAS and not "intermediate analog substrate" has been described elsewhere [14].

The monomer design of using β-CF of two glucose unit structures was important. When β-glucosyl fluoride of a monosaccharide structure was used, its polymerization by cellulase catalysis produced only cello-oligomers. This indicates that the monosaccharide fluoride is not appropriate enough as a starting monomer compared with β-CF. The effectiveness of the disaccharide monomer structure β-CF was considered as a possibile explanation of the cellulose biosynthesis mechanism [33–35].

In relation to the substrate specificity of cellulase catalysis, α-cellobiosyl fluoride (α-CF, the anomer of β-CF) was not reacted under reaction conditions similar to those of β-CF polymerization, α-CF being recovered unchanged (Scheme 8). This suggests that α-CF cannot be recognized by cel-

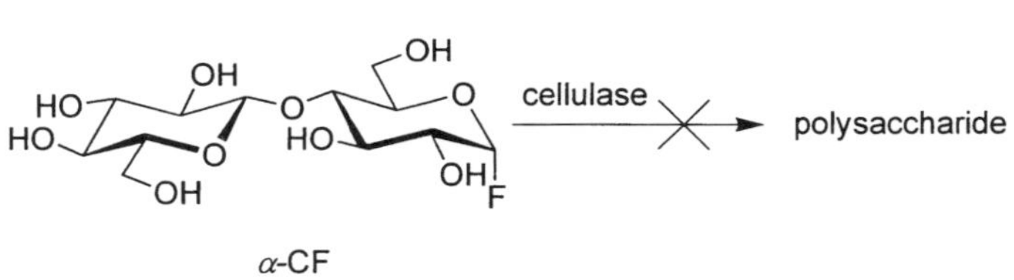

α-CF

Scheme 8

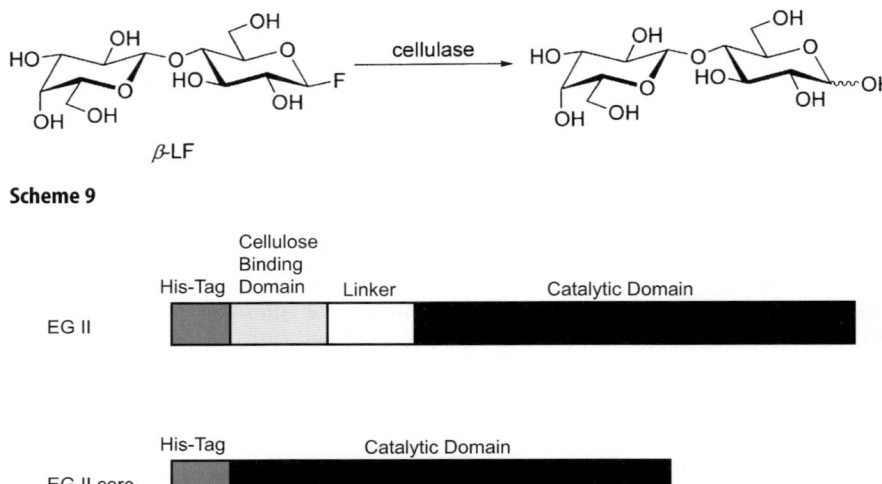

Fig. 6 Schematic illustrations of endoglucanase II fusion protein (EG II) expressed by *Saccaromyces sereviside* and the mutant endoglucanase (EG II core)

lulase. The cellulase-catalyzed reaction of β-lactosyl fluoride (β-LF) quantitatively yielded the corresponding hydrolysis product of lactose (Scheme 9). This result indicates that β-LF is recognized at the donor site of cellulase and activated to react with water, but not at the acceptor site [31].

Very recent results gave important information on the role of the cellulase function [36]. Cellulase (endoglucanase, EG) is constituted from three domains: cellulose binding domain (CBD), linker domain, and catalytic domain. In hydrolysis, CBD first binds cellulose molecules and then the catalytic domain catalyzes their hydrolysis, the linker domain linking these two domains. Using biotechnology, two types of protein enzymes were prepared from yeast, the one having all domains (EG II, molecular weight 5.0×10^4) and the other having only the catalytic domain lacking CBD and the linker domain (EG II core, molecular weight 3.8×10^4) (Fig. 6). Both enzymes showed high polymerization activity for monomer β-CF, giving synthetic cellulose. With time, the cellulose produced gradually disappeared with EG II due to hydrolysis, but the product cellulose hardly changed with the EG II core. These results suggest that the CBD plays an important role in the hydrolysis of the product, but not for the polymerization of β-CF.

3.2.4
High-Order Molecular Structures

In living systems, cellulose forms high-ordered crystalline structures of different allomorphs. Primarily two allomorphs are typical: cellulose I showing a parallel structure of glucan chains and cellulose II showing an antiparallel structure (Fig. 7) [37, 38]. Cellulose I is a thermodynamically metastable form, which had long been believed to be produced only in living systems. Cellulose II, in contrast, is thermodynamically more stable; therefore, once cellulose I is converted, for example, via Mercelization, to cellulose II, it never returns to cellulose I.

Utilizing the characteristics of the polymerization that produces synthetic cellulose via a one-step reaction, the in situ polymerization system by TEM was directly observed. Polymerization of β-CF by a crude cellulase catalyst in acetonitrile/buffer (5 : 1) solution yielded the product of irregular rodlets characteristic of a cellulose II allomorph [39]. Using a partially purified cellulase as catalyst, the assembly of synthetic cellulose I was accomplished in an optimized acetonitrile/buffer ratio (2 : 1) via "choroselective" polymerization as illustrated in Fig. 7. The cellulose I allomorph was characterized by TEM (Fig. 8), electron diffraction, and cellobiohydrolase I-colloidal gold binding [40]. Formation of a crystalline synthetic cellulose I allomorph was

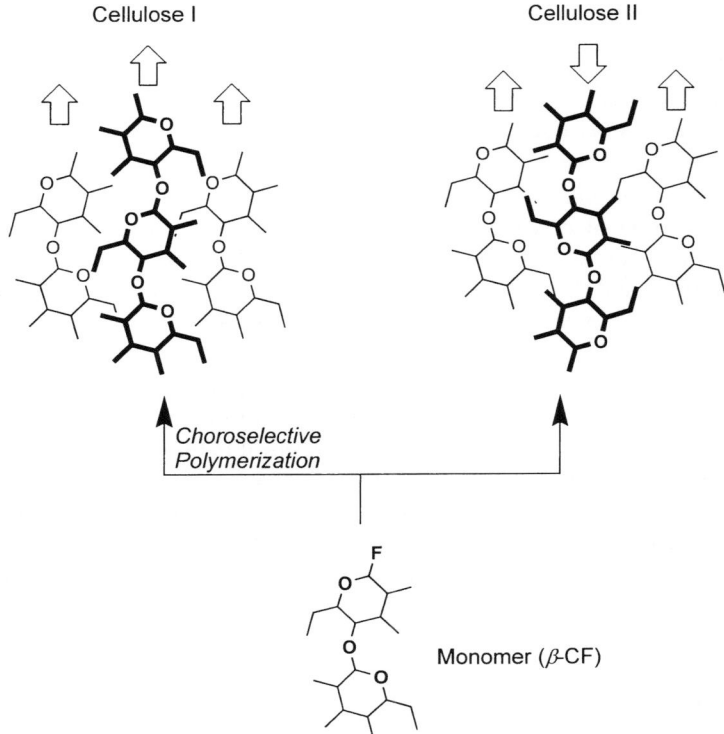

Fig. 7 Two allomorphs of cellulose I (parallel) and cellulose II (anti-parallel) structures. Choroselective polymerization of β-CF produced cellulose I or cellulose II

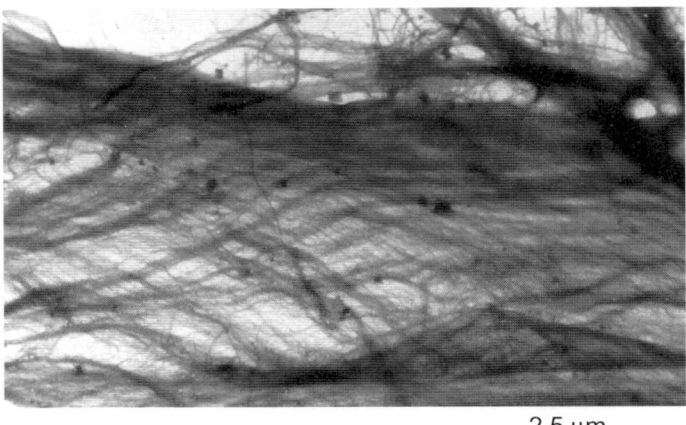

Fig. 8 Microfibrils of synthetic cellulose I observed by TEM

achieved due to (i) purification and enrichment of the enzyme protein responsible for the polymerization and (ii) micelle formation owing to the optimized acetonitrile/buffer ratio. There was an indication of microscopic micellar aggregates to organize catalytic subunits for assembling synthetic cellulose I [40–42]. This is the first successful instance of cellulose I allomorph formation in a nonbiosynthetic way. These findings may provide clues for further understanding of in vivo biosynthetic mechanisms leading to native cellulose I formation.

For this type of control in high-order molecular assembly during polymerization, a new concept called "choroselectivity" was introduced [41, 42], which represents the *intermolecular relationship* generated when macromolecules are formed in spatially oriented alignment due to supramolecular interactions. In other words, when polymer chains have direction, e.g., cellulose molecules having reducing and nonreducing ends, a parallel or antiparallel spatial directional relationship due to noncovalent, intermolecular interactions between the polymer chains comes into existence. If the polymer chain showing the one preferential spatial direction of the allomorph over the other is formed during polymerization, the reaction is to be defined as "choroselective." This selectivity is interestingly compared with the hitherto-known selectivities, i.e., chemo-, stereo-, and regioselectivities, all of which are concerned with the *intramolecular relationship* generated when the covalent bond is formed between two reaction centers.

In situ observation of the polymerization process of β-CF by polarization optical microscopy (POM), scanning electron microscopy (SEM), and TEM provided information on the molecular orientation within the spherulites of synthetic cellulose [43]. POM observations revealed that two types of spherulites were formed: one displaying a well-defined Maltese cross identified

Enzymatic Polymerization to Polysaccharides

Fig. 9 SEM observations of synthetic cellulose spherulites

as positive type under the microscope with a sensitive color plate and another of negative type. The negative-type spherulites were dominant, and the shape was three-dimensionally round with platelike single crystals of the synthetic cellulose originating radially from the center as observed by SEM (Fig. 9).

This structure is completely different from the 2D structure of the spherulite composed of fibrous bacterial cellulose. TEM observations demonstrated that the negative-type spherulites are composed of thin crystal plates approximately 8 nm thick, in which the cellulose chains are aligned vertically to the plate, having a cellulose I allomorph. These high-order structures of cellulose are formed only via the polymerization of β-CF.

Formation of the high-order molecular structure was also examined by time-resolved small-angle neutron scattering (SANS) measurement [44]. In situ SANS observations of the polymerization gave information on the self-assembling process of the synthesized cellulose with a space scale of approximately 20–200 nm. The results revealed that the newly synthesized cellulose self-assembled into aggregates whose surfaces are characterized by a fractal surface with the fractal surface dimension (D_s) increasing from 2 to 2.4 and the scattering intensity at a low scattering vector (q) region also increasing by about 100 times with time.

3.2.5
Extension to Cellulose-Related Saccharides

Using a substrate-specific function of an enzyme, glycosylation with a sugar fluoride was extended for stepwise elongation for a cell-oligomer synthesis via two enzymatic reactions (Scheme 10) [45, 46]. First, β-LF was glycosidated as a donor with cellulase catalysis by methyl β-cellobioside as an acceptor to form the $\beta(1 \rightarrow 4)$ linkage regioselectively. β-LF did not act as an acceptor. Second, the galactosyl group was selectively removed with β-galactosidase

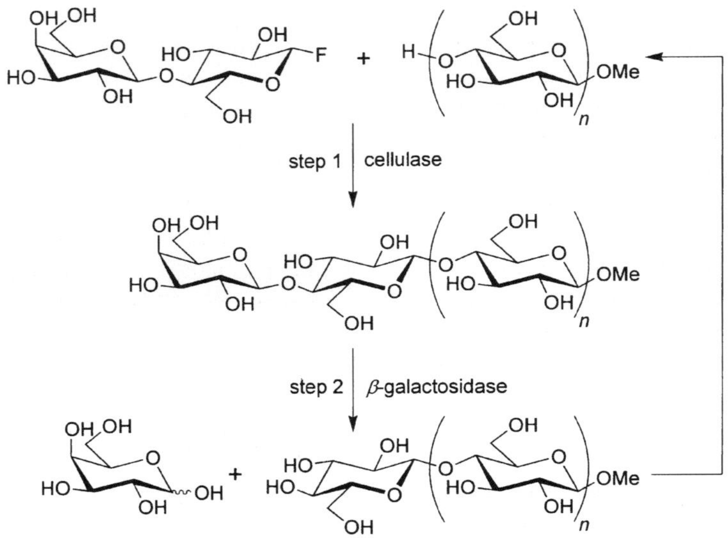

Scheme 10

catalysis from the first product to give methyl β-cellotrioside. Thus, repeating the lactosylation and degalactosylation produces a one-glucose-unit elongated cello-oligomer. This is a convenient general method of elongation to produce an oligomer with the desired degree of polymerization, without protection and deprotection procedures.

Starting from naturally occurring xyloglucans, a trisaccharide of xyopyranosylcellobiose was obtained. Fluorine atom was introduced to the anomeric carbon of the trisaccharide according to the known procedures. The trisaccharide fluoride acted as a monomer for endo-1,4-glucanase (from *Trichoderma reesei*) in a similar reaction condition as cellulose synthesis to produce synthetic xyloglucan oligomers (Scheme 11) [47].

A mutant enzyme was prepared from the *Agrobacterium* sp. β-glucosidase/galactosidase that catalyzes hydrolysis and transglycosylation of β(1 → 4) saccharide compounds. A glutamic acid unit at 358 of the enzyme was replaced by alanine to give the Glu358Ala mutant, where a catalytic nu-

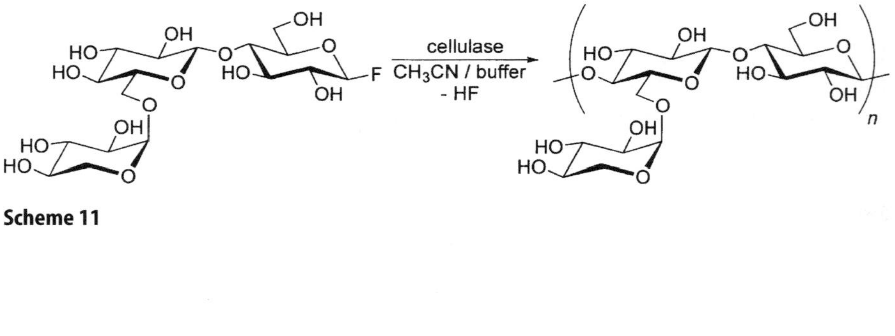

Scheme 11

cleophile residue (CO_2H group) was replaced by a nonnucleophile residue (CH_3 group) ("glycosynthase") [48]. α-glycosyl fluorides (α-galactosyl and α-glucosyl fluorides) acted as good donors for the mutant-catalyzed synthesis of oligosaccharides. For example, α-galactosyl fluoride and an acceptor of Glc-β(1 → 4)-Glc-p-nitrophenyl yielded the product trisaccharide Gal-β(1 → 4)-Glc-β(1 → 4)-Glc-p-nitrophenyl in 92% yields. With α-glucosyl fluoride as donor, Glc-p-nitrophenyl as acceptor reacted to give the product disaccharide Glc-β(1 → 4)-Glc-p-nitrophenyl in 48% yields and also the corresponding trisaccharide in 34% yields. In the case of α-glucosyl fluoride, glycosylation took place twice for trisaccharide formation. All reactions proceeded in a β(1 → 4) mode involving the inversion of the C1 configuration as shown in Fig. 10. The hydrolysis mechanism by the wild-type enzyme having two carboxyl groups is considered similar to that shown in Fig. 5A.

Sugar fluorides were extensively used for enzymatic synthesis of oligo- and polysaccharides. The enzyme used was a "glycosynthase" mutant of Cel7B from *Humicola insolens*, where glutamic acid 197 was replaced by alanine (Glu197Ala) [49]. Glycosidation of α-lactosyl fluoride (α-LF) as a donor with mono- and disaccharide acceptors by Glu197Ala catalyst gave the products in high yields. For example, the reaction of α-LF with benzyl β-glucoside produced the corresponding trisaccharide in 83% yields (Scheme 12). As acceptors, disaccharides were better than monosaccharides. It is to be stressed that several β(1 → 4) and β(1 → 3)-disaccharides acted as acceptors and pro-

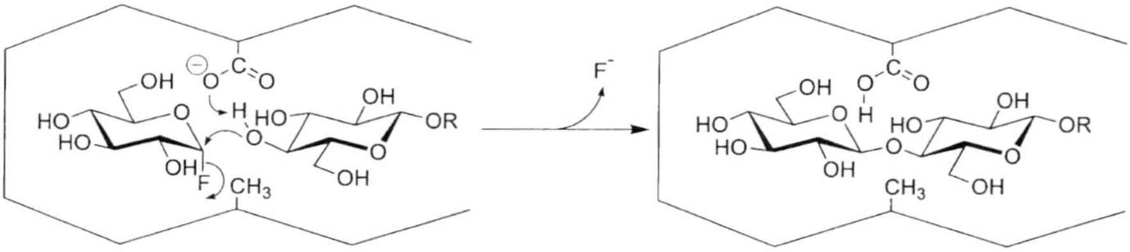

Fig. 10 Proposed mechanism of glycosynthase Glu358Ala catalysis

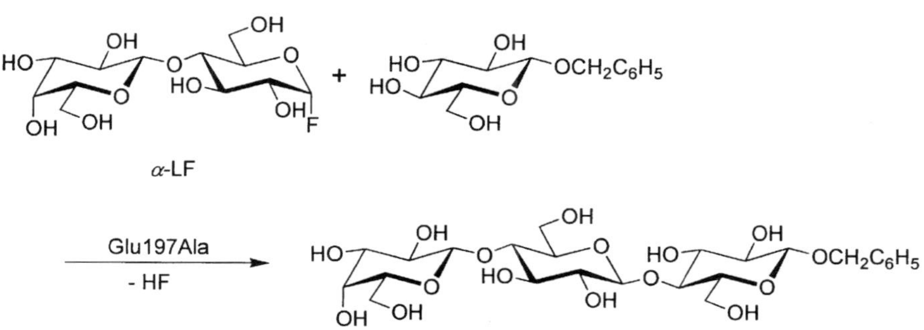

Scheme 12

duced tetrasaccharides after reacting with α-LF. In all reactions only $\beta(1 \to 4)$ linkages were formed.

As to the structure of Glu197Ala, there were no interactions with the protein involving the 6-OH at both the −2 and +1 glucosyl residues. The −2 subsite 6-OH is extremely solvent exposed and often disordered. It was thus speculated that the mutant Glu197Ala could not only be used as a "glycosynthase" for the synthesis of natural oligo- and polysaccharides, but that the lack of steric restriction on the C6-OH interactions could be exploited for the synthesis of substituted $\beta(1 \to 4)$ glucans.

The expectation was realized in the enzymatic polymerization of α-cellobiosyl fluoride (α-CF) and various derivatives (α-CFa – α-CFd) catalyzed by the mutant Glu197Ala giving rise to cellulose and its derivatives (Scheme 13). The reaction was carried out in phosphate buffer (0.1 M, pH 7.0) at 40 °C for 24 h. As the reaction proceeded, the product of a white crystalline precipitate was formed, which was characterized as shown in Scheme 13. The product from α-CF was a cello-oligomer showing a cellulose II type. It is interesting that the glycosylation was repeated involving the inversion of the C1 carbon configuration.

Arabinoxylans and $\beta(1 \to 3, 1 \to 4)$-D-glucans are the major noncellulosic cell wall components of the higher plants belonging to the Poaceae family. These β-glucans are linear polymers, and their DP reaches 1200. The proportion of $(1 \to 3)$ and $(1 \to 4)$ linkages differ from one another. The amount of $(1 \to 3)$ linkages does not exceed 25–30%, which means that the molecules consist essentially of $(1 \to 4)$-linked cello-oligosaccharides of DP 2–4 [50, 51]. The $\beta(1 \to 3, 1 \to 4)$-D-glucans in the cell walls are hydrolytically

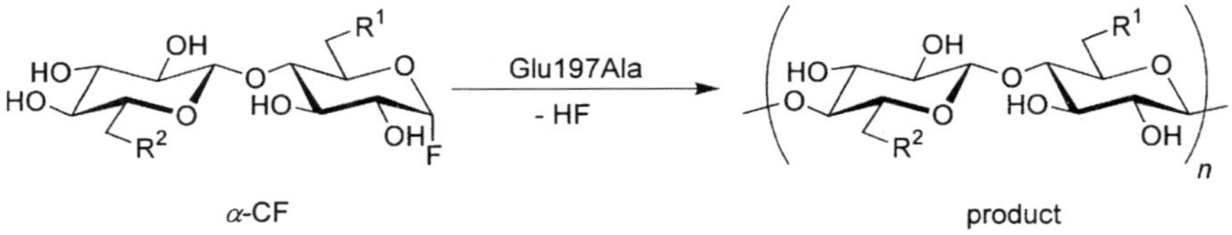

monomer	R^1	R^2	product	yield (%)	n value
α-CF	OH	OH	cello-oligomer	81	–
α-CFa	OH	Br	c. d.	68	>2
α-CFb	OH	NH$_2$	c. d.	–	–
α-CFc	OH	X	c. d.	72	6~9
α-CFd	Br	Br	–	0	–

X: α-S-xylopyranosyl
c. d.: cellulose derivative

Scheme 13

degraded by the $\beta(1 \to 3, 1 \to 4)$-glucanases found in a number of bacteria and fungi due to an energy source of the microorganisms. For in vitro synthesis of a $\beta(1 \to 3, 1 \to 4)$-D-glucan, the $\beta(1 \to 3, 1 \to 4)$-D-glucanase from *Bacillus* sp. was mutated by replacing Glu 134 with Ala to give a glycosynthase (Glu134Ala) [52]. It was shown that the self-condensation of the monomer α-laminaribiosyl fluoride was catalyzed by the Glu134Ala glycosynthase (Scheme 14). The polymerization of the monomer was carried out at 35 °C at pH 7.2 (phosphate buffer) for 3 d using Glu134Ala catalyst. A gel product was obtained in 80% yields, showing that the product has a $\beta(1 \to 3)$ and $\beta(1 \to 4)$ alternating structure with DP of 12 by electron-impact mass spectroscopic analysis and DP values ranging from 6 to 12 by MALDI-TOF mass spectroscopy. The reaction involved the exclusive $\beta(1 \to 4)$ linkage formations.

Light microscopy observations revealed that the product consists of spherulites from 5 to 50 μm in diameter composed of lamellar elements, or platelet crystals. SEM studies showed a porous surface structure of the spherulites. An X-ray diffraction diagram contained two strong diffraction rings whose d-spacings were calculated as 0.587 nm and 0.417 nm, leading to a monoclinic unit cell with $a = 0.834$ nm, $b = 0.825$ nm, $c = 2.04$ nm, and $\gamma = 90.5°$. The dimensions of the ab plane are similar to those of cellulose Iβ, but the length of the c-axis is nearly twice that of cellulose I.

Enzymatic polymerizations mentioned thus far belong in principle to a kinetic approach, where an activated glycosyl donor such as a glycosyl fluoride,

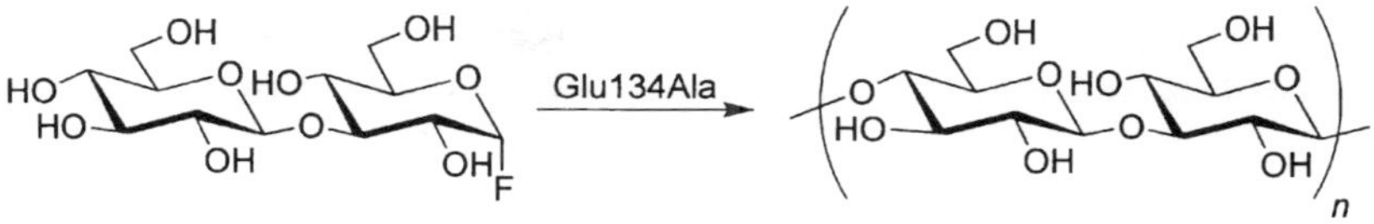

Scheme 14

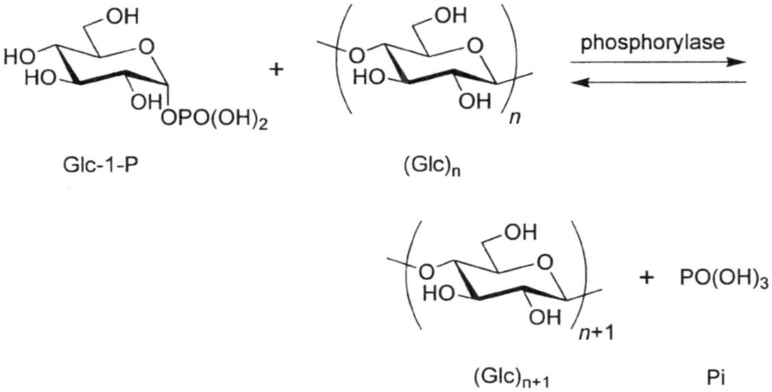

Scheme 15

designed as a TSAS, forms rapidly the glycosyl-enzyme complex to lead to the product poly- or oligo-saccharide (kinetic-controlled synthesis) [12]. On the other hand, the following phosphorylase method belongs to the equilibrium-controlled synthesis (Scheme 15) [53]. The phosphorylase enzyme was isolated from *Clostridium thermocellum* and used as catalyst for the reaction of α-glucosyl phosphate (Glc-1-P) as glycosyl donor in a large excess amount for an acceptor like cellobiose ($(Glc)_n$, $n \geq 2$) to shift the equilibrium to the right side. The chain elongation took place to produce cello-oligomers with DP of about 8, showing a cellulose II allomorph. As acceptors, 4-thiocellobiose, methyl β-cellobioside, and methyl 4-thio-α-cellobioside were used with the same efficiency as cellobiose to allow the introduction of these groups at the end of the cello-oligomers.

3.3
Amylose

Following the same principle for cellulose synthesis, α-maltosyl fluoride (α-MF) was employed for the synthesis of amylose by α-amylase catalysis (Scheme 16) [54]. Again, the monomer α-MF acted both as donor and acceptor; it is a polycondensation type to produce malto-oligosaccharides via $\alpha(1 \rightarrow 4)$ bond formation. The reaction was examined in various organic/water mixed solvents. The best solvent mixture to afford the highest oligomers was a methanol/phosphate buffer ($2/1\ v/v$).

Phosphorylases are key enzymes of carbohydrate metabolism and catalyze the phosphorolysis of glycosidic bonds in oligo- and polysaccharide substrates. This type of reaction afforded amylose via in vitro polymerization of D-glucosyl

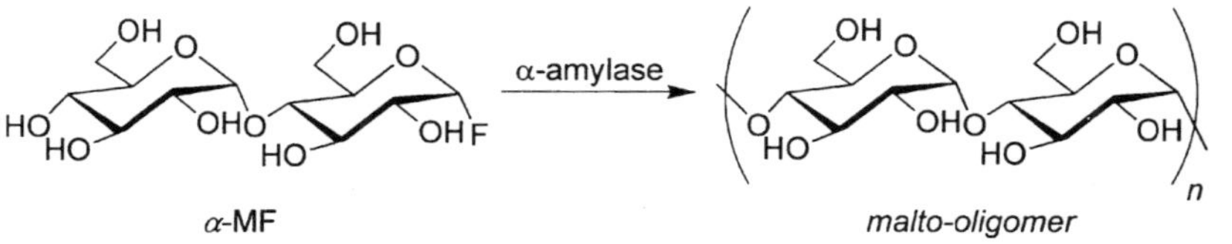

Scheme 16

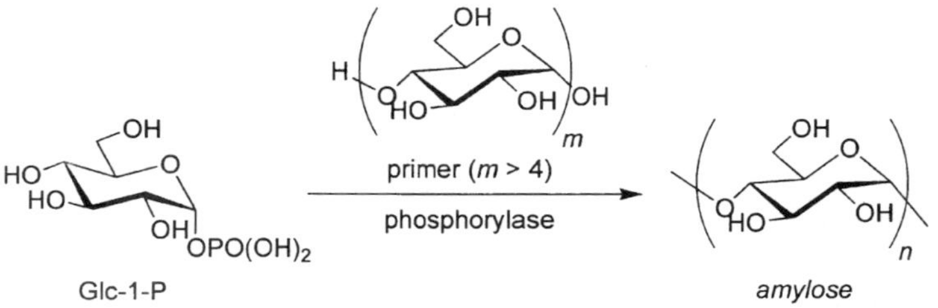

Scheme 17

phosphate (Glc-1-P) catalyzed by potato phosphorylase (Scheme 17) [55]. The reaction belongs to a biosynthetic pathway, and hence it is not within the context of the defined enzymatic polymerization in this article. However, it is cited here for reference. This approach is an equilibrium-controlled synthesis, and therefore the reaction required a large excess amount of Glc-1-P as also observed in Scheme 15. In addition, the reaction required a primer of malto-oligomer with minimum length of tetramer and proceeded in living-like polymerization to form amylose with uniform chain length.

3.4
Xylan

Xylan, a xylose polymer with $\beta(1 \rightarrow 4)$ glycosidic linkage in the main chain, is the important component of hemicellulose in plant cell walls. It interacts supramolecularly with cellulose to form a composite. Naturally occurring xylan normally contains 4-O-methylglucronic acid, L-arabinose, etc., as a minor unit in the side chain. Synthetic xylan consisting exclusively of xylose units was prepared by the crude cellulase (containing xylanase) polymerization of β-xylobiosyl fluoride (β-XF) as monomer in a mixed solvent of acetonitrile/acetate buffer (Scheme 18) [56]. The reaction proceeded by giving a perfect $\beta(1 \rightarrow 4)$ glycosidic bond of synthetic xylan having a DP of 23.

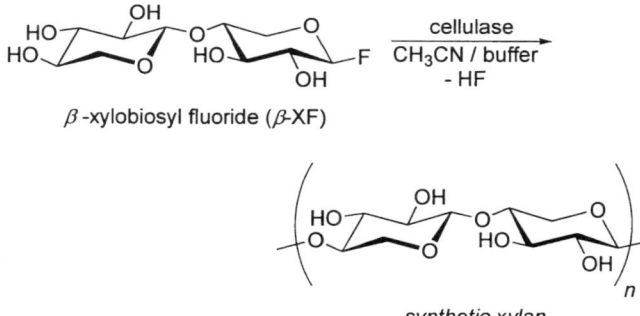

Scheme 18

3.5
Chitin

3.5.1
Synthesis Via Enzymatic Polymerization

Chitin is the most abundant organic substance in the animal world. Its quantity is estimated as approximately 1% of that of cellulose. It attracts much interest in a number of scientific and application areas as a multifunctional substance [57]. Taking the chitin structure into account, a repeating unit was hypothesized to be a chitobiose (disaccharide unit) rather than N-acetylglucosamine (monosaccharide). Then, a chitobiose oxazoline derivative (Chi-oxa) was designed as "TSAS monomer" and prepared for the catalysis of chitinase, a hydrolysis enzyme of chitin. Synthetic chitin was thus prepared for the first time via a ring-opening polyaddition [58–61], where the methyloxazoline acted as a latent N-acetyl group with a cationic ring opening (Scheme 19) [62]. The chitinase used was a retaining glycanase. The monomer oxazoline derivative has a distorted conformation of an α-anomer configuration, which mimics the transition-state structure of the chitinase hydrolysis. During polymerization, the $\beta(1 \rightarrow 4)$ glycosidic linkage was repeatedly achieved involving the inversion of the configuration of the C1 carbon to form synthetic chitin with a perfectly controlled structure.

Chi-oxa monomer in phosphate buffer (0.01 M, pH 10.6) was incubated in the presence of chitinase (*Bacillus* sp., 1 wt % for the monomer) for 50 h at 25 °C. As the reaction progressed, precipitation of the product synthetic chitin started, which was isolated in a quantitative yield. The $\beta(1 \rightarrow 4)$ structure was definitely determined by CP/MAS solid ^{13}C NMR spectroscopy. The viscosity-average molecular weight (M_v) of synthetic chitin was determined as a value higher than 10^4 according to the Mark–Houwink–Sakurada equation, $[\eta] = KM^\alpha$. However, later the molecular weight value was evaluated as less than that value. For the synthesis of chitin, a monomer oxazoline de-

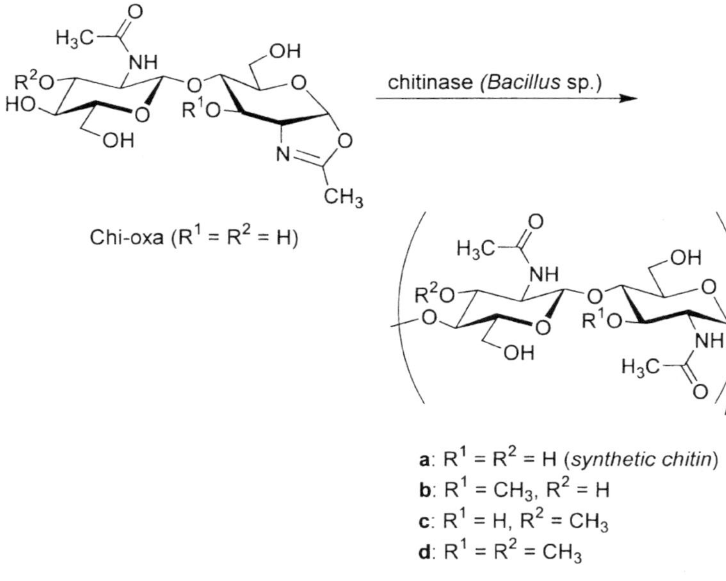

Scheme 19

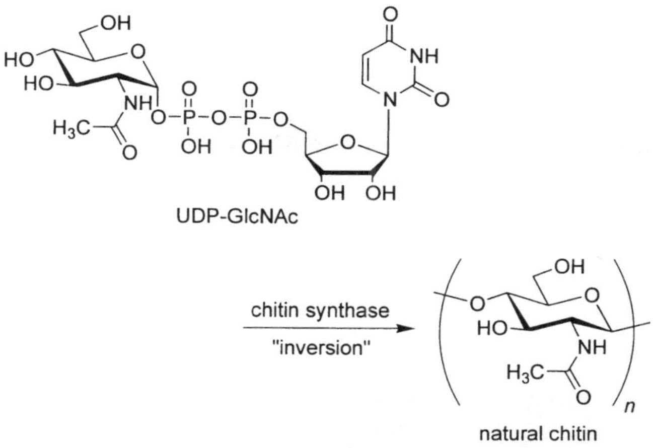

Scheme 20

rived from *N*-acetylglucosamine (monosaccharide) gave only a mixture of chito-oligomers up to a pentamer. For reference, in vivo chitin synthesis via a biosynthetic pathway is given in Scheme 20, which is quite different from the in vitro synthesis.

In nature, a bacterial cell-wall peptidoglycan is well known as an alternatingly 3-*O*-modified derivative of chitin and as a natural substrate of lysozyme enzymes. Synthesis of such chitin derivatives was attempted by extending the reaction to form synthetic chitin (Scheme 19) [63]. The ring-opening polyaddition reactivity of the monomers having a methyl group at 3, 3′ and 3,3′ positions was very low due to the steric hindrance of the substituent; the product was 3-*O*-methylated chitin oligomers up to an octamer.

3.5.2
Mechanistic Aspects

The monomer design of in vitro chitin synthesis with chitinase catalysis was performed on the basis of the new concept of TSAS, which stemmed from a speculated transition-state structure involved in chitin hydrolysis. The concept led to Chi-oxa having an oxazoline structure, which was actually prepared and orally presented in 1995 [58–60].

On the other hand, much information on the hydrolysis mechanism of chitin in vivo by a chitinase enzyme has accumulated, particularly since 1995. There are mainly two mechanistic aspects in terms of the nature of the reaction intermediate: (1) an oxazolinium intermediate [64–67] and (2) an oxocarbenium ion intermediate or a covalent glycosyl carboxylate intermediate (Scheme 21) [68]. An X-ray crystallographic analysis of chitinase A was performed that suggested an oxazolinium intermediate (Fig. 11) [65].

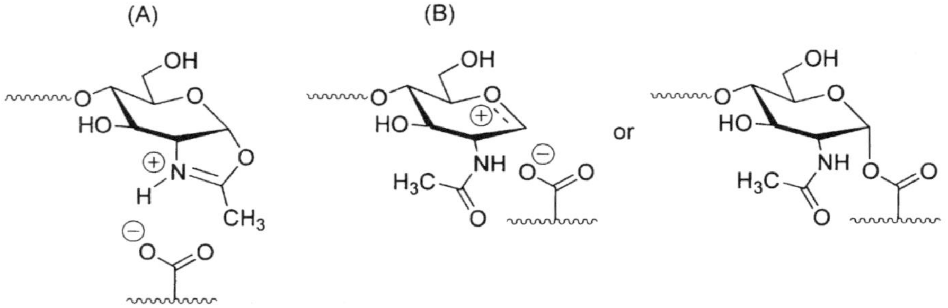

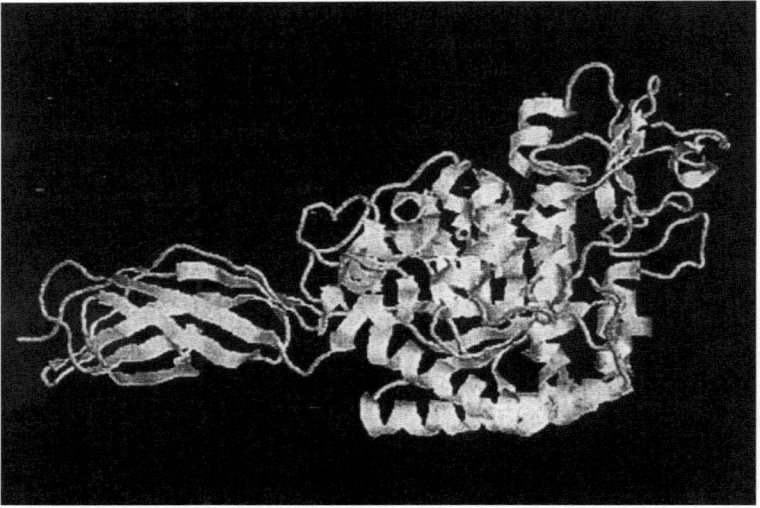

Scheme 21

Fig. 11 Crystal structure of chitinase A

Chitinases from families 18 and 20 are understood to involve a protonation on glycosidic oxygen by one of the carboxyl groups via a general acid-base mode and to form an oxazolinium intermediate. A positive charge at C1 in a transition state is stabilized intramolecularly by a carbonyl oxygen of the N-acetyl group at C2 by cleaving the C1 – O glycosidic bond as indicated from stage a to b in the hydrolysis of chitin in vivo (Fig. 12A) [64]. The anchimeric stabilization was estimated as 38 kcal/mol [66]. Then, a water molecule nucleophilically attacks the oxazolinium ion to open the ring, giving rise to a hydrolysis product from stage b to c. The whole reaction involves the inversion of the C1 configuration two times.

On the other hand, in the in vitro polymerization (B), the Chi-oxa monomer has a distorted α-C1 structure, which was established by ^{1}H NMR analysis, and is readily recognized by family 18 chitinase due to the TSAS. The polymeriza-

Enzymatic Polymerization to Polysaccharides 187

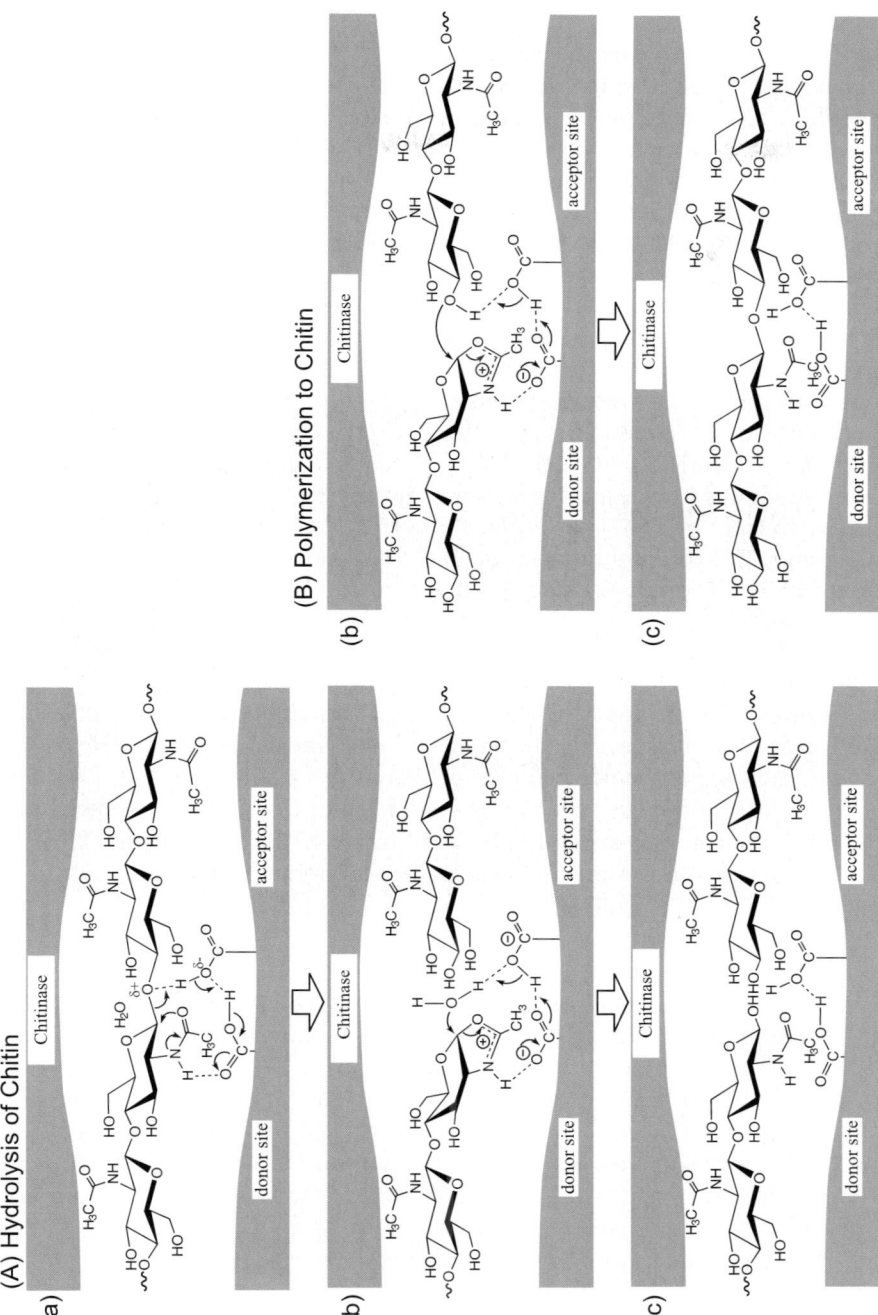

Fig. 12 Possible reaction mechanisms of chitinase catalysis. **A** Hydrolysis of chitin. **B** Polymerization of Chi-oxa monomer to synthetic chitin

tion lacks a protonation step on the glycosidic oxygen of the hydrolysis but only needs the protonation on the oxazoline nitrogen to form the oxazolinium ion, which takes place under less acidic conditions as shown at stage b. Then, a nucleophilic attack of the 4′-hydroxyl group on the C1 carbon of the oxazolinium opens the ring by inverting the C1 carbon configuration from α to β, forming a new glycosidic bond from stage b to c. The product shifts to the right side, and another monomer comes in at the donor site. Repetition of this glycosidic bond formation eventually gives synthetic chitin. It should be noted that the oxazolinium ion species is involved as a common intermediate in both hydrolysis of chitin in vivo and polymerization to chitin in vitro. The transition state of both reactions must lie in the ring opening of the oxazolinium species from stage b to stage c, which comes later in comparison with the case of cellulose synthesis. In other words, the polymerization of fluoride monomers involves a relatively early transition state whereas that of oxazoline monomers a relatively late transition state. This view accords with the fact that normally the former requires a medium with a lower pH value for the reaction to proceed.

The concept of TSAS was further proven to be valid by the following experiments. Chi-oxa monomer was subjected to polymerization by enzymatic catalysis. The hydrolysis enzymes used were chitinase (family 18) involving an oxazolinium intermediate and lysozyme involving an oxocarbenium ion intermediate (Fig. 13). With the former enzyme, synthetic chitin was quantitatively obtained after 50 h at pH 10.6, whereas with the latter no chitin was produced after 165 h of reaction [69]. This implies that the oxazoline monomer could not be a substrate for the lysozyme enzyme.

It is known that the optimal pH for hydrolysis by chitinase is 7.8 [70]. The effects of pH on the ring-opening polyaddition of the monomer showed that the optimal value was around pH 10.6 in terms of the chitin yield (Table 3); at pH 8.0 the reaction time for the complete monomer conversion took 2.5 h; however, the yield of synthetic chitin was 38%. At pH 10.6 it took 50 h and the

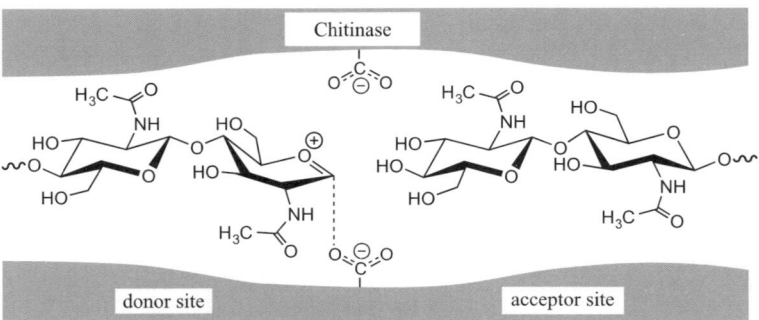

Fig. 13 An oxocarbenium ion intermediate involved in the hydrolysis of chitin by lysozyme or chitinase (Family 19)

Table 3 Effect of pH on the ring-opening polyaddition of Chi-oxa monomer [a]

pH	Reaction time / h [b]	Yields of synthetic chitin / %
8.0	2.5	38
9.0	20	62
10.6	50	~100
11.4	150	87
12.9	300	0

[a] Reaction conditions; chitinase 1 wt % for monomer at r.t.
[b] Reaction time for complete monomer consumption

chitin was obtained quantitatively [69]. This means that the polymerization was very rapid at around pH 8, but the product chitin was also hydrolyzed rapidly; at pH 10.6 polymerization was exclusively promoted, suppressing the hydrolytic activity significantly. These pH effects are shown roughly in Fig. 14. Polymerization is less sensitive to pH, probably because it does not require protonation at glycosidic oxygen. Therefore, polymerization is to be carried out at a pH where the difference in the rate of polymerization and hydrolysis is maximal.

Based on the above discussion (Fig. 12) it was speculated that polymerization may need only one of the carboxyl groups at the active site for the protonation of the monomer. To prove this, wild-type chitinase A1 was mutated, where aspartic acid 202 (D202) was left and glutamic acid 204 was replaced by glutamine (E204Q) [71, 72]. As anticipated, the mutant chitinase showed only polymerization (oligomerization) activity and no hydrolysis abil-

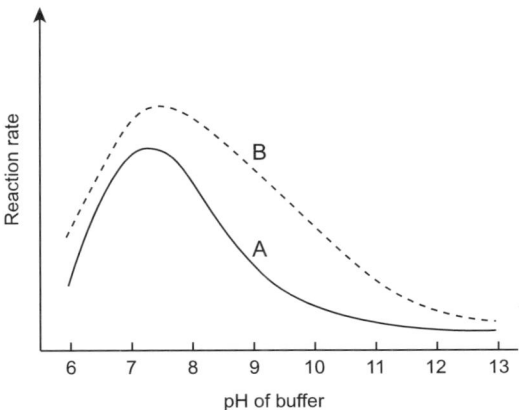

Fig. 14 A rough sketch of pH effects on the rate of chitinase-catalyzed reactions: **A** Hydrolysis activity. **B** Polymerization

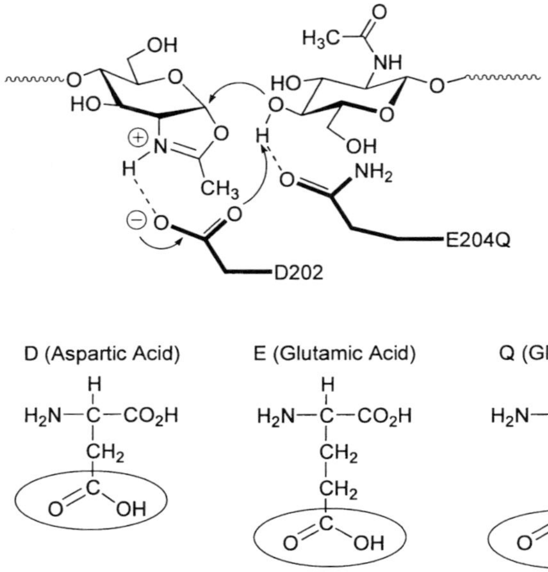

Scheme 22

ity, although the overall activity was much reduced from that of the wild chitinase [73]. The transition state of the glycosylation (propagation) by the mutant is shown in Scheme 22.

3.5.3
High-Order Molecular Structures

X-ray diffraction measurements of synthetic chitin showed a high-order structure of α-chitin having an antiparallel chain alignment and a thermo-

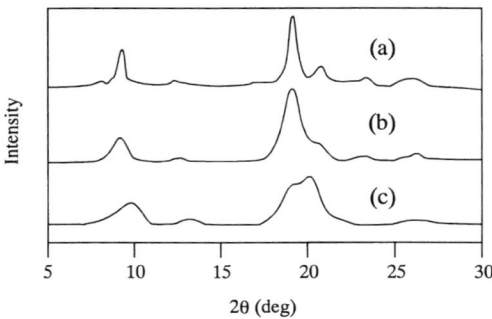

Fig. 15 X-ray diffractograms of **a** artificial chitin, **b** natural α-chitin from queen crab, and **c** natural β-chitin from squid pens

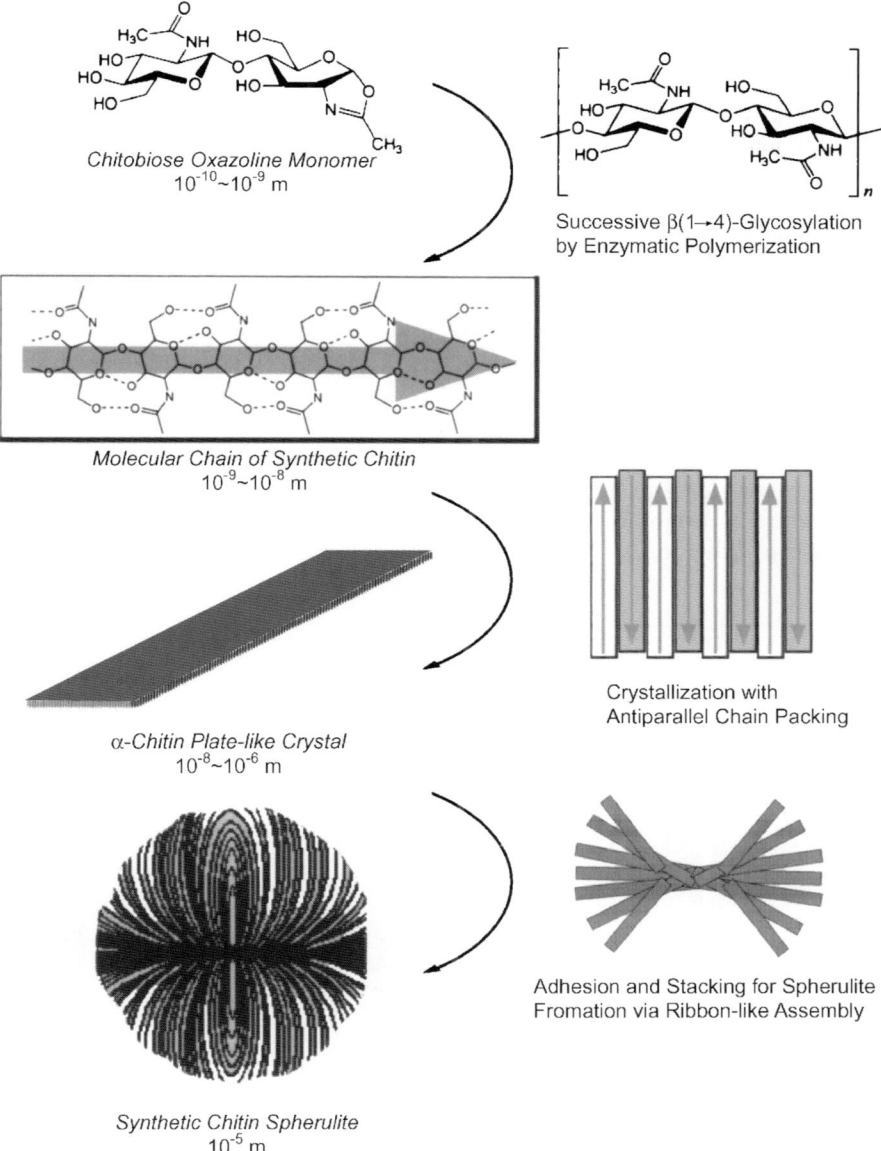

Fig. 16 Hierarchy structures in vitro formed by synthetic chitin via enzymatic polymerization

dynamically more stable form [74] and β-chitin having a parallel metastable chain structure (Fig. 15) [75].

Self-organization of polymers into high-order structures such as lamellas, spherulites, and liquid crystals is known [76, 77]. Those structures were reproduced by acid hydrosates of cellulose and chitin [78, 79]. The ring-opening polyaddition of Chi-oxa produced platelike single crystals of synthetic α-chitin and induced organization of the ribbons assembled from the crystals into the spherulites for the first time [80]. These high-order hierarchy structures are illustrated in Fig. 16.

3.5.4
Chitin Oligomer Derivatives

Chitinase-catalyzed polymerization using an oxazoline monomer was extended to a stepwise elongation to prepare chitin oligomer derivatives (Scheme 23) [81]. The reaction involves a chitinase-catalyzed transglycosylation of the oxazoline derivative of N-acetyllactosamine as donor to an oligosaccharide as acceptor (step 1) and a β-galactosidase-catalyzed cleavage of the galactose unit from the nonreducing end of the product (step 2). Thus, the N-acetylglucosamine unit can be added stepwise to the nonreducing end of the starting oligosaccharide, as observed in the stepwise cello-oligomer synthesis of Scheme 10.

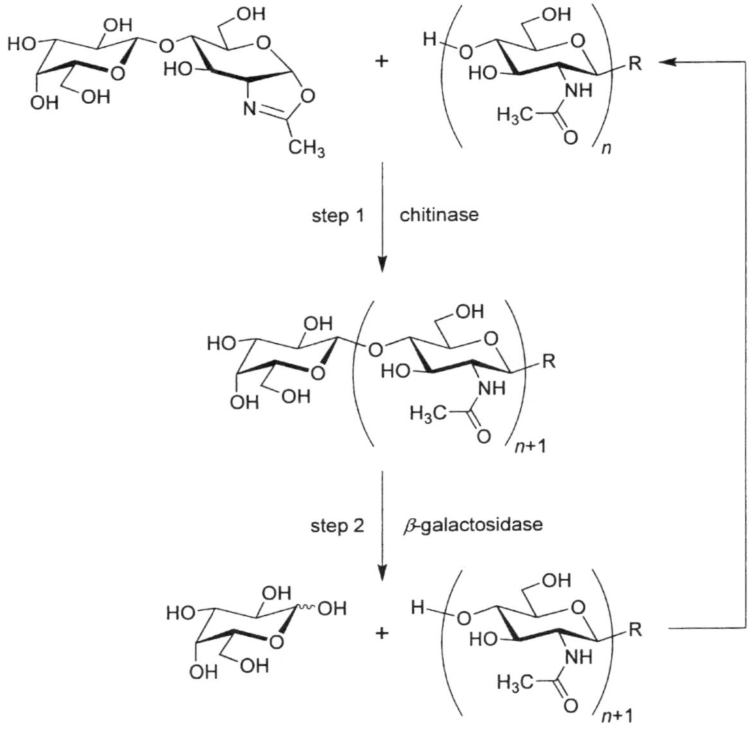

Scheme 23

3.6
Glycosaminoglycans

Extracellular matrices (ECMs) are substances found between living cells (Fig. 17) [82, 83]. ECMs are constituted from fibronectins, collagens, proteoglycans, etc. Proteoglycans are formed from core proteins to which many polysaccharide chains are linked. These are all linear heteropolysaccharides called glycosaminoglycans (GAGs), formerly designated as "mucopolysaccharides". GAGs include seven biomacromolecules: hyaluronan (hyaluronic acid, HA), chondroitin (Ch), chondroitin sulfate (ChS) (Scheme 1), heparin, heperan sulfate, dermatan sulfate, and keratan sulfate (Scheme 24) [84].

HA, Ch, and ChS are the main components of GAGs, particularly in dermis and cartilage, in which large molecular complexes are formed by association of these molecules [84]. GAGs play crucial roles in differentiation and proliferation of cells, tissue morphogenesis, and wound healing via signaling by interactions with growth factors and morphogens [82, 85–92]. They have a single repeating structural motif of a disaccharide unit containing a hexosamine and show a great deal of structural diversity causing discrete structural forms generated by complex patterns of deacetylation, sulfation, and epimerization,

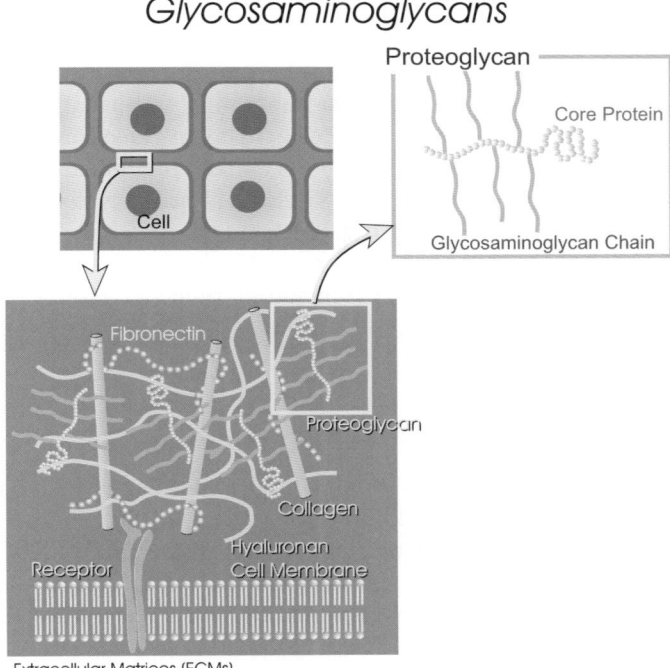

Fig. 17 Illustrations of glycosaminoglycans found in extracellular matrices

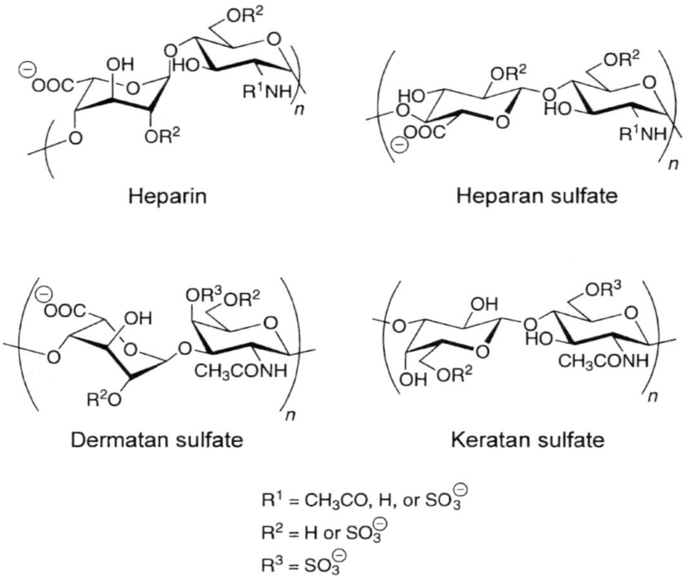

Scheme 24

$R^1 = CH_3CO, H, \text{ or } SO_3^{\ominus}$
$R^2 = H \text{ or } SO_3^{\ominus}$
$R^3 = SO_3^{\ominus}$

which are found in particular tissues and influenced by disease and aging [93, 94]. GAGs containing glucosamines as a hexosamine constituent are designated as glucosaminoglycans, and those containing galactosamines as galactosaminoglycans [84]. HA and Ch are the best known glucosaminoglycan and galactosaminoglycan, composed of $\beta(1 \rightarrow 4)$ linked β-D-glucuronyl-$(1 \rightarrow 3)$-N-acetyl-D-glucosamine (GlcA$\beta(1 \rightarrow 3)$GlcNAc; N-acetylhyalobiuronate) and β-D-glucuronyl-$(1 \rightarrow 3)$-N-acetyl-D-galactosamine (GlcA$\beta(1 \rightarrow 3)$GalNAc; N-acetylchondrosine) repeating units, respectively. This section focuses on the precision synthesis of HA and Ch including their derivatives [95–98] via enzymatic polymerization.

3.6.1
Hyaluronan and Its Derivatives

HA was first found in 1934, which is a high-molecular-weight biopolysaccharide contained in the vitreous humor of cattle eyes [99]. In vivo synthesis of HA is performed in the plasma membrane by the catalysis of HA synthases through the alternate addition of GlcNAc and GlcA in a $\beta(1 \rightarrow 4)$ and $\beta(1 \rightarrow 3)$ fashion, respectively, employing the corresponding uridine-5′-diphospho (UDP)-sugar substrates (Fig. 18) [100].

Some bacteria produce HA as a capsular polysaccharide by the HA synthase, which facilitates evasion of the immune response by host organ-

Enzymatic Polymerization to Polysaccharides

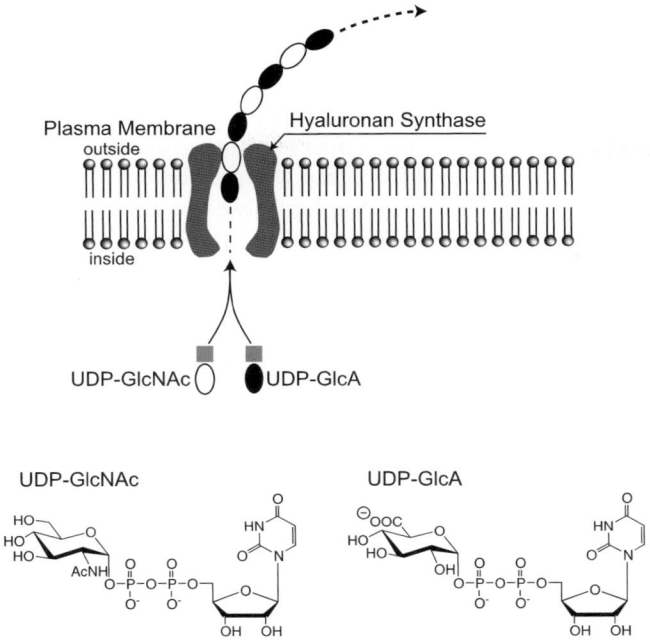

Fig. 18 Biosynthesis of HA by HA synthase with using UDP-sugar substrates

isms [101]. The molecular weight value of naturally occurring HA often reaches several million daltons. This large molecule of HA interacts with a number of other ECM molecules, resulting in the formation of a strong network of viscoelastic ECMs [102, 103]. HA exhibits critical bioactivities such as matrix formation around cumulus cells during ovulation and fertilization [104–106], mediation of immune cell adhesion [107], and activation of intracellular signaling [108]. Furthermore, intracellular HA was proved to exist in many kinds of proliferating cells [109, 110], which is believed to exert some influences on cell mitosis. Such multifunctions of HA have been utilized in the medical and pharmaceutical fields [111]. Therefore, HA has been frequently modified for the development of biocompatible materials. Precise control of the modification of HA is normally difficult due to lower reactivity of functional groups on a high-molecular weight HA molecule. Further, unexpected side reactions, like bond cleavage of the HA chain causing reduction of the molecular weight, often occur during modification under severe reaction conditions [112, 113]. HA derivatives with a perfectly controlled structure bear potentials to regulate their biological activities as well as their chemical and physical properties. A number of chemists have attempted to synthesize such an indispensable molecule of HA [114]; however, an octasaccharide derivative is the largest molecule prepared by organic chemistry to date [115].

3.6.1.1
Synthesis of Hyaluronan Via Enzymatic Polymerization

As illustrated in Fig. 19, HA has two kinds of glycosidic linkages, i.e., (1 → 4)-β-GlcNAc and (1 → 3)-β-GlcA linkages. Hyaluronoglucosaminidase (EC 3.2.1.35), generally called hyaluronidase (HAase) belonging to glycoside hydrolase family 56, participates in the hydrolysis of the β(1 → 4) glycosidic linkage, and hyaluronoglucuronidase (EC 3.2.1.36) hydrolyzes the β(1 → 3) glycoside [116]. These views allow for the design of two types of substrate monomers: the oxazoline-type monomer for HAase catalysis and the β-fluoride-type monomer for hyaluronoglucuronidase.

The anomeric carbon atoms at the reducing end of both monomers are activated by the structure of [2,1-d]-oxazoline and by the electronegativity of fluorine atoms. HAase is involved in the catabolism of (1 → 4)-β-N-acetylglucosaminide linkage through a "substrate-assisted" mechanism [117], which contains an oxazolinium ion species as a transition state of enzymatic hydrolysis. The structure of the oxazoline monomer is close to oxazolinium,

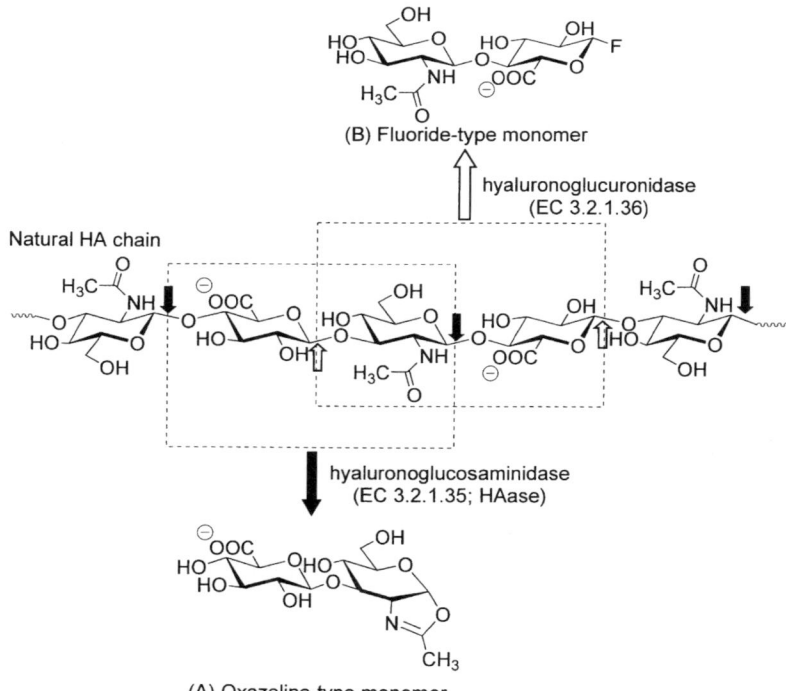

Fig. 19 Cleaving points on HA by HAase (*black arrows*) and hyaluronoglucronidase (*white arrows*) and possible monomer structures for the enzymes: **A** Oxazoline-type monomer for HAase catalysis. **B** Fluoride-type monomer for the catalysis of hyaluronoglucronidase

which is a TSAS monomer for HAase. HAase is a commercially available enzyme, whereas hyaluronoglucuronidase is difficult to obtain; the latter has been found in leeches and small sea crustaceans [116]. Thus, the combination of the oxazoline monomer and HAase is feasible and selected for the synthesis of HA [95, 96] (Scheme 25).

The substrate monomer was synthesized via conventional chemical methods [96]. During enzymatic reaction, three kinds of reactions are possible: polymerization of the monomer to synthetic HA, enzymatic and nonenzymatic hydrolysis of the monomer via the oxazoline ring opening, leading to the GlcAβ(1 → 3)GlcNAc disaccharide (Scheme 26).

The monomer was effectively recognized and consumed by HAase from ovine testes (OTH) at pH 7.5, giving rise to synthetic HA in good yields with high molecular weight values (M_n) via ring-opening polyaddition under total

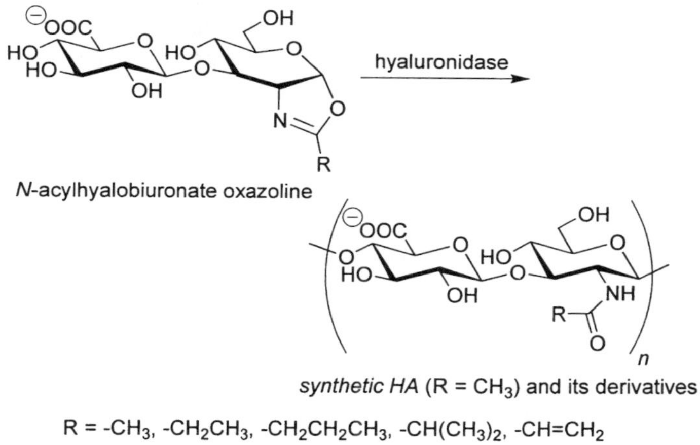

Scheme 25

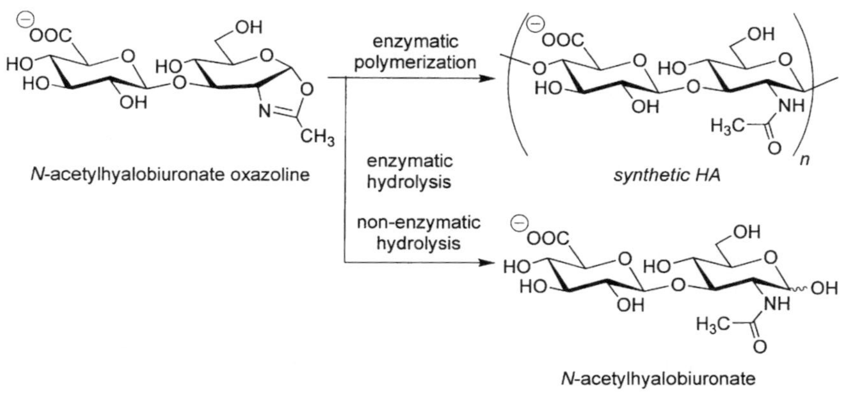

R = -CH$_3$, -CH$_2$CH$_3$, -CH$_2$CH$_2$CH$_3$, -CH(CH$_3$)$_2$, -CH=CH$_2$

Scheme 26

control of regioselectivity and stereochemistry. Table 4 indicates the results of polymerization under various reaction conditions. The pH strongly affected both the yield and the molecular weight of synthetic HA. Neutral conditions at around pH 7.5 allow for effective polymer production prior to hydrolysis of the monomer. OTH has the optimum activity for the hydrolysis of HA at pH 4.5–6.5 [116]. Therefore, acidic pH conditions accompanied a preceding hydrolysis of the monomer with rapid consumption rate. Weak alkaline conditions resulted in a slower rate of monomer consumption, lower yields, and molecular weight values of the product HA due to a reduction in enzyme activity. Bovine testicular HAase (BTH) also produced synthetic HA effectively, and the molecular weight was up to 17 700 (88–90 saccharide units). However, bee venom containing HAase exhibited no activity for the polymerization, although the reaction mechanism of HA hydrolysis is very similar to that of OTH and BTH [117].

The catalysis mechanism of HAase is similar to that of chitinase, which belongs to glycoside hydrolase family 18 described in Sect. 3.5.2 (Fig. 12). HAase belonging to glycoside hydrolase family 56 has a carboxyl group as a catalytic proton donor during the hydrolysis reaction but no catalytic nucleophiles; carbonyl oxygen of the C2 acetamido group in GlcNAc attacks the neighbor-

Table 4 HAase-catalyzed polymerization to synthetic HA and its derivatives

Entry	Polymerization[a] Monomer	Enzyme[b]	pH	Time/h	Synthetic HA or HA derivative Yield/%	M_n[c]	M_w[c]
1	2-methyl	OTH	6.0	8	19	1300	1400
3	2-methyl	OTH	7.0	48	72	4400	9800
4	2-methyl	OTH	7.5	52	78	5500	13 800
	2-methyl				53[d]	13 300	22 000
5	2-methyl	OTH	8.0	60	73	5700	14 900
6	2-methyl	OTH	9.0	72	65	4600	11 300
7	2-methyl	OTH	9.5	168	0	–	–
9	2-methyl	BTH	7.5	60	53	7800	17 600
					34[d]	17 700	25 000
10	2-methyl	bee venom	7.5	72	trace	–	–
11	2-methyl	H – OTH	7.5	6	69	6800	19 500
12	2-ethyl	OTH	7.5	48	65[d]	6900	17 500
13	2-n-propyl	OTH	7.5	60	47[d]	4500	13 500
14	2-vinyl	OTH	7.5	48	50[d]	5900	16 800

[a] In a carbonate buffer (50 mM) at 30 °C; monomer conc., 0.1 M; amount of enzyme, 10 wt % for monomer
[b] OTH, ovine testicular HAase (560 units/mg); BTH, bovine testicular HAase (500 units/mg)
[c] Determined by SEC calibrated with hyaluronan standards. [d] Isolated yields after purification

ing C1 atom to cleave the glycosidic bond, resulting in the formation of an oxazolinium ion. Thus, the oxazoline monomer is a TSAS monomer, which is readily recognized and polymerized by HAase, leading to synthetic HA in a regio- and stereoselective manner.

3.6.1.2
Enzymatic Polymerization to Hyaluronan Derivatives

Synthesis of HA derivatives was performed via HAase-catalyzed polymerization of 2-ethyl, 2-*n*-propyl, 2-isopropyl, 2-phenyl, 2-vinyl, and 2-isopropenyl oxazoline monomers (Scheme 25). Polymerization of these monomers enables the single-step production of HA derivatives with a well-defined structure bearing a corresponding *N*-acyl group via the oxazoline ring opening, leading to the development of biomaterials having novel physiological activities. Among the monomers, 2-ethyl, 2-*n*-propyl, and 2-vinyl monomers gave the corresponding HA derivatives bearing *N*-propionyl, *N*-butyryl, and *N*-acryloyl groups in all glucosamine units (Table 4). The 2-isopropyl oxazoline derivative was little consumed by the enzyme, providing a trace amount of the *N*-isobutyrylated HA oligomers. The 2-phenyl and 2-isopropenyl oxazoline derivatives were not catalyzed by OTH. Similar results were obtained using BTH as catalyst.

The monomers of 2-methyl, 2-ethyl, 2-*n*-propyl, and 2-vinyl oxazoline derivatives were copolymerized by HAase, providing HA derivatives having *N*-acyl groups in various proportions (Scheme 27) [118]. Composition of the *N*-acyl groups in a polymer chain was controllable by varying the

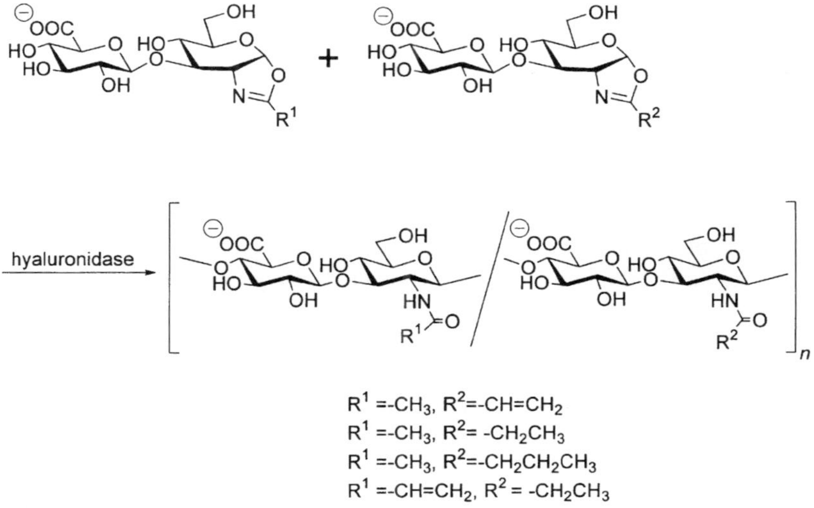

R^1 =-CH$_3$, R^2=-CH=CH$_2$
R^1 =-CH$_3$, R^2= -CH$_2$CH$_3$
R^1 =-CH$_3$, R^2=-CH$_2$CH$_2$CH$_3$
R^1 =-CH=CH$_2$, R^2 = -CH$_2$CH$_3$

Scheme 27

comonomer feed ratio. This is the first successful example of enzymatic copolymerization for the synthesis of polysaccharides.

Thus the catalysis allowed the production of not only natural-type HA but also HA derivatives having N-propionyl, N-butyryl, and N-acryloyl groups in various proportions in a polymer chain. This implies that natural HAase possesses enough space at the active site for these sterically larger substituents compared with N-acetyl group, or that HAase is dynamic during polymerization as a host with the larger group guests. Isopropyl, phenyl, and isopropenyl groups on the monomer were sterically too large for HAase catalysis.

3.6.2
Chondroitin and Its Derivatives

Chondroitin (Ch) exists predominantly as a carbohydrate part of proteoglycans in *Caenorhabditis elegans* [119] or in the higher organisms as a precursor of ChS, mainly found in cartilage, cornea, and brain matrices [120]. Molecular weight values (M_n) of naturally occurring ChS range widely from 0.5×10^4 to 35×10^4. A number of reports describe the biological functions of Ch and ChS, for example, maintenance of cartilage elasticity [121], regulation of tissue morphogenesis, and promotion of neurite outgrowth [122] and neuronal migration [123]. Biosynthesis of Ch is performed in the Golgi apparatus by the catalysis of chondroitin synthase [124, 125] and other glycosyltransferases [126, 127] employing UDP-GlcA and UDP-GalNAc as substrates. Ch is further sulfated by several kinds of sulfotransferases, giving rise to ChS [128]. Like the bacterial HA, Ch is also synthesized by some bacteria as a capsular polysaccharide [129, 130]. Detailed mechanisms for the production of Ch in *C. elegans* have been fully elucidated [131]; however, those of Ch and ChS in the higher organisms have been unclear.

3.6.2.1
Synthesis of Chondroitin Via Enzymatic Polymerization

From the chemical structure of Ch, two possible monomers for Ch synthesis were designed based on the function of hydrolysis enzymes: the oxazoline-type monomer (N-acetylchondrosine oxazoline) for HAase catalysis responsible for the hydrolysis of $(1 \rightarrow 4)$-β-N-acetylgalactosaminide linkages in Ch, and the fluoride-type monomer of GalNAc$\beta(1 \rightarrow 4)$GlcAβ-F for endo-β-glucuronidase involved in Ch degradation through bond cleavage of $(1 \rightarrow 3)$-β-glucuronide linkages. However, the glucuronidase found in rabbit liver [132] is not a commercially available enzyme. Therefore, the combination of the oxazoline-type monomer and HAase was selected for the Ch synthesis (Scheme 28).

The oxazoline monomer was synthesized from D-galactose and glucurono-6,3-lactone as starting materials via conventional chemical methods [97]. The

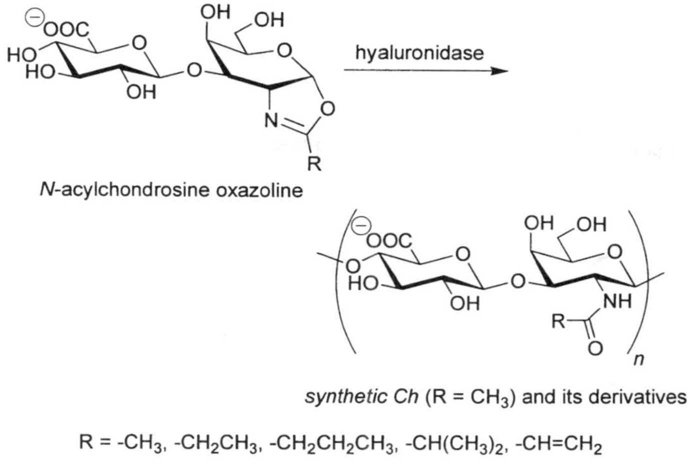

Scheme 28

monomer was recognized and catalyzed by OTH, giving rise to synthetic Ch under total control of regioselectivity and stereochemistry. Table 5 summarizes polymerization results with varying reaction conditions. Like HA synthesis, polymerization behaviors greatly depend on the reaction conditions of pH and the enzyme origin; Ch was effectively produced by OTH at a pH of around 7.5. The M_n value of synthetic Ch reached 4600, which corresponds to that of naturally occurring Ch.

3.6.2.2
Enzymatic Polymerization to Chondroitin Derivatives

In contrast to HA derivatives, synthesis of Ch derivatives has been less widely reported. Therefore, precise production of Ch derivatives is very important for the investigation of Ch functions and for the development of Ch-related biomaterials. Monomers with 2-ethyl, 2-*n*-propyl, 2-isopropyl, 2-phenyl, and 2-vinyl groups were designed and synthesized [97]. Polymerization of the 2-ethyl and the 2-vinyl oxazoline monomers effectively proceeded by OTH, providing the corresponding *N*-propionyl and *N*-acryloyl derivatives of Ch in good yields (Table 5). The polymerization reactivity of these monomers depends mainly on the steric bulkiness of the 2-substituent in the oxazoline, i.e., 2-ethyl > 2-vinyl ≫ 2-*n*-propyl > 2-isopropyl ≫ 2-phenyl.

Synthesis of ChS with a well-defined structure was achieved by HAase using sulfated *N*-acetylchondrosine oxazoline monomers [133]. Monomers bearing a sulfate group at C4 in GalNAc were effectively polymerized by HAase, giving rise to ChS having a sulfate group at C4 of all GalNAc units with an M_n value of over 10 000.

Table 5 HAase-catalyzed polymerization to synthetic Ch and its derivatives

	Polymerization[a]				Synthetic Ch or Ch derivative		
Entry	Monomer	Enzyme[b]	pH	Time/h	Yield/%	M_n[c]	M_w[c]
1	2-methyl	H-OTH	6.0	1	28	1600	1700
2	2-methyl	H-OTH	7.0	9	47	1900	2200
3	2-methyl	H-OTH	7.5	23	50	2100	2500
4	2-methyl	H-OTH	8.0	33	49	2200	2700
5	2-methyl	H-OTH	8.5	6	9	4600	6500
6	2-methyl	H-OTH	9.0	72	0	–	–
7	2-methyl	OTH	7.5	23	35	2500	3200
8	2-methyl	H-BTH	7.5	40	29	2600	3400
9	2-methyl	BTH	7.5	40	10	2800	3600
10	2-ethyl	H-OTH	7.5	35	46	2700	3600
11	2-vinyl	H-OTH	7.5	24	19	3400	4600

[a] In a phosphate buffer (50 mM) at 30 °C; concentration of monomer, 0.1 M; amount of enzyme, 10 wt % for monomer
[b] OTH, ovine testicular hyaluronidase (560 units/mg); BTH, bovine testicular hyaluronidase (330 units/mg); H-OTH, ovine testicular hyaluronidase (2160 units/mg); H-BTH, bovine testicular hyaluronidase (1010 units/mg)
[c] Determined by SEC calibrated with hyaluronan standards.

3.6.3
Synthesis of Glycosaminoglycans by Other Methods

Another method of GAGs synthesis utilizing HAase has been reported. HA oligomers up to 22 sugar units were synthesized by enzymatic reconstruction of HA chains by HAase (Scheme 29) [134].

The reaction harnessed the transglycosylation activity of HAase, which transfers an N-acetylhyalobiuronate unit from the donor HA molecule successively to the acceptor 2-pyridylamino (PA)-HA hexasaccharide, yielding PA-HA oligomers. Like the synthesis of PA-HA, various kinds of PA-ChS oligosaccharide libraries were prepared [134, 135].

On the other hand, there are some reports describing the synthesis of GAGs using glycosyltransferases via biosynthetic pathways [136], which exclusively catalyze glycosidic bond formation. The following are examples of GAGs synthesis using glycosyltransferases: HA synthase from *Streptococcus equisimilis* was employed for milligram-scale synthesis of HA [137]. In this reaction system, UDP-sugars (UDP-GlcA and UDP-GlcNAc) were effectively regenerated by the catalyses of several kinds of enzymes, giving rise to synthetic HA in a 90% yield with Mn 5.5×10^5. In addition, mutated HA synthase was recently utilized for the stepwise synthesis of HA, which has a monodispersed molecular mass of up to 20 sugar units [138].

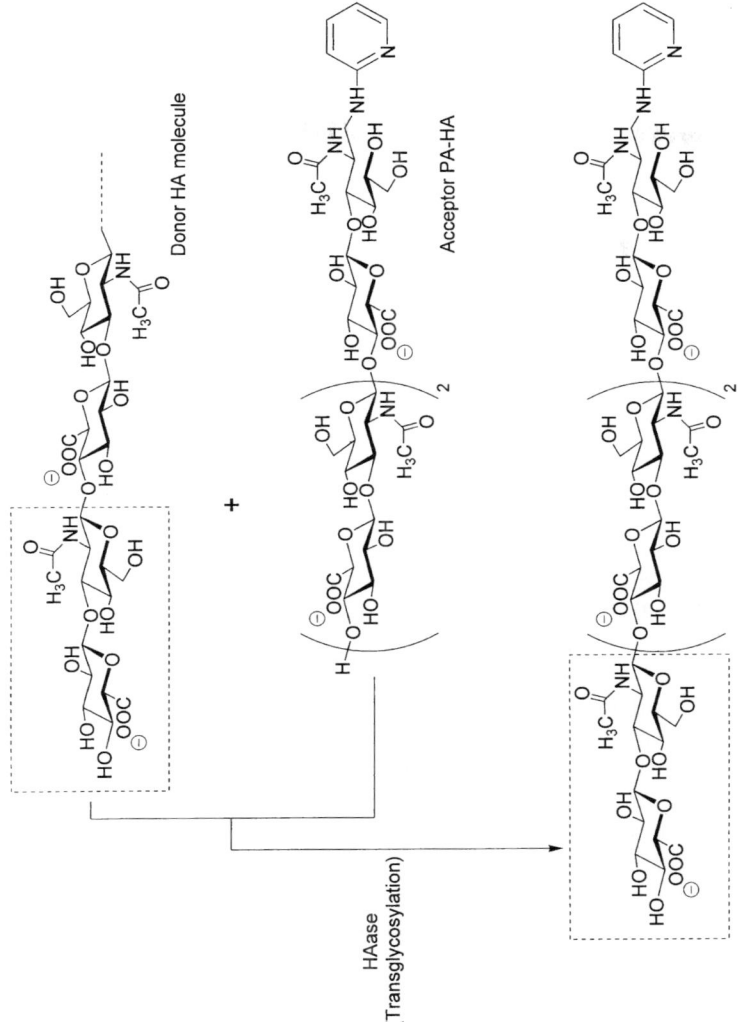

Scheme 29

3.7
Unnatural Polysaccharides

An enzyme is often dynamic and allows for the kind of catalysis required for the formation of unnatural products, if an artificial substrate is appropriately designed. Then, the reaction follows the circle of Fig. 2B. For instance, a 6-O-methyl-β-CF monomer was successfully polymerized by the catalysis of cellulase from *Trichoderma viride* in a regio- and stereoselective manner, giving rise to alternatingly 6-O-methylated cellulose derivatives over tetradecasac-

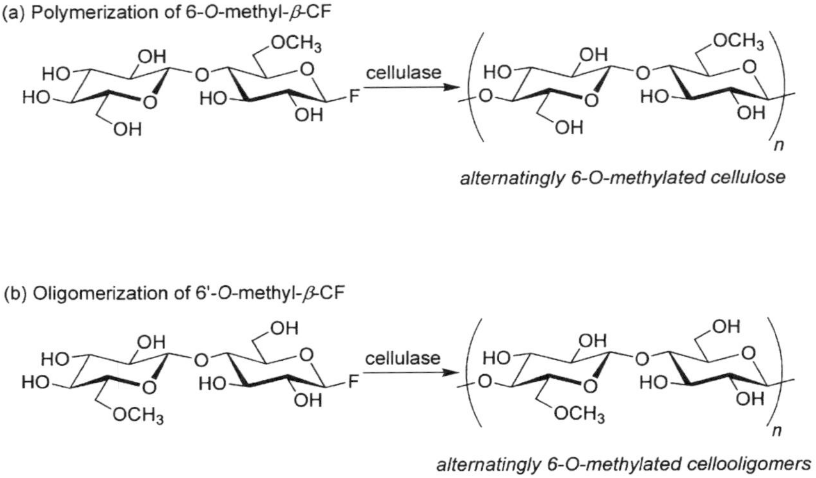

(a) Polymerization of 6-O-methyl-β-CF

alternatingly 6-O-methylated cellulose

(b) Oligomerization of 6'-O-methyl-β-CF

alternatingly 6-O-methylated cellooligomers

Scheme 30

charide (Scheme 30a). In contrast, 6'-O-methyl-β-CF was less catalyzed by the enzyme, affording mainly the tetrasaccharide with an alternating methylated structure (Scheme 30b) [139]. These results provided important suggestions about the kind of enzyme catalysis that requires the recognition of 6-OH at the nonreducing end but not at the reducing end of the monomer.

These findings were extended to the chitinase-catalyzed glycosidation of sugar oxazoline derivatives with a carboxylate group as glycosyl donors [140]. The glycosyl donors having a carboxylate group at the nonreducing end effectively coupled with acceptor chito-oligosaccharides through $\beta(1 \rightarrow 4)$ glycosidic linkage. It is intriguing that alternatingly 6-O-carboxymethylated chitotetraose was selectively produced by the catalysis of chitinase from *Streptomyces griseus* using a 6'-O-carboxymethyl-Chi-oxa derivative as substrate [141].

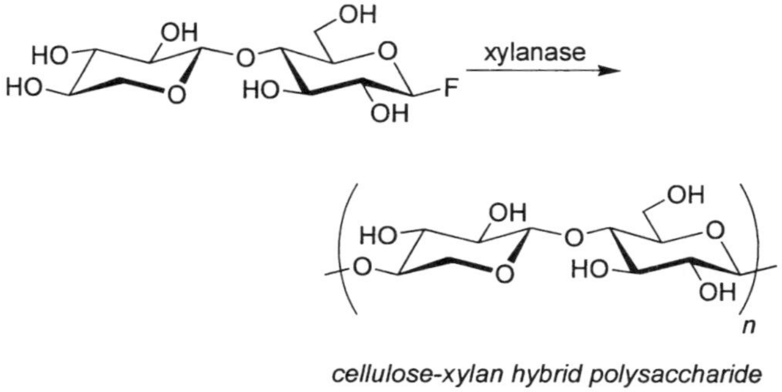

cellulose-xylan hybrid polysaccharide

Scheme 31

A hybrid-type polysaccharide was produced using a monomer consisting of different sugar moieties; β-D-xylosyl-(1 → 4)-β-D-glucosyl fluoride was successfully polymerized by the catalysis of xylanase from *Trichoderma viride*, giving rise to a cellulose–xylan hybrid polysaccharide (Scheme 31) [142]. This hybrid polymer has the alternating structure of (1 → 4)-β-D-xyloside and (1 → 4)-β-D-glucoside; therefore, both natures of cellulose and xylan are expected to be present in a molecule.

A sugar fluoride monomer was further used for the synthesis of thio-oligosaccharides, which can be an inhibitor of cellulase and are useful for the mechanistic study of cellulase function [143]. Hemithiocello-oligomers from tetraose to tetradecaose were obtained via enzymatic polymerization using 4-thio-β-cellobiosyl fluoride as monomer by cellulase catalyst [144].

4
Concluding Remarks

This review was focused on polysaccharide synthesis via enzymatic polymerization. The reaction using a hydrolase enzyme as catalyst and an appropriately designed monomer brought about the first chemical synthesis of several natural and unnatural polysaccharides, all of which had been difficult to synthesize by conventional methods. Stereochemistry and regioselectivity were perfectly controlled during polymerization to give synthetic polysaccharides with a precise structure. As a contribution to basic science, the enzymatic polymerizations to synthetic cellulose and synthetic chitin provided fundamental information on biosynthetic mechanisms and supramolecular high-order molecular structure formation. Some synthetic polysaccharides were functionalized by introducing a reactive group, for example, acryloylated hyaluronan and chondroitin were readily obtained. These techniques can be further applied to developing new telechelics, macromonomers, end-functionalized polymers, gel materials, bioactive materials [145], and polymeric drugs derived from these poly- and oligosaccharides. In addition, preparation of a mutant enzyme is a new technique to modify the catalyst function, which is an important future research direction.

We would like to note again enzymatic catalysis and antibody enzyme (abzyme) catalysis, both for in vitro reactions [8]. In the enzymatic polymerization of the present article, the key point is the appropriate design of monomers so that they can form an enzyme-substrate complex for the reaction to proceed, using an isolated enzyme as catalyst (*substrate design*). In the abzyme catalysis, the catalyst enzyme is designed in living cells using the immune system of an antigen-antibody relationship (*catalyst design*) [146, 147]. Figure 20 shows a typical example of abzyme design for an ester hydrolysis catalyst, where a tetracoordinate phosphorus compound (hapten) is used

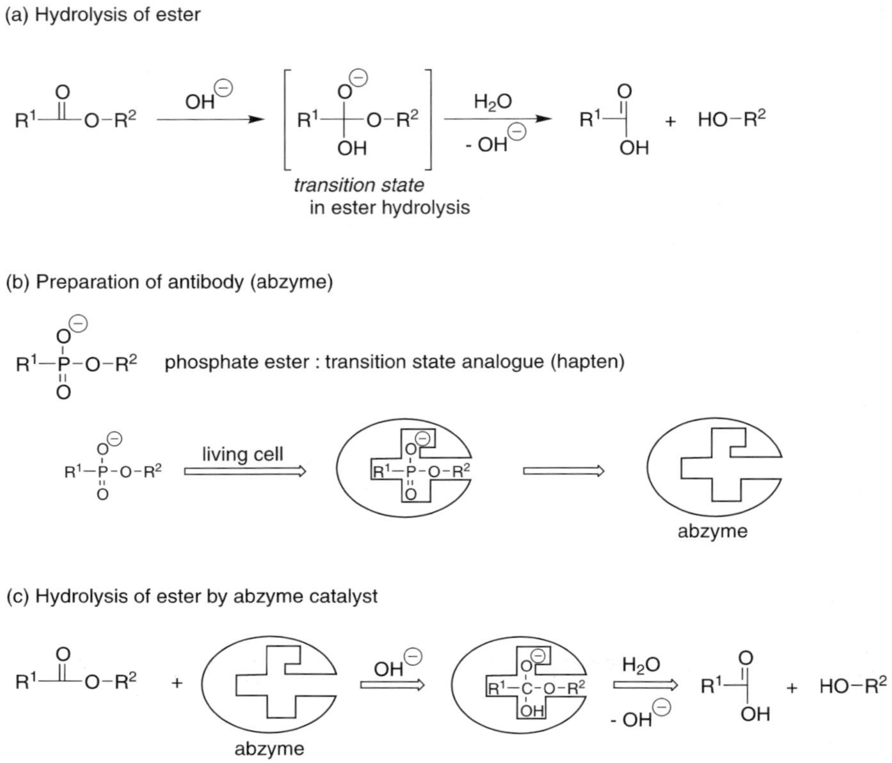

(a) Hydrolysis of ester

[reaction scheme: R^1-C(=O)-O-R^2 + OH^- → [transition state: R^1-C(O$^-$)(OH)-O-R^2] + H_2O / $-OH^-$ → R^1-C(=O)-OH + HO-R^2]

transition state
in ester hydrolysis

(b) Preparation of antibody (abzyme)

R^1-P(=O)(O$^-$)-O-R^2 phosphate ester : transition state analogue (hapten)

[scheme: phosphate ester → living cell → cell containing hapten → abzyme]

(c) Hydrolysis of ester by abzyme catalyst

[scheme: R^1-C(=O)-O-R^2 + abzyme → abzyme-bound tetrahedral intermediate R^1-C(O$^-$)(OH)-O-R^2 → R^1-COOH + HO-R^2]

Fig. 20 Illustration of the abzyme design and catalysis in the hydrolysis of ester

as the model of a transition state involved in the base-catalyzed hydrolysis of the ester. The living cells recognize the hapten as an antigen to produce an antibody. The antibody thus formed is isolated and utilized as catalyst for the ester hydrolysis. This idea spurred many chemists to study abzyme catalysis extensively. The results looked interesting; the rate enhancement by abzyme catalysis was 10^3- to 10^6-fold depending on the reaction [148], which is to be compared by enzyme catalysis with $\sim 10^{12}$-fold and even a value reaching 10^{20}-fold as recently described [149]. The key, therefore, in catalyst design when living cells produce an antibody by complexing with a hapten at the ground state is the extent to which the antibody reflects the 3D dynamic structure of the transition state of the reaction. In addition to the rate acceleration, the catalysis involves other factors of controlling reaction selectivities such as stereochemistry and regioselectivity. Furthermore, it was found recently that a natural antibody shows catalytic activity in rare cases; such an antibody is called an "antigenase" [150–152]. These results interrelate the catalytic functions of enzyme and antibody, which seems very significant.

References

1. Kobayashi S (1997) Introduction. In: Kobayashi S (ed) Catalysis in precision polymerization. Wiley, Chichester, UK, p 1
2. Kobayashi S, Shoda S, Uyama H (1995) Adv Polym Sci 121:1
3. Kobayashi S, Shoda S, Uyama H (1996) Enzymatic polymerization. In: Salamone JC (ed) The polymeric materials encyclopedia. CRC Press, Roca Raton, FL, p 2102
4. Kobayashi S, Shoda S, Uyama H (1997) Enzymatic catalysis. In: Kobayashi S (ed) Catalysis in precision polymerization. Wiley, Chichester, UK, p 417
5. Ritter H (1997) In: Arshady R (ed) Desk reference of functional polymers, synthesis and applications. American Chemical Society, Washington, DC, p 103
6. Kobayashi S, Uyama H (1998) Biocatalytical routes to polymers. In: Schlueter AD (ed) Material science and technology-synthesis of polymers. Wiley, Weinheim, Germany, p 549
7. ACS Symposium Series (1998) Gross RA, Kaplan DL, Swift G (eds) American Chemical Society, Washington, DC, Vol 684
8. Kobayashi S (1999) J Polym Sci Part A Polym Chem 37:3041
9. Kobayashi S, Uyama H, Ohmae M (2001) Bull Chem Soc Jpn 74:613
10. Kobayashi S, Uyama H, Kimura S (2001) Chem Rev 101:3793
11. Gross RA, Kumar A, Karla B (2001) Chem Rev 101:2097
12. Kobayashi S, Sakamoto J, Kimura S (2001) Prog Polym Sci 26:1525
13. Kobayashi S, Uyama H (2002) Enzymatic polymerization. In: Kroschwitz JI (ed) Encyclopedia of polymer science and technology, 3rd edn Web site. Wiley, New York
14. Kobayashi S (2005) J Polym Sci Part A Polym Chem 43:693
15. Fischer E (1894) Ber Chem Dtsch Chem Ges 27:2985
16. Pauling L (1946) Chem Eng News 24:1375
17. Kollman PA, Kuhn B, Donini O, Perakyla M, Santon R, Bakowies D (2001) Acc Chem Res 34:72
18. Kennedy JF, White CA (1979) In: Barton D, Ollis WD, Haslam E (eds) Comprehensive organic chemistry. Pergamon, Oxford 5:755
19. Koenig W, Knorr E (1901) Ber Chem Dtsch Chem Ges 34:957
20. Schmidt RR, Michel J (1980) Angew Chem Int Ed Engl 19:731
21. Schmidt RR (1986) Angew Chem Int Ed Engl 25:212
22. Mark H (1980) Cellul Chem Technol 14:569
23. Klemm D, Heinze T, Wagenknecht W (1997) Acta Polym 48:277
24. Heinze T (1998) Macromol Chem Phys 199:2341
25. Kobayashi S, Uyama H (2003) Macromol Chem Phys 204:235
26. Schlubach HH, Luehrs (1941) Liebigs Ann Chem 547:73
27. Huseman E, Mueller GJM (1966) Macromol Chem 91:212
28. Kochtkov NK (1987) Tetrahedron 43:2389
29. Uryu T, Yamaguchi C, Morikawa K, Terui K, Kanai T, Matsuzaki K (1985) Macromolecules 18:599
30. Nakatsubo F, Takano T, Kawada T, Murakami K (1989) In: Kennedy JF, Phillips GO, Williams PA (eds) Cellulose, structural and functional aspects. Ellis Horwood, Sussex, UK, p 201
31. Kobayashi S, Kashiwa K, Kawasaki T, Shoda S (1991) J Am Chem Soc 113:3079
32. Davies GJ, Dauter M, Brozowski AM, Børnvad ME, Andersen KV, Schuelein M (1998) Biochemistry 37:1926
33. Saxena IM, Brown RM Jr, Fevre M, Geremia RA, Henrissat B (1995) J Bacteriol p 1419
34. Brown RM Jr, Saxena IM, Kudlicka K (1996) Trends Plant Sci 1:149

35. Delmer DP (1999) Annu Rev Plant Physiol Plant Microbiol 50:245
36. Nakamura I, Yoneda H, Maeda T, Makino A, Ohmae M, Sugiyama J, Ueda M, Kobayashi S, Kimura S (2005) Macromol Biosci 5:623
37. Gardner KH, Blackwell J (1974) Biopolymers 13:1975
38. Kolpak FJ, Blackwell J (1976) Macromolecules 9:273
39. Kobayashi S, Shoda S, Lee J, Okuda K, Brown PM Jr, Kuga S (1994) Macromol Chem Phys 195:1319
40. Lee JH, Brown RM Jr, Kuga S, Shoda S, Kobayashi S (1994) Proc Natl Acad Sci USA 91:7425
41. Kobayashi S, Okamoto E, Wen X, Shoda S (1996) J Macromol Sci Pure Appl Chem 33:1375
42. Kobayashi S, Shoda S, Wen X, Okamoto E, Kiyosada T (1997) J Macromol Sci Pure Appl Chem 34:2135
43. Kobayashi S, Hobson LJ, Sakamoto J, Kimura S, Sugiyama J, Imai T, Ito T (2000) Biomacromolecules 1:168,509
44. Tanaka H, Hashimoto T, Kurosaki K, Ohmae M, Kobayashi S, Koizumi S (2003) Polym Prepr Jpn 52:1873
45. Kobayashi S, Kawasaki T, Obata K, Shoda S (1993) Chem Lett 685
46. Karthaus O, Shoda S, Takano H, Obata K, Kobayashi S (1994) J Chem Soc Perkin Trans I 1851
47. Shoda S, Chigira Y, Mitsuishi (1999) Glycoconjugate J 16:S18
48. Mackenzie LF, Wang Q, Warren RAJ, Withers SG (1998) J Am Chem Soc 120:5583
49. Fort S, Boyer V, Greffe L, Davies GJ, Morez O, Christiansen L, Schuelein M, Cottaz S, Bringuez H (2000) J Am Chem Soc 122:5429
50. Carpita NC, Gibeaut DM (1993) Plant J 3:1
51. Woodward JR, Fincher GB, Stone BA (1983) Carbohydr Polym 3:207
52. Faijes M, Imai T, Bulone V, Planas A (2004) Biochem J 380:635
53. Samain E, Lancelon-Pin C, Ferigo F, Moreau V, Chanzy H, Heyraud A, Driguez H (1995) Carbohydr Res 271:217
54. Kobayashi S, Shimada J, Kashiwa K, Shoda S (1992) Macromolecules 25:3237
55. Ziegast G, Pfannemueller B (1987) Carbohydr Res 160:185
56. Kobayashi S, Wen X, Shoda S (1996) Macromolecules 29:2698
57. Muzzarelli RAA, Jeuniaux C, Gooday G (1986) Chitin in nature and technology. Plenum, New York
58. Kiyosada T, Takada E, Shoda S, Kobayashi S (1995) Polym Prepr Jpn 44:660
59. Kiyosada T, Shoda S, Kobayashi S (1995) Polym Prepr Jpn 44:1230
60. Kobayashi S, Kiyosada T, Shoda S (1996) J Am Chem Soc 118:13113
61. Sato H, Mizutani S, Tsuge S, Ohtani H, Aoi K, Okada M, Kobayashi S, Kiyosada T, Shoda S (1998) Anal Chem 70:7
62. Kobayashi S (1990) Prog Polym Sci 15:751
63. Sakamoto J, Kobayashi S (2004) Chem Lett 33:698
64. Terwisscha van Scheltinga AC, Armand S, Kalk KH, Isogai A, Henrissat B, Dijkstra BW (1995) Biochemistry 34:15619
65. Tews I, Terwisscha van Scheltinga AC, Perrakis A, Wilson KS, Dijkstra BW (1997) J Am Chem Soc 119:7954
66. Brameld KA, Shrader WD, Imperiali B, Goddard III WA (1998) J Mol Biol 280:913
67. Brameld KA, Goddard III WA (1998) J Am Chem Soc 120:3571
68. Bramed KA, Goddard III WA (1998) Proc Natl Acad Sci USA 95:4276
69. Kiyosada T (1998) PhD Thesis, Tohoku University
70. Tominaga Y, Tsujisaka Y (1976) Agric Biol Chem 40:2325
71. Watanabe T, Suzuki K, Oyanagi W, Ohnishi K, Tanaka H (1990) J Biol Chem 265:15659

72. Watanabe T, Kobori K, Miyashita K, Fujii T, Sakai H, Uchida M, Tanaka H (1993) J Biol Chem 268:1856
73. Sakamoto J, Watanabe T, Ariga Y, Kobayashi S (2001) Chem Lett 1180
74. Minke R, Blackwell J (1978) J Mol Biol 120:167
75. Dweltz N (1961) Biochim Biophys Acta 51:283
76. Keller A, Waring JRS (1955) J Polym Sci 17:447
77. Keith HD, Padden FJ Jr (1959) J Polym Sci 39:123
78. Marchessault RH, Morehead FF, Walter NM (1959) Nature 184:632
79. Revol JF, Marchessault RH (1993) Int J Biol Macromol 15:329
80. Sakamoto J, Sugiyama J, Kimura S, Imai T, Itoh T, Watanabe T, Kobayashi S (2000) Macromolecules 33:4155,4982
81. Shoda S, Fujita M, Lohavisavapanichi C, Misawa Y, Ushizaki K, Tawata Y, Kuriyama M, Kohri M, Kuwata H (2002) Helv Chim Acta 85:3919
82. Iozzo RV (1998) Annu Rev Biochem 67:609
83. Bernfield M, Götte M, Park PW, Reizes O, Fitzgerald ML, Lincecum J, Zako M (1999) Annu Rev Biochem 68:729
84. Prydz K, Dalen KT (2000) J Cell Sci 113:193
85. Bullock SL, Fletcher JM, Beddington RSP, Wilson VA (1998) Genes Dev 12:1894
86. Perrimon N, Bernfield M (2000) Nature 404:725
87. Lin X, Wei G, Shi Z, Dryer L, Esko JD, Wells DE, Matzuk MM (2000) Dev Biol 224:299
88. Lander AD, Selleck SB (2000) J Cell Biol 148:227
89. Chudo H, Toyoda H (2001) Seikagaku 73:449
90. Habuchi O (2001) Cell Technol 20:204
91. Maeda N (2001) Cell Technol 20:1074
92. Selleck SB (2000) Trend Genet 16:206
93. Maccarana M, Sakura Y, Tawada A, Yoshida K, Lindahl U (1996) J Biol Chem 271:17804
94. Feyzi E, Saldeen T, Larsson E, Lindhal U, Salmivirta M (1998) J Biol Chem 273:13395
95. Kobayashi S, Morii H, Itoh R, Kimura S, Ohmae M (2001) J Am Chem Soc 123:11825
96. Ochiai H, Mori T, Ohmae M, Kobayashi S (2005) Biomacromolecules 6:1068
97. Kobayashi S, Fujikawa S, Ohmae M (2003) J Am Chem Soc 125:14357
98. Kobayashi S, Ohmae M, Fujikawa S, Ochiai H (2005) Macromol Symp 226:147
99. Meyer K, Palmer JW (1934) J Biol Chem 107:629
100. Itano N, Kimata K (2002) IUBMB Life 54:195
101. Weigel PH (2002) IUBMB Life 54:201
102. Selleck SB (2000) Trend Genet 16:206
103. Doege K, Sasaki M, Horigan E, Hassell JR, Yamada Y (1987) J Biol Chem 262:17757
104. Zhuo L, Kimata K (2001) Cell Struct Funct 26:189
105. Salustri A, Yanagishita M, Underhill CB, Laurent TC, Hascall VC (1992) Dev Biol 151:541
106. Fulop C, Salustri A, Hascall VC (1997) Arch Biochem Biophys 337:261
107. de la Motte CA, Hascall VC, Calabro A, Yen-Lieberman B, Strong SA (1999) J Biol Chem 274:30747
108. Thome RF, Legg JW, Isacke CM (2004) J Cell Sci 117:373
109. Evanko SP, Wight TN (1999) J Histochem Cytochem 47:1331
110. Tammi R, Rilla K, Pienimäki J-P, MacCallum DK, Hogg M, Luukkonen M, Hascall VC, Tammi M (2001) J Biol Chem 276:35111
111. Lapčík L Jr, Lapčík L (1998) Chem Rev 98:2663
112. Dahl LB, Laurent TC, Smedsrod B (1988) Anal Biochem 175:397
113. Curvall M, Lindberg B, Lonngren J (1975) Carbohydr Res 41:235
114. Karst NA, Linhardt RJ (2003) Curr Med Chem 10:1993
115. Blatter G, Jaquinet JC (1996) Carbohydr Res 288:109

116. Frost GI, Csóka T, Stern R (1996) Trends Glycosci Glycotechnol 8:419
117. Marković-Housley Z, Miglierini G, Soldatova L, Rizkallah PJ, Müller U, Schirmer T (2000) Structure 8:1025
118. Ochiai H, Mori T, Ohmae M, Kobayashi S (2004) Polym Prepr Jpn 53:1882
119. Bulik DA, Wei G, Toyoda H, Kinoshita-Toyoda A, Waldrip WR, Esko JD, Robbins PW, Selleck SB (2000) Proc Natl Acad Sci USA 97:10838
120. Faissner A, Clement A, Lochter A, Streit A, Mandl C, Schachner M (1994) J Cell Biol 126:783
121. Watanabe H, Kimata K, Line S, Strong D, Gao LY, Kozak CA, Yamada Y (1994) Nat Genet 7:154
122. Clement AM, Nadanaka S, Masayama K, Mandl C, Sugahara K, Faissner A (1998) J Biol Chem 273:28444
123. Maeda N, Noda M (1998) J Cell Biol 142:203
124. Kitagawa H, Uyama T, Sugahara K (2001) J Biol Chem 276:38721
125. Kitagawa H, Izumikawa T, Uyama T, Sugahara K (2003) J Biol Chem 278:23666
126. Sato T, Gotoh M, Kiyohara K, Akashima T, Iwasaki H, Kameyama A, Mochizuki H, Yada T, Inaba N, Togayachi A, Kudo T, Asada M, Watanabe H, Imamura T, Kimata K, Narimatsu H (2003) J Biol Chem 278:3063
127. Uyama T, Kitagawa H, Tanaka J, Tamura J, Ogawa T, Sugahara K (2003) J Biol Chem 278:3072
128. Kusche-Gullberg M, Kjellén L (2003) Curr Opin Struc Biol 13:605
129. DeAngelis PL, Padgett-McCue AJ (2000) J Biol Chem 275:24124
130. Ninomiya T, Sugiura N, Tawada A, Sugimoto K, Watanabe H, Kimata K (2002) J Biol Chem 277:21567
131. Sugahara K, Mikami T, Uyama T, Mizuguchi S, Nomura K, Kitagawa H (2003) Curr Opin Struct Biol 13:612
132. Takagaki K, Nakamura T, Majima M, Endo M (1988) J Biol Chem 263:7000
133. Fujikawa S, Ohmae M, Kobayashi S (2004) Polym Prepr Jpn 53:1880
134. Saitoh H, Takagaki K, Majima M, Nakamura T, Matsuki A, Kasai M, Narita H, Endo M (1995) J Biol Chem 270:3741
135. Takagaki K, Ishido K (2000) Trends Glycosci Glycotechnol 12:295
136. For example, see the following review and cited references therein: DeAngelis PL (2002) Anat Rec 268:317
137. De Luca C, Lansing M, Martin I, Crescenzi F, Shen GJ, O'Regan M, Wong CH (1995) J Am Chem Soc 117:5869
138. DeAngelis PL (2003) J Biol Chem 278:35199
139. Okamoto E, Kiyosada T, Shoda S, Kobayashi S (1997) Cellulose 4:161
140. Ochiai H, Ohmae M, Kobayashi S (2004) Carbohydr Res 339:2769
141. Ochiai H, Ohmae M, Kobayashi S (2004) Chem Lett 33:694
142. Fujita M, Shoda S, Kobayashi S (1998) J Am Chem Soc 120:6411
143. Driguez H (1997) Top Curr Chem 187:86
144. Moreau V, Driguez H (1996) J Chem Soc Perkin Trans 1 p 525
145. Uryu T (1993) Prog Polym Sci 18:717
146. Tramontano A, Kim DJ, Lerner RA (1986) Science 234:1566
147. Pollack SJ, Jacobs JW, Schultz PG (1986) Science 234:1570
148. Lerner RA, Benkovic SJ, Schultz PG (1991) Science 252:659
149. Borman S (2004) Chem Eng News February 23:35
150. Matsuura K, Yamamoto K, Shinohara H (1994) Biochem Biophys Res Commun 204:57
151. Hifumi E, Okamoto Y, Uda T (1999) J Biosci Bioeng 88:323
152. Uda T, Hifumi E (2004) J Biosci Bioeng 97:143

In Vitro Enzyme-Induced Vinyl Polymerization

Amarjit Singh[1] · David L. Kaplan[2] (✉)

[1]Department of Radiology, Harvard Medical School, Boston, MA 02115, USA

[2]Departments of Chemical & Biological Engineering and Biomedical Engineering, Bioengineering & Biotechnology Center, Tufts University, Medford, MA 02155, USA
david.kaplan@tufts.edu

1	Introduction	211
2	Polymerization of Acrylic Acid Based Monomers	212
3	Polymerization of Styrene Based Monomers	215
4	Vitamin C Functionalized Vinyl Polymers	218
5	Kinetics and Mechanism	220
6	Conclusions	222
	References	222

Abstract The in vitro enzyme-mediated polymerization of vinyl monomers is reviewed with a scope covering enzymatic polymerization of vitamin C functionalized vinyl monomers, styrene, derivatives of styrene, acrylates, and acrylamide in water and water-miscible cosolvents. Vitamin C functionalized polymers were synthesized via a two-step biocatalytic approach where vitamin C was first regioselectively coupled to vinyl monomers and then subsequently polymerized. The analysis of this enzymatic cascade approach to functionalized vinyl polymers showed that the vitamin C in polymeric form retained its antioxidant property. Kinetic and mechanistic studies revealed that a ternary system (horseradish peroxidase, H_2O_2, initiator β-diketone) was required for efficient polymerization and that the initiator controls the characteristics of the polymer. The main attributes of enzymatic approaches to vinyl polymerization when compared with more traditional synthetic approaches include facile ambient reaction environments of temperature and pressure, aqueous conditions, and direct control of selectivity to generate functionalized materials as described for the ascorbic acid modified polymers.

Keywords Enzymatic polymerization · Vinyl monomer · Vitamin C · Biocatalysis

1
Introduction

Nature is responsible for the synthesis of a diverse set of polymers in biosystems with sophisticated structural control. In vivo these polymers are synthesized by enzymes. When these enzymes are isolated from biological systems and used in vitro, they retain catalytic properties to carry out biotransformations in a selective fashion in unnatural environments. These selective transformations provide advantages over traditional synthetic methods with respect to stereo-, regio-, and enantioselective control to produce functionalized polymers, biomaterials, agrochemicals, and pharmaceuticals. Studies on enzyme-based polymerization reactions have expanded for the past 15 years. Part of this expansion is due to the variety of families of enzymes that can be utilized for transformations of their natural substrates as well as a wide range of synthetic compounds. Enzymatic, chemoenzymatic, and multienzymatic approaches have been used to exploit enzymatic technology for polymer synthesis [1–5]. The main target macromolecules for enzymatic polymerization are polysaccharides, polyesters, poly(amino acids), and polyaromatics. Lipases and peroxidases have shown remarkable utility toward these targets. Lipase-catalyzed selective modification of monomers, ring-opening polymerization, polycondensation of alcohols with acids, and polycarbonate synthesis have been reported. Peroxidases catalyze the oxidation of a variety of substrates such as phenols, aromatic amines, and vinyl monomers in the presence of hydrogen peroxide, leading to polymerization in aqueous, non-aqueous, and interfacial systems [6–14].

In this chapter, the focus is on in vitro enzyme catalysis for vinyl polymerization. To the best of our knowledge, prior to the work of Derango et al. (1992) there is a single short report showing the formation of low molecular weight vinyl polymers when studied in a suspension of *Escherichia coli* in the presence of methyl methacrylate [15, 16]. Unlike polyaromatics, vinyl polymerization offers better control of polymer characteristics, as has been demonstrated with ternary systems (enzyme, oxidant, and initiator such as β-diketone). The number of different vinyl monomer chemistries investigated for susceptibility toward enzymatic polymerization (**1–12**) is fewer than reported aromatics, as is the extent of literature covering these types of syntheses. In addition, the discovery of multienzymatic approaches for the synthesis of antioxidant-functionalized vinyl polymers provides new impetus for the use of enzymatic methods related to vinyl polymers.

2
Polymerization of Acrylic Acid Based Monomers

Resins of poly(sodium acrylate) are used as water absorbent materials for cleaning surfaces, and in water and oil conditioning, personal care products, and disposable materials for medical applications [17–20]. The large amount of lignin produced as a by-product of the pulp and paper industry has little commercial utility. Fungal laccase-catalyzed grafting of lignin with acrylamide and other vinyl monomers is an approach being used to convert this "waste" lignin into a new class of marketable engineering plastics [21–23]. In 1951 Parravano reported for the first time that enzymatic systems such as *E. coli* in formic acid and xanthine oxidase in formaldehyde can initiate polymerization of methyl methacrylate in aqueous medium free of oxygen [15]. There was no significant activity in the study of enzymatic vinyl polymerization until 1992 when Derango et al. reported the polymerization of vinyl monomers such as acrylamide, methyl acrylate, hydroxyethyl methacrylate, acrylic acid, and methyl methacrylate in the presence of various enzymes such as xanthine oxidase, horseradish peroxidase, alcohol oxidase, and chloroperoxidase [16].

The authors reported the generation of polymers in the presence of a large excess of oxidant (monomer : oxidant 1.6 : 1.0 vol/vol) without any information on the mechanism of polymerization, polymer molecular weight, or role and influence of oxidant on the enzyme or reaction kinetics. Subsequently, the laccase-catalyzed polymerization of acrylamide in water without any initiator (β-diketone) at temperatures ranging from 50 to 80 °C was reported; however, the reactions were efficiently carried out at room temperature with the use of laccase/β-diketone (2,4-pentanedione) initiator to produce polyacrylamide in 97% yield ($M_n = 2.3 \times 10^5$) [24, 25]. The molecular weight of the polymer in these studies was determined by size exclusion chromatography. At 50, 65, and 80 °C when the polymerization was carried out for 4 h, polyacrylamides with $M_n = 9.4, 9.2,$ and 10×10^5, respectively, were produced with the highest yield of 70% at 65 °C.

In later studies on enzymatic vinyl polymerization, the same authors used a ternary system (horseradish peroxidase, H_2O_2, β-diketone) to generate free radicals in the oxidoreductive pathway. Teixeira et al. [26] studied the role of eight β-diketones (2,4-pentanedione, 2-acetylcyclohexanone, 2-acetylcyclopentanone, 1,3-cyclohexanedione, 5,5-dimethyl-1,3-cyclohexanedione, 2-methyl-1,3-cyclohexanedione, 1,3-cyclopentanedione, 2-methyl-1,3-cyclopentanedione) in the polymerization of acrylamide. The results demonstrated that β-diketones are key compounds in the reactions, and by changing the initiator the molecular weight ($M_n = 5100$ to $124\,000$), polydispersity (PD = 2.5 to 4.4), and yield (38 to 93%) of vinyl polymers were greatly influenced. Kalra et al. [27] performed the horseradish peroxidase-mediated polymerization of acrylamide in aqueous medium with and without the addition of surfactant. The surfactants bis(2-ethylhexyl)sodium sulfosuccinate and cetyltrimethylammonium bromide influenced the reaction yield with time. Reactions were completed in 60 and 75 min, respectively, compared to 180 min without surfactant, to produce polymer with the same yield (94 to 99%). Concentrated emulsions prepared from sodium monooleate were also used as a medium for acrylamide free radical polymerization. Polyacrylamide in 99% yield was produced in 75 min; the polydispersity was > 4, higher than that for the aqueous system without surfactant. The horseradish peroxidase/H_2O_2 system was extended to the chemoselective polymerization of 2-(4-hydroxyphenyl)ethyl methacrylate and polymerization of 2-phenylethyl methacrylate (Fig. 1) [28].

Kobayashi et al. reported the laccase-catalyzed polymerization of acrylamide without the addition of hydrogen peroxide. Polymerization occurs at 50 °C and higher temperatures, but no polymerization was observed at room temperature. In the presence of initiator such as 2,4-pentanedione, laccase catalyzed the vinyl polymerization at room temperature. The highest molecular weight ($M_n = 6 \times 10^5$, PD = 1.99) was achieved when the reaction was carried at 65 °C. In the polymerization at this temperature the polymer yield gradually increased with time, e.g., 37, 70, and 81% in 1, 4, and 24 h, respectively. No polymerization was observed when the reaction was carried out under air. In terms of yield of the reaction, water was a better solvent than N,N-dimethylformamide. The horseradish peroxidase-catalyzed polymerization of the phenolic moiety in the presence of the methacrylic group was carried out using hydrogen peroxide as oxidizing agent in acetone/acetate buffer (pH 7; 50 : 50 vol %) at room temperature under air for 24 h. Polymer with a pendant acrylic group was produced in 77% yield.

Kalra et al. [29] reported that horseradish peroxidase type II gave the best yield, polydispersity, and tacticity in the enzyme-mediated free radical polymerization of methyl methacrylate in mixtures of water and water-miscible cosolvents, when compared with horseradish peroxidase type I, soybean peroxidase, and *Arythomyces ramosus* peroxidase. The polymerization of methyl methacrylate has predominantly shown the formation of syndiotactic polymers. Changing the enzyme concentration from 70 to 80 mg/mL in a given

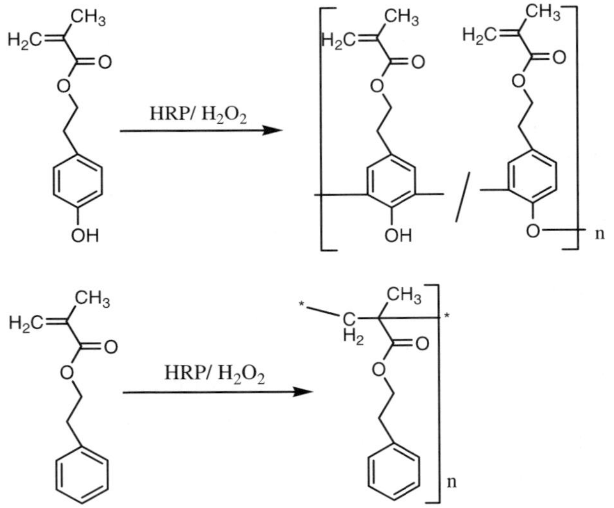

Fig. 1 Chemoselective polymerization of phenol derivative by horseradish peroxidase

reaction resulted in an increase in the *syn*-dyad fraction from 0.84 to 0.85 and an increase in polymer yield from 46 to 56%. On increasing the hydrogen peroxide concentration from 0.056 to 0.110 mmol, the yield decreased from 65 to 13% and M_n decreased from 91 000 to 13 000. The concentration of 2,4-pentanedione greatly influenced the polymer yield, which increased from 18 to 88% when the concentration was increased from 0.097 to 0.116 mmol, whereas further increases in concentration to 0.136 and 0.115 mmol resulted in a decrease in product yield (45 and 14%, respectively).

3
Polymerization of Styrene Based Monomers

Polystyrene is used in a wide range of commercial applications as packaging material, injection molded parts, and UV screening agents [30, 31]. An enzymatic strategy was used for the synthesis of polymers from styrene, 4-methylstyrene, 2-vinylnaphthalene, and sodium styrenesulfonate [32, 33]. Molecular weight and yield of polystyrene were influenced by solvent, concentration of hydrogen peroxide, and initiator. Five solvents, on the basis of different polarity, were investigated in these reactions to explore monomer and polymer compatibility. The highest polymer yield was in aqueous tetrahydrofuran. With high water content (H_2O/THF (v/v)> 5) the polymer yield was low. This may be due to insufficient tetrahydrofuran to dissolve the styrene monomer and polymer, leading to phase separation. As the tetrahydrofuran content was increased, polystyrene yield increased to 21% (H_2O/THF = 3, v/v). Polystyrene yield was low in aqueous DMF because this solvent was not miscible in water and thus decreased the initiation of the reaction. The polystyrene yield in methanol was also low, perhaps because styrene was soluble in methanol but polystyrene had poor solubility, thus oligomers precipitated and terminated polymerization. Polymer yield in dioxane was also low, despite a solvent polarity comparable to that of tetrahydrofuran. This low yield may be due to the effect of dioxane on enzyme activity. Polydispersity was similar in tetrahydrofuran and dioxane. As a result of these studies, tetrahydrofuran was selected as a suitable solvent for subsequent polymerization reactions due to the combination of high yield and high molecular weight of polymer formed. Polystyrene molecular weight increased with reaction time while polydispersity also increased, presumably due to the free radical process. The influence of six different initiators (2,4-pentanedione, dibenzoylmethane, benzoylacetone, tetronic acid, 4-hydroxycoumarin, and 1,3-cyclopentanedione) was examined for the horseradish peroxidase-catalyzed polymerization of styrene. The results showed that molecular weights (M_n = 26 923 to 96 504) and yields (14 to 60%) were effected by the initiator.

Polymers of *p*-styrenesulfonic acid are used in many applications including ion-exchange membranes and resins [34–38], in biomaterials to influence cell adhesion [39–43], and as inhibitors of virus infections [44–51]. Horseradish peroxidase, hematin, and pegylated hematin (PEG-hematin) were used for the polymerization of sodium styrenesulfonate in aqueous systems (Fig. 2). PEG-hematin was synthesized to maintain solubility and function at neutral pH, which was not possible with unmodified hematin due to the limitations in solubility at neutral pH. The molecular weight of poly(sodium styrenesulfonate) increased (from M_n 65 882 to 163 280) with reaction time while the polydispersity also increased (2.5 to 3.4), presumably due to the free radical process. More than 80% of the sodium styrenesulfonate was converted to poly(sodium styrenesulfonate) in 4 h. Experiments were run for 15, 30, 60, 120, and 240 min with three different ratios of hydrogen peroxide and 2,4-pentanedione (H_2O_2 : dione): (a) 0.056 : 0.082, (b) 0.056 : 0.224, and (c) 0.224 : 0.082 mmol. When the ratio of 2,4-pentanedione was 1.46 times the molarity of hydrogen peroxide (condition "a"), a steady increase in molecular weight (from M_n 37 795 to 90 389) and yield (7.0 to 82.2%) was observed. When the concentration of 2,4-pentanedione was increased four times (condition "b"), a decrease in molecular weight (from M_n 88 106 to 49 407) and increase in polydispersity (2.3 to 3.8) was observed. Reaction "b" was slow initially but yielded the same amount of polymer as in case "a" over time. In comparison, with an increased ratio of hydrogen peroxide (condition "c") no product was detected for the reaction run of 15 or 30 min, while after 240 min with same amount of hydrogen peroxide a 56% yield of polymer (M_n = 137 098, PD = 2.8) was obtained. The higher molecular weight achieved when the initiator was 1.46 times the hydrogen peroxide used, on a molar basis, can be explained on the basis of the mechanistic cycle for horseradish peroxidase-catalyzed polymerization of vinyl monomers (Fig. 3). One molecule of hydrogen peroxide produces two catalytically active forms of horseradish peroxidase, each of which oxidizes the initiator to produce a radical. The initiator radical generates

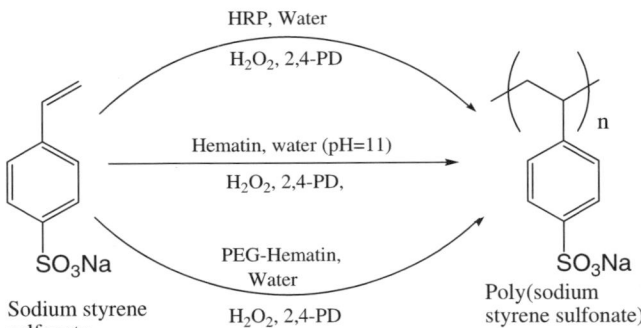

Fig. 2 Schematic of sodium styrenesulfonate polymerization by the different catalysts. HRP: horseradish peroxidase; 2,4-PD: 2,4-pentanedione; H_2O_2: hydrogen peroxide

monomer styrene radicals, which react with other radicals. The initiator radical also generates monomer styrene radicals that react with other monomers to produce polymer. Therefore, based on the mechanism and optimal stoichiometry, one molecule of hydrogen peroxide requires two molecules of 2,4-pentanedione.

When 2,4-pentanedione was present at four times the level of hydrogen peroxide, many initiator radicals were formed early in the reaction leading to the formation of many polymer chains and thus lower molecular weight polymer. The higher concentration of hydrogen peroxide likely inhibited the horseradish peroxidase activity in the short time frame (up to 30 min) leading to low overall yield after 240 min.

To evaluate the influence of hydrogen peroxide and 2,4-pentanedione on the molecular weight and polydispersity of poly(sodium styrenesulfonate), stepwise changes in the ratio of these two components were studied, with each reaction run for 4 h at room temperature. The increase in hydrogen peroxide resulted in increased molecular weight of polymer (M_n = 80 925 to 137 098), while the increase in 2,4-pentanedione resulted in decreased molecular weight and increased polydispersity (M_n = 152 251 to 44 245; PD = 3.2 to 4.0). The yield was influenced by the amount of water, with the low water content yielding the most polymer (94%) and the highest water content the lowest yield (58%). PEG-hematin yielded poly(sodium styrenesulfonate) with the highest molecular weight (M_n = 223 520, PD = 3.48) versus hematin and horseradish peroxidase. More than 98% of the sodium acrylate was polymerized with hematin at pH 11.0 to form poly(sodium acrylate) at room temperature in water.

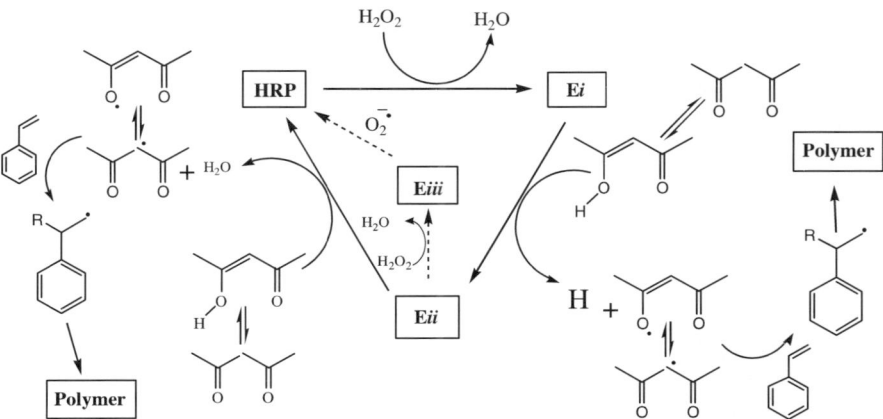

Fig. 3 Catalytic cycle of horseradish peroxidase showing the possible mechanism for free radical polymerization of vinyl monomer (styrene) with β-diketone initiator (2,4-pentanedione). HRP: horseradish peroxidase; E*i*, E*ii*, and E*iii*: oxidized states of horseradish peroxidase; R: 2,4-pentanedione]

4
Vitamin C Functionalized Vinyl Polymers

Antioxidant compounds are added to protect the polymers during and after processing due to thermal and oxidative degradation [52]. The performance of an antioxidant compound depends on its ability to scavenge and retard the radical chain oxidation process. The properties of a good antioxidant candidate include volatility, compatibility, and solubility with the matrix to avoid aggregation and heterogeneous distribution. Chemical methods were used to produce macromolecules of antioxidant compounds in order to reduce volatility. The antioxidant compounds were bound to polymers for more homogeneous distribution. Commonly used antioxidant compounds include synthetic phenolic compounds. These phenolic compounds are added in large quantities to protect polymers from degradation but can also leach into foods and consumer-related products. Since the antioxidants must be added in large quantities, this can also result in discoloration and off-flavors during processing. Furthermore, this loading with antioxidants results in the exposure of consumers to excess levels of such compounds. There has been growing interest in the substitution of synthetic food antioxidants like butylated hydroxyanisole (BHA) and butylated hydroxytoluene (BHT) by natural antioxidants because they are suspected carcinogens and the U.S. Food and Drug Administration (FDA) has therefore restricted their use [53]. There are numerous natural antioxidants and many are polyphenolic compounds including flavonoids such as quercetin, various tannins, caffeic acid, extracts of spices and tea, tocopherols (vitamin E), β-carotene, and vitamin C. Most of these natural antioxidants are not good candidates for foods and other consumer-related applications due to their poor lipid solubility, high production cost when compared to BHA and BHT, and instability with time leading to discoloration of foods. When ascorbic acid is used in these types of consumer-related applications, it is prone to moisture-induced degradation which leads to the formation of brown coloration [54–56]. In addition, the hydrophilic behavior of ascorbic acid prevents its application in cosmetics or use in the presence of fats and oils. At present, 6-0-palmitoyl ascorbic acid is produced by chemical means to resolve these issues. The chemical synthesis to generate this derivative of ascorbic acid results in the formation of a mixture of products with a preponderance of 0-6 substitution. This requires further purification, leading to lower yields [57, 58]. To overcome these problems enzymatic modifications of ascorbic acid to ascorbyl palmitate have been pursued [59, 60].

Recently a new strategy was developed whereby a mild and highly selective enzymatic method was used to covalently couple the primary hydroxyl group of ascorbic acid with styrene and methyl acrylate monomers, followed by a second enzymatic reaction catalyzed by horseradish peroxidase to polymerize the styrene and acrylate monomers yielding vitamin C functionalized polystyrene and poly(methyl methacrylate), respectively (Fig. 4) [61, 62]. The coupling approach was conducted with a *Candida antarctica* lipase-mediated transesterification reaction where the primary hydroxyl group of ascorbic acid was regioselectively acylated with trifluoroethyl 4-vinylbenzoate and 2,2,2-trifluoroethyl methacrylate. In a general procedure for polymerization of L-ascorbyl methyl methacrylate, 5 mL water and 10 mL tetrahydrofuran were flushed with nitrogen for 15 min. L-Ascorbyl methyl methacrylate (2.0 g, 8.2 mM) was added to the reaction mixture. Horseradish peroxidase (16 mg) was dissolved in 200 µL water. Hydrogen peroxide (0.82 mM, 93 µL) and 2,4-pentanedione (1.64 mM, 177 µL) were added simultaneously after the addition of the enzyme, and polymerization was conducted for 24 h with continuous stirring. The polymerization of L-ascorbyl 4-vinylbenzoate was carried out in a mixture of water and methanol. The antioxidant activity of ascorbic acid, ascorbic acid functionalized polystyrene, and poly(methyl methacrylate) were compared by using the DPPH (2,2-diphenyl-1-picrylhydrazyl radical) method [63–65]. The test compound (ascorbic acid, L-ascorbyl methyl methacrylate, or polymer of L-ascorbyl methyl methacrylate) and DPPH (final conc. 0.2 mM) predissolved in methanol were assessed spectrophotometrically. Methanol without DPPH was used as reference. Vitamin C functionalized polystyrene, when used at a concentration of 238 µM, fully scavenged DPPH free radicals (0.2 mM). In another case, ascorbic acid, L-ascorbyl

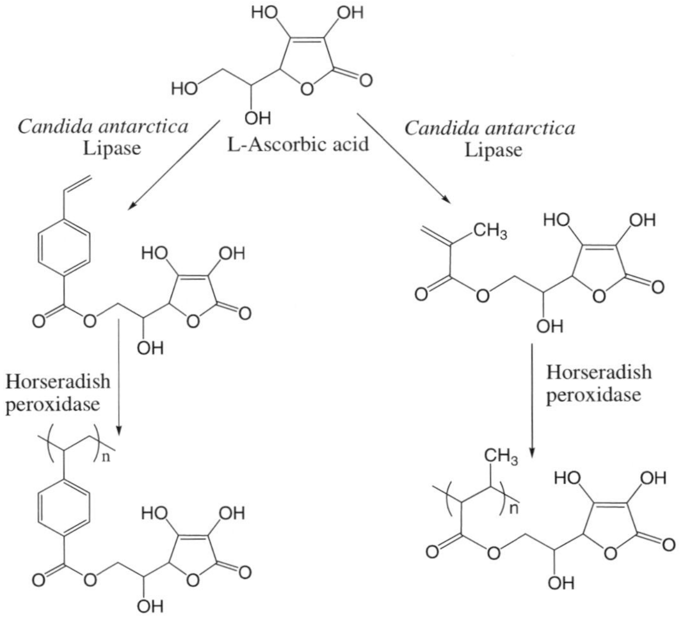

Fig. 4 Enzymatic synthesis of vitamin C functionalized polymers

methyl methacrylate, and polymerized L-ascorbyl methyl methacrylate, when used at a concentration up to 133 μM, fully scavenged DPPH free radicals (0.2 mM). This analysis showed that the formation of polymers with active pendant antioxidant compounds was feasible, since the vitamin C retained its ability to scavenge radicals while in polymeric form.

5
Kinetics and Mechanism

A molar equivalent of hydrogen peroxide to monomer and horseradish peroxidase is a well-known redox system that catalyzes the free radical polymerization of phenol, anilines, and their derivatives [6–14]. Horseradish peroxidase-mediated polymerization of styrene and methyl methacrylate, with a monomer (styrene or methyl methacrylate) to hydrogen peroxide ratio of 40 : 1, did not occur in the absence of 2,4-pentanedione. Therefore, it is likely that this compound is involved in the initiation of free radical formation. A reasonable hypothesis for the horseradish peroxidase-catalyzed polymerization of vinyl monomers is that the enzyme is oxidized by hydrogen peroxide and passes from its native state through two catalytically active forms (Ei and Eii). Each of these active forms oxidizes the initiator (β-diketone, 2,4-pentanedione) while the enzyme returns to the native form. The Eii state of enzyme is oxidized by hydrogen peroxide to produce inactive enzyme, Eiii, which spontaneously reverts to the native form of enzyme. The free radicals produced from the initiator generate radicals in the vinyl monomer to form polymer (Fig. 2).

Durand et al. [66, 67] performed a detailed study on optimization conditions for acrylamide polymerization and the role of the concentration of hydrogen peroxide, 2,4-pentanedione, and horseradish peroxidase in the initial stages of the reaction. Hydrogen peroxide is involved in three processes: (a) the catalytic oxidation of 2,4-pentanedione, (b) degradation reactions of horseradish peroxidase through the formation of an inactive form of enzyme (Eiii), and (c) chemical oxidation of 2,4-pentanedione (cycloperoxidation). For a successful polymerization reaction hydrogen peroxide concentrations ranging from 10^{-4} to 10^{-2} mol L^{-1} with respect to horseradish peroxidase 1.9 g L^{-1} were required. The hydrogen peroxide concentration did not have a great influence on the molecular weight except when it exceeded upper and lower limits. The minimum concentration of 2,4-pentanedione required in a reaction was 4 mmol L^{-1}, as no acrylamide polymer was recovered below this limit. Unlike hydrogen peroxide, there is no upper limit for 2,4-pentanedione as it is involved in the production of radicals and not in the degradation of horseradish peroxidase; molecular weight was influenced by its concentration. Experiments on acrylamide polymerization with a fixed concentration of hydrogen peroxide (7×10^{-4}) and various concentrations of horseradish peroxidase (0.06 to 12 g L^{-1}) showed that the higher horseradish peroxidase concentration resulted in reduced polymer molecular weight. A concentration higher than 0.49 g L^{-1} did have a significant effect on molecular weight. Lalot et al. studied the free radical polymerization of acrylamide in water at room temperature with catalysis by horseradish peroxidase to produce radical species from hydrogen peroxide and 2,4-pentanedione through an oxidative pathway [68]. The initial period of polymerization inhibition was observed and residual oxygen was assumed to be the inhibitor. The polymerization process was a first-order reaction with respect to monomer in the 70–75% conversion range. The deviation of the reaction from first order could be due to a decrease of the propagating radical concentrations arising from the increase of the polyacrylamide solution viscosity.

Initiators greatly influenced styrene polymerization in terms of molecular weight and yield. The role of the initiator in vinyl polymerization is complex. According to the mechanistic hypothesis, it is the enol form of the initiator (β-diketone) that is used for production of radicals and not the keto form. Based on this mechanism, the initial concentration of the initiator will depend on the equilibrium in the β-diketone (keto ↔ enol forms) and it could differ from one initiator to another in the same polymerization medium. In addition, the actual concentration of initiator would be reduced by the chemical oxidation of β-diketone to cyclic peroxide, which will depend on the nature of the β-diketone (cyclic or acyclic) and amount of hydrogen peroxide in the reaction. Two additional issues that could be considered responsible for the control of molecular weight along with the change of the β-diketone are (a) chain transfer reactions and (b) reactivity of enzyme with the β-diketone. The production of radicals first occurs on the enol form and chain transfer to the β-diketone could affect the molecular weight of polymer. The use of cyclic β-diketones resulted in dramatic changes in the molecular weight of styrene and acrylamide. In order to check if these variations came from chain transfer, acrylamide polymerization was carried out using the chemical catalyst 4,4′-azodicyanovaleric acid as initiator in the presence of β-diketones (2,4-pentanedione, 2-acetylcyclopentanone, 2-methylcyclohexane-1,3-dione, and 2-methylcyclopentane-1,3-dione) and the results were compared with those without β-diketones. The presence of β-diketones decreased the molecular weight, which could be due to chain transfer taking place to the β-diketones. In the chemical polymerization initiated by 4,4′-azodicyanovaleric acid the yield remained near 100%, while it varied between 38 and 93% for the horseradish peroxidase-mediated ternary system. The results demonstrated that chain transfer is not the only parameter to influence the output of the polymerization reaction, but in addition reactivity of the enzyme with β-diketone could be partly responsible. This may also help explain the results of styrene polymerization where changing the β-diketone resulted in changes in yield and molecular weight; however, when 4-hydroxycoumarin with a stable enol form was used as initiator the yield was low (14%). Coumarins may have lower reactivity with horseradish peroxidase, and compounds like cyclic β-diketones are good substrates for horseradish peroxidase. This result has been further supported by the fact that cyclic β-diketones, such as 5,5-dimethyl-1,3-cyclohexanedione, are substrates for chloroperoxidases, which belong to the same subclass of enzymes as horseradish peroxidase [69].

6
Conclusions

There are tremendous opportunities for enzymatic polymerization reactions, including: (1) most of the enzymes used in polymer synthesis are commercially available and genetic engineering can be used to enhance selective reactivities; (2) a pool of heterocyclic compounds like coumarins, flavones, and chalcones are well known to chemists and provide fertile ground for further exploration; (3) a more diverse set of monomers can be synthesized using chemoenzymatic methods; (4) most of the target polymers (synthetic) are formed in living organisms with better structural control by using enzymes to further validate the concept of enzymatic polymerization, and much can be done in parallel to natural polymers in terms of control of stereoregularity; and (5) in living organisms these polymers are synthesized for specific purposes like energy storage, architectural control, and information transfer. In vitro polymers are synthesized for academic purposes. If more specific controls can be achieved, enzymatic polymerization can meet challenges in a variety of technological applications, including controlled release of pharmaceuticals, tissue engineering, the food industry, and other consumer-related needs. Vitamin C functionalized vinyl polymers have shown an ability to scavenge free radicals which can be used to improve the shelf life of labile components in foods and pharmaceuticals, while also reducing human exposure to commonly used antioxidants such as BHA and BHT.

References

1. Kaplan DL, Dordick J, Gross RA, Swift G (1998) ACS Symp Ser 684:2
2. Singh A, Kaplan DL (2002) J Polym Environ 10:85
3. Kobayashi S, Uyama H, Kimura S (2001) Chem Rev 101:3793
4. Gross RA, Kumar A, Kalra B (2001) Chem Rev 101:2097
5. Kobayashi S, Uyama H (2001) Adv Biochem Eng Biotechnol 71:241
6. Dordick JS, Marletta MA, Klibanov AM (1987) Biotechnol Bioeng 30:31
7. Akkara JA, Senecal KJ, Kaplan DL (1991) J Polym Sci A-1 29:1561
8. Rao AM, John VT, Gonzalez RD, Akkara JA, Kaplan DL (1993) Biotechnol Bioeng 41:531
9. Karayigitoglu CF, Kommareddi N, Gozalez RD, John VT, McPherson GL, Akkara JA, Kaplan DL (1995) Mater Sci Eng C2:165

10. Akkara JA, Ayyagari M, Bruno F, Samuelson L, John VT, Karayigitoglu C, Tripathy S, Marx KA, Rao DVGLN, Kaplan DL (1994) Biomimetics 2:331
11. Ayyagari MS, Marx KA, Tripathy SK, Akkara JA, Kaplan DL (1995) Macromolecules 28:5192
12. Ayyagari M, Akkara JA, Kaplan DL (1996) Mater Sci Eng C4:169
13. Akkara JA, Kaplan DL, John VT, Tripathy SK (1996) Polym Mater Encyc 3:2115
14. Banerjee S, John VT, McPherson GL, O'Conor CJ, Buisson YSL, Akkara JA, Kaplan DL (1997) Colloid Polym Sci 275:930
15. Parravano G (1951) J Am Chem Soc 73:183
16. Derango RA, Chiang L, Dowbenko R, Lasch JG (1992) Biotechnol Lett 6:523
17. Xue C, Chen X (1997) Sichuan Lianhe Daxue Xuebao, Gongcheng Kexueban 1:41
18. Totsuka F, Sakakibara N, Osawa T (1997) JP 09157531 A2
19. Zhao X, Zheng Y, She Y, Zhang Y (1995) Jianghan Shiyou Xueyuan Xuebao 17:68
20. Suzuki T (1985) JP 60149609 A2
21. Mai C, Milstein O, Huttermann A (1999) Appl Microbiol Biotechnol 51:517
22. Mai C, Milstein O, Huttermann A (2000) J Biotechnol 79:173
23. Huttermann A, Kharazipour A (1996) VTT Symp 163:143
24. Ikeda R, Tanaka H, Uyama H, Kobayashi S (1998) Macromol Rapid Commun 19:423
25. Emery O, Lalot T, Brigodiot M, Marechal E (1997) J Polym Sci A-1 35:3331
26. Teixeira D, Lalot T, Brigodiot M, Marechal E (1999) Macromolecules 32:70
27. Kalra B, Gross RA (2002) Green Chem 4:174
28. Uyama H, Lohavisavapanich C, Ikeda R, Kobayashi S (1998) Macromolecules 31:554
29. Kalra B, Gross R (2000) Biomacromolecules 1:501
30. Miller RL, Nielson LE (1961) J Polym Sci 55:643
31. Eichhorn J (1964) J Appl Polym Sci 8:2497
32. Singh A, Ma D, Kaplan DL (2000) Biomacromolecules 1:592
33. Singh A, Roy S, Samuelson L, Bruno F, Nagarajan R, Kumar J, John V, Kaplan DL (2001) J Macromol Sci Pure Appl Chem A38:1219
34. Wei Z, Wang W, Jin D, Yang D, Tartakovskaya L (1997) J Appl Polym Sci 64:1893
35. Carretta N, Tricoli V, Picchioni F (2000) J Memb Sci 166:189
36. Lam-Leung SY, Li HR (1995) J Appl Polym Sci 57:1373
37. Goldammer EV, Conway BE (1973) J Polym Sci B 11:2767
38. Gryte CC, Gregor HP (1976) J Polym Sci B 14:1839
39. Kowalczynska HM, Nowak-Wyrzkowska M (1999) Cell Biol Int 23:359
40. Lee JH, Lee JW, Khang G, Lee HB (1997) Biomaterials 18:351
41. Detrait E, Lhoest JB, Bertrand P, Van den Bosch de Aguilar P (1999) J Biomed Mater Res 45:404
42. Charef S, Tapon-Bretaudiere J, Fischer A, Flunger F, Jozefowicz M, Labarre D (1996) Biomaterials 17:903
43. Liekens S, Leali D, Neyts J, Esnouf R, Rusnati M, Dellera P, Maudgal PC, De Clercq E, Presta M (1999) Mol Pharmacol 56:204
44. Liekens S, Neyts J, Degreve B, De Clercq E (1997) Oncol Res 9:173
45. Zeitlin L, Whaley KJ, Hegarty TA, Moench TR, Cone RA (1997) Contraception 56:329
46. Mohan P, Schols D, Baba M, De Clercq E (1992) Antiviral Res 18:139
47. Tan GT, Wickramasinghe A, Verma S, Hughes SH, Pezzuto JM, Baba M, Mohan P (1993) Biochim Biophys Acta 1181:183
48. Ikeda S, Neyts J, Verma S, Wickramasinghe A, Mohan P, De Clercq E (1994) Antimicrobial Agents Chemother 38:256
49. Anderson RA, Feathergill K, Diao X, Cooper M, Kirkpatrick R, Spear P, Waller DP, Chany C, Doncel GF, Herold B, Zaneveld LJD (2000) J Androl 21:862

50. Herold BC, Bourne N, Marcellino D, Kirkpatrick R, Strauss DM, Zaneveld LJD, Waller DP, Anderson RA, Chany CJ, Barham BJ, Stanberry LR, Cooper MD (2000) J Infect Dis 181:770
51. Luscher-Mattli M (2000) Antivir Chem Chemother 11:249
52. Breese KD, Lamethe JF, DeArmitt C (2000) Polym Degrad Stab 70:89 and references cited therein
53. Code of Federal Regulations, 21CFR172.110, Title 21, vol 3, parts 170–199, pp 28–29
54. Shephard AB, Nichols SC, Braithwaite A (1999) Talanta 48:585
55. Shephard AB, Nichols SC, Braithwaite A (1999) Talanta 48:595
56. Shephard AB, Nichols SC, Braithwaite A (1999) Talanta 48:607
57. Nickels H, Hackenberger A (1984) Ger Offen DE 3308922
58. Pauling H, Wehrli C (1998) Eur Pat Appl EP 296483
59. Watanabe Y, Adachi S, Nakanishi K, Matsuno R (2001) JAOCS 78:823
60. Luhong T, Hao Z, Shehate MM, Yunfei HA (2000) Biotechnol Appl Biochem 32:35
61. Singh A, Kaplan DL (2003) Adv Mater 15:1291
62. Singh A, Kaplan DL (2004) J Macromol Sci Pure Appl Chem 41:1377–1386
63. Chen Y, Wang M, Rosen RT, Ho C (1999) J Agric Food Chem 47:2226
64. Masuda T, Yonemori S, Oyama Y, Takeda Y, Tanaka T, Andoh T, Shinohara A, Nakata M (1999) J Agric Food Chem 47:1749
65. Duh P, Yen G, Yen W, Chang L (2001) J Agric Food Chem 49:1455
66. Durand A, Lalot T, Brigodiot M, Marechal E (2000) Polymer 41:8183
67. Durand A, Lalot T, Brigodiot M, Marechal E (2001) Polymer 42:5515
68. Lalot T, Brigodiot M, Marechal E (1999) Polym Int 48:288
69. Ashley PL, Griffin BW (1981) Arch Biochem Biophys 210:167

Author Index Volumes 101–194

Author Index Volumes 1–100 see Volume 100

de Abajo, J. and *de la Campa, J. G.*: Processable Aromatic Polyimides. Vol. 140, pp. 23–60.
Abe, A., Furuya, H., Zhou, Z., Hiejima, T. and *Kobayashi, Y.*: Stepwise Phase Transitions of Chain Molecules: Crystallization/Melting via a Nematic Liquid-Crystalline Phase. Vol. 181, pp. 121–152.
Abetz, V. and *Simon, P. F. W.*: Phase Behaviour and Morphologies of Block Copolymers. Vol. 189, pp. 125–212.
Abetz, V. see Förster, S.: Vol. 166, pp. 173–210.
Adolf, D. B. see Ediger, M. D.: Vol. 116, pp. 73–110.
Aharoni, S. M. and *Edwards, S. F.*: Rigid Polymer Networks. Vol. 118, pp. 1–231.
Alakhov, V. Y. see Kabanov, A. V.: Vol. 193, pp. 173–198.
Albertsson, A.-C. and *Varma, I. K.*: Aliphatic Polyesters: Synthesis, Properties and Applications. Vol. 157, pp. 99–138.
Albertsson, A.-C. see Edlund, U.: Vol. 157, pp. 53–98.
Albertsson, A.-C. see Söderqvist Lindblad, M.: Vol. 157, pp. 139–161.
Albertsson, A.-C. see Stridsberg, K. M.: Vol. 157, pp. 27–51.
Albertsson, A.-C. see Al-Malaika, S.: Vol. 169, pp. 177–199.
Allegra, G. and *Meille, S. V.*: Pre-Crystalline, High-Entropy Aggregates: A Role in Polymer Crystallization? Vol. 191, pp. 87–135.
Allen, S. see Ellis, J. S.: Vol. 193, pp. 123–172.
Al-Malaika, S.: Perspectives in Stabilisation of Polyolefins. Vol. 169, pp. 121–150.
Altstädt, V.: The Influence of Molecular Variables on Fatigue Resistance in Stress Cracking Environments. Vol. 188, pp. 105–152.
Améduri, B., Boutevin, B. and *Gramain, P.*: Synthesis of Block Copolymers by Radical Polymerization and Telomerization. Vol. 127, pp. 87–142.
Améduri, B. and *Boutevin, B.*: Synthesis and Properties of Fluorinated Telechelic Monodispersed Compounds. Vol. 102, pp. 133–170.
Ameduri, B. see Taguet, A.: Vol. 184, pp. 127–211.
Amir, R. J. and *Shabat, D.*: Domino Dendrimers. Vol. 192, pp. 59–94.
Amselem, S. see Domb, A. J.: Vol. 107, pp. 93–142.
Anantawaraskul, S., Soares, J. B. P. and *Wood-Adams, P. M.*: Fractionation of Semicrystalline Polymers by Crystallization Analysis Fractionation and Temperature Rising Elution Fractionation. Vol. 182, pp. 1–54.
Andrady, A. L.: Wavelenght Sensitivity in Polymer Photodegradation. Vol. 128, pp. 47–94.
Andreis, M. and *Koenig, J. L.*: Application of Nitrogen-15 NMR to Polymers. Vol. 124, pp. 191–238.
Angiolini, L. see Carlini, C.: Vol. 123, pp. 127–214.
Anjum, N. see Gupta, B.: Vol. 162, pp. 37–63.
Anseth, K. S., Newman, S. M. and *Bowman, C. N.*: Polymeric Dental Composites: Properties and Reaction Behavior of Multimethacrylate Dental Restorations. Vol. 122, pp. 177–218.

Antonietti, M. see *Cölfen, H.:* Vol. 150, pp. 67–187.
Aoki, H. see *Ito, S.:* Vol. 182, pp. 131–170.
Armitage, B. A. see *O'Brien, D. F.:* Vol. 126, pp. 53–58.
Arnal, M. L. see *Müller, A. J.:* Vol. 190, pp. 1–63.
Arndt, M. see *Kaminski, W.:* Vol. 127, pp. 143–187.
Arnold, A. and *Holm, C.:* Efficient Methods to Compute Long-Range Interactions for Soft Matter Systems. Vol. 185, pp. 59–109.
Arnold Jr., F. E. and *Arnold, F. E.:* Rigid-Rod Polymers and Molecular Composites. Vol. 117, pp. 257–296.
Arora, M. see *Kumar, M. N. V. R.:* Vol. 160, pp. 45–118.
Arshady, R.: Polymer Synthesis via Activated Esters: A New Dimension of Creativity in Macromolecular Chemistry. Vol. 111, pp. 1–42.
Auer, S. and *Frenkel, D.:* Numerical Simulation of Crystal Nucleation in Colloids. Vol. 173, pp. 149–208.
Auriemma, F., de Rosa, C. and *Corradini, P.:* Solid Mesophases in Semicrystalline Polymers: Structural Analysis by Diffraction Techniques. Vol. 181, pp. 1–74.

Bahar, I., Erman, B. and *Monnerie, L.:* Effect of Molecular Structure on Local Chain Dynamics: Analytical Approaches and Computational Methods. Vol. 116, pp. 145–206.
Baietto-Dubourg, M. C. see *Chateauminois, A.:* Vol. 188, pp. 153–193.
Ballauff, M. see *Dingenouts, N.:* Vol. 144, pp. 1–48.
Ballauff, M. see *Holm, C.:* Vol. 166, pp. 1–27.
Ballauff, M. see *Rühe, J.:* Vol. 165, pp. 79–150.
Balsamo, V. see *Müller, A. J.:* Vol. 190, pp. 1–63.
Baltá-Calleja, F. J., González Arche, A., Ezquerra, T. A., Santa Cruz, C., Batallón, F., Frick, B. and *López Cabarcos, E.:* Structure and Properties of Ferroelectric Copolymers of Poly(vinylidene) Fluoride. Vol. 108, pp. 1–48.
Baltussen, J. J. M. see *Northolt, M. G.:* Vol. 178, pp. 1–108.
Barnes, M. D. see *Otaigbe, J. U.:* Vol. 154, pp. 1–86.
Barnes, C. M. see *Satchi-Fainaro, R.:* Vol. 193, pp. 1–65.
Barsett, H. see *Paulsen, S. B.:* Vol. 186, pp. 69–101.
Barshtein, G. R. and *Sabsai, O. Y.:* Compositions with Mineralorganic Fillers. Vol. 101, pp. 1–28.
Barton, J. see *Hunkeler, D.:* Vol. 112, pp. 115–134.
Baschnagel, J., Binder, K., Doruker, P., Gusev, A. A., Hahn, O., Kremer, K., Mattice, W. L., Müller-Plathe, F., Murat, M., Paul, W., Santos, S., Sutter, U. W. and *Tries, V.:* Bridging the Gap Between Atomistic and Coarse-Grained Models of Polymers: Status and Perspectives. Vol. 152, pp. 41–156.
Bassett, D. C.: On the Role of the Hexagonal Phase in the Crystallization of Polyethylene. Vol. 180, pp. 1–16.
Batallán, F. see *Baltá-Calleja, F. J.:* Vol. 108, pp. 1–48.
Batog, A. E., Pet'ko, I. P. and *Penczek, P.:* Aliphatic-Cycloaliphatic Epoxy Compounds and Polymers. Vol. 144, pp. 49–114.
Batrakova, E. V. see *Kabanov, A. V.:* Vol. 193, pp. 173–198.
Baughman, T. W. and *Wagener, K. B.:* Recent Advances in ADMET Polymerization. Vol. 176, pp. 1–42.
Becker, O. and *Simon, G. P.:* Epoxy Layered Silicate Nanocomposites. Vol. 179, pp. 29–82.
Bell, C. L. and *Peppas, N. A.:* Biomedical Membranes from Hydrogels and Interpolymer Complexes. Vol. 122, pp. 125–176.
Bellon-Maurel, A. see *Calmon-Decriaud, A.:* Vol. 135, pp. 207–226.

Bennett, D. E. see O'Brien, D. F.: Vol. 126, pp. 53–84.
Berry, G. C.: Static and Dynamic Light Scattering on Moderately Concentraded Solutions: Isotropic Solutions of Flexible and Rodlike Chains and Nematic Solutions of Rodlike Chains. Vol. 114, pp. 233–290.
Bershtein, V. A. and *Ryzhov, V. A.*: Far Infrared Spectroscopy of Polymers. Vol. 114, pp. 43–122.
Bhargava, R., Wang, S.-Q. and *Koenig, J. L*: FTIR Microspectroscopy of Polymeric Systems. Vol. 163, pp. 137–191.
Biesalski, M. see Rühe, J.: Vol. 165, pp. 79–150.
Bigg, D. M.: Thermal Conductivity of Heterophase Polymer Compositions. Vol. 119, pp. 1–30.
Binder, K.: Phase Transitions in Polymer Blends and Block Copolymer Melts: Some Recent Developments. Vol. 112, pp. 115–134.
Binder, K.: Phase Transitions of Polymer Blends and Block Copolymer Melts in Thin Films. Vol. 138, pp. 1–90.
Binder, K. see Baschnagel, J.: Vol. 152, pp. 41–156.
Binder, K., Müller, M., Virnau, P. and *González MacDowell, L.*: Polymer+Solvent Systems: Phase Diagrams, Interface Free Energies, and Nucleation. Vol. 173, pp. 1–104.
Bird, R. B. see Curtiss, C. F.: Vol. 125, pp. 1–102.
Biswas, M. and *Mukherjee, A.*: Synthesis and Evaluation of Metal-Containing Polymers. Vol. 115, pp. 89–124.
Biswas, M. and *Sinha Ray, S.*: Recent Progress in Synthesis and Evaluation of Polymer-Montmorillonite Nanocomposites. Vol. 155, pp. 167–221.
Blankenburg, L. see Klemm, E.: Vol. 177, pp. 53–90.
Blumen, A. see Gurtovenko, A. A.: Vol. 182, pp. 171–282.
Bogdal, D., Penczek, P., Pielichowski, J. and *Prociak, A.*: Microwave Assisted Synthesis, Crosslinking, and Processing of Polymeric Materials. Vol. 163, pp. 193–263.
Bohrisch, J., Eisenbach, C. D., Jaeger, W., Mori, H., Müller, A. H. E., Rehahn, M., Schaller, C., Traser, S. and *Wittmeyer, P.*: New Polyelectrolyte Architectures. Vol. 165, pp. 1–41.
Bolze, J. see Dingenouts, N.: Vol. 144, pp. 1–48.
Bosshard, C.: see Gubler, U.: Vol. 158, pp. 123–190.
Boutevin, B. and *Robin, J. J.*: Synthesis and Properties of Fluorinated Diols. Vol. 102, pp. 105–132.
Boutevin, B. see Améduri, B.: Vol. 102, pp. 133–170.
Boutevin, B. see Améduri, B.: Vol. 127, pp. 87–142.
Boutevin, B. see Guida-Pietrasanta, F.: Vol. 179, pp. 1–27.
Boutevin, B. see Taguet, A.: Vol. 184, pp. 127–211.
Bowman, C. N. see Anseth, K. S.: Vol. 122, pp. 177–218.
Boyd, R. H.: Prediction of Polymer Crystal Structures and Properties. Vol. 116, pp. 1–26.
Bracco, S. see Sozzani, P.: Vol. 181, pp. 153–177.
Briber, R. M. see Hedrick, J. L.: Vol. 141, pp. 1–44.
Bronnikov, S. V., Vettegren, V. I. and *Frenkel, S. Y.*: Kinetics of Deformation and Relaxation in Highly Oriented Polymers. Vol. 125, pp. 103–146.
Brown, H. R. see Creton, C.: Vol. 156, pp. 53–135.
Bruza, K. J. see Kirchhoff, R. A.: Vol. 117, pp. 1–66.
Buchmeiser, M. R.: Regioselective Polymerization of 1-Alkynes and Stereoselective Cyclopolymerization of a, w-Heptadiynes. Vol. 176, pp. 89–119.
Budkowski, A.: Interfacial Phenomena in Thin Polymer Films: Phase Coexistence and Segregation. Vol. 148, pp. 1–112.
Bunz, U. H. F.: Synthesis and Structure of PAEs. Vol. 177, pp. 1–52.

Burban, J. H. see *Cussler, E. L.*: Vol. 110, pp. 67–80.
Burchard, W.: Solution Properties of Branched Macromolecules. Vol. 143, pp. 113–194.
Butté, A. see *Schork, F. J.*: Vol. 175, pp. 129–255.

Calmon-Decriaud, A., Bellon-Maurel, V., Silvestre, F.: Standard Methods for Testing the Aerobic Biodegradation of Polymeric Materials. Vol. 135, pp. 207–226.
Cameron, N. R. and *Sherrington, D. C.*: High Internal Phase Emulsions (HIPEs)-Structure, Properties and Use in Polymer Preparation. Vol. 126, pp. 163–214.
de la Campa, J. G. see *de Abajo, J.*: Vol. 140, pp. 23–60.
Candau, F. see *Hunkeler, D.*: Vol. 112, pp. 115–134.
Canelas, D. A. and *DeSimone, J. M.*: Polymerizations in Liquid and Supercritical Carbon Dioxide. Vol. 133, pp. 103–140.
Canva, M. and *Stegeman, G. I.*: Quadratic Parametric Interactions in Organic Waveguides. Vol. 158, pp. 87–121.
Capek, I.: Kinetics of the Free-Radical Emulsion Polymerization of Vinyl Chloride. Vol. 120, pp. 135–206.
Capek, I.: Radical Polymerization of Polyoxyethylene Macromonomers in Disperse Systems. Vol. 145, pp. 1–56.
Capek, I. and *Chern, C.-S.*: Radical Polymerization in Direct Mini-Emulsion Systems. Vol. 155, pp. 101–166.
Cappella, B. see *Munz, M.*: Vol. 164, pp. 87–210.
Carlesso, G. see *Prokop, A.*: Vol. 160, pp. 119–174.
Carlini, C. and *Angiolini, L.*: Polymers as Free Radical Photoinitiators. Vol. 123, pp. 127–214.
Carter, K. R. see *Hedrick, J. L.*: Vol. 141, pp. 1–44.
Casas-Vazquez, J. see *Jou, D.*: Vol. 120, pp. 207–266.
Chan, C.-M. and *Li, L.*: Direct Observation of the Growth of Lamellae and Spherulites by AFM. Vol. 188, pp. 1–41.
Chandrasekhar, V.: Polymer Solid Electrolytes: Synthesis and Structure. Vol. 135, pp. 139–206.
Chang, J. Y. see *Han, M. J.*: Vol. 153, pp. 1–36.
Chang, T.: Recent Advances in Liquid Chromatography Analysis of Synthetic Polymers. Vol. 163, pp. 1–60.
Charleux, B. and *Faust, R.*: Synthesis of Branched Polymers by Cationic Polymerization. Vol. 142, pp. 1–70.
Chateauminois, A. and *Baietto-Dubourg, M. C.*: Fracture of Glassy Polymers Within Sliding Contacts. Vol. 188, pp. 153–193.
Chen, P. see *Jaffe, M.*: Vol. 117, pp. 297–328.
Chern, C.-S. see *Capek, I.*: Vol. 155, pp. 101–166.
Chevolot, Y. see *Mathieu, H. J.*: Vol. 162, pp. 1–35.
Chim, Y. T. A. see *Ellis, J. S.*: Vol. 193, pp. 123–172.
Choe, E.-W. see *Jaffe, M.*: Vol. 117, pp. 297–328.
Chow, P. Y. and *Gan, L. M.*: Microemulsion Polymerizations and Reactions. Vol. 175, pp. 257–298.
Chow, T. S.: Glassy State Relaxation and Deformation in Polymers. Vol. 103, pp. 149–190.
Chujo, Y. see *Uemura, T.*: Vol. 167, pp. 81–106.
Chung, S.-J. see *Lin, T.-C.*: Vol. 161, pp. 157–193.
Chung, T.-S. see *Jaffe, M.*: Vol. 117, pp. 297–328.
Clarke, N.: Effect of Shear Flow on Polymer Blends. Vol. 183, pp. 127–173.
Coenjarts, C. see *Li, M.*: Vol. 190, pp. 183–226.

Cölfen, H. and *Antonietti, M.*: Field-Flow Fractionation Techniques for Polymer and Colloid Analysis. Vol. 150, pp. 67–187.
Colmenero, J. see Richter, D.: Vol. 174, pp. 1–221.
Comanita, B. see Roovers, J.: Vol. 142, pp. 179–228.
Comotti, A. see Sozzani, P.: Vol. 181, pp. 153–177.
Connell, J. W. see Hergenrother, P. M.: Vol. 117, pp. 67–110.
Corradini, P. see Auriemma, F.: Vol. 181, pp. 1–74.
Creton, C., Kramer, E. J., Brown, H. R. and *Hui, C.-Y.*: Adhesion and Fracture of Interfaces Between Immiscible Polymers: From the Molecular to the Continuum Scale. Vol. 156, pp. 53–135.
Criado-Sancho, M. see Jou, D.: Vol. 120, pp. 207–266.
Curro, J. G. see Schweizer, K. S.: Vol. 116, pp. 319–378.
Curtiss, C. F. and *Bird, R. B.*: Statistical Mechanics of Transport Phenomena: Polymeric Liquid Mixtures. Vol. 125, pp. 1–102.
Cussler, E. L., Wang, K. L. and *Burban, J. H.*: Hydrogels as Separation Agents. Vol. 110, pp. 67–80.
Czub, P. see Penczek, P.: Vol. 184, pp. 1–95.

Dalton, L.: Nonlinear Optical Polymeric Materials: From Chromophore Design to Commercial Applications. Vol. 158, pp. 1–86.
Dautzenberg, H. see Holm, C.: Vol. 166, pp. 113–171.
Davidson, J. M. see Prokop, A.: Vol. 160, pp. 119–174.
Davies, M. C. see Ellis, J. S.: Vol. 193, pp. 123–172.
Den Decker, M. G. see Northolt, M. G.: Vol. 178, pp. 1–108.
Desai, S. M. and *Singh, R. P.*: Surface Modification of Polyethylene. Vol. 169, pp. 231–293.
DeSimone, J. M. see Canelas, D. A.: Vol. 133, pp. 103–140.
DeSimone, J. M. see Kennedy, K. A.: Vol. 175, pp. 329–346.
Dhal, P. K., Holmes-Farley, S. R., Huval, C. C. and *Jozefiak, T. H.*: Polymers as Drugs. Vol. 192, pp. 9–58.
DiMari, S. see Prokop, A.: Vol. 136, pp. 1–52.
Dimonie, M. V. see Hunkeler, D.: Vol. 112, pp. 115–134.
Dingenouts, N., Bolze, J., Pötschke, D. and *Ballauf, M.*: Analysis of Polymer Latexes by Small-Angle X-Ray Scattering. Vol. 144, pp. 1–48.
Dodd, L. R. and *Theodorou, D. N.*: Atomistic Monte Carlo Simulation and Continuum Mean Field Theory of the Structure and Equation of State Properties of Alkane and Polymer Melts. Vol. 116, pp. 249–282.
Doelker, E.: Cellulose Derivatives. Vol. 107, pp. 199–266.
Dolden, J. G.: Calculation of a Mesogenic Index with Emphasis Upon LC-Polyimides. Vol. 141, pp. 189–245.
Domb, A. J., Amselem, S., Shah, J. and *Maniar, M.*: Polyanhydrides: Synthesis and Characterization. Vol. 107, pp. 93–142.
Domb, A. J. see Kumar, M. N. V. R.: Vol. 160, pp. 45–118.
Doruker, P. see Baschnagel, J.: Vol. 152, pp. 41–156.
Dubois, P. see Mecerreyes, D.: Vol. 147, pp. 1–60.
Dubrovskii, S. A. see Kazanskii, K. S.: Vol. 104, pp. 97–134.
Dudowicz, J. see Freed, K. F.: Vol. 183, pp. 63–126.
Duncan, R., Ringsdorf, H. and *Satchi-Fainaro, R.*: Polymer Therapeutics: Polymers as Drugs, Drug and Protein Conjugates and Gene Delivery Systems: Past, Present and Future Opportunities. Vol. 192, pp. 1–8.
Duncan, R. see Satchi-Fainaro, R.: Vol. 193, pp. 1–65.

Dunkin, I. R. see *Steinke, J.*: Vol. 123, pp. 81–126.
Dunson, D. L. see *McGrath, J. E.*: Vol. 140, pp. 61–106.
Dziezok, P. see *Rühe, J.*: Vol. 165, pp. 79–150.

Eastmond, G. C.: Poly(e-caprolactone) Blends. Vol. 149, pp. 59–223.
Ebringerová, A., Hromádková, Z. and *Heinze, T.*: Hemicellulose. Vol. 186, pp. 1–67.
Economy, J. and *Goranov, K.*: Thermotropic Liquid Crystalline Polymers for High Performance Applications. Vol. 117, pp. 221–256.
Ediger, M. D. and *Adolf, D. B.*: Brownian Dynamics Simulations of Local Polymer Dynamics. Vol. 116, pp. 73–110.
Edlund, U. and *Albertsson, A.-C.*: Degradable Polymer Microspheres for Controlled Drug Delivery. Vol. 157, pp. 53–98.
Edwards, S. F. see *Aharoni, S. M.*: Vol. 118, pp. 1–231.
Eisenbach, C. D. see *Bohrisch, J.*: Vol. 165, pp. 1–41.
Ellis, J. S., Allen, S., Chim, Y. T. A., Roberts, C. J., Tendler, S. J. B. and *Davies, M. C.*: Molecular-Scale Studies on Biopolymers Using Atomic Force Microscopy. Vol. 193, pp. 123–172.
Endo, T. see *Yagci, Y.*: Vol. 127, pp. 59–86.
Engelhardt, H. and *Grosche, O.*: Capillary Electrophoresis in Polymer Analysis. Vol. 150, pp. 189–217.
Engelhardt, H. and *Martin, H.*: Characterization of Synthetic Polyelectrolytes by Capillary Electrophoretic Methods. Vol. 165, pp. 211–247.
Eriksson, P. see *Jacobson, K.*: Vol. 169, pp. 151–176.
Erman, B. see *Bahar, I.*: Vol. 116, pp. 145–206.
Eschner, M. see *Spange, S.*: Vol. 165, pp. 43–78.
Estel, K. see *Spange, S.*: Vol. 165, pp. 43–78.
Estevez, R. and *Van der Giessen, E.*: Modeling and Computational Analysis of Fracture of Glassy Polymers. Vol. 188, pp. 195–234.
Ewen, B. and *Richter, D.*: Neutron Spin Echo Investigations on the Segmental Dynamics of Polymers in Melts, Networks and Solutions. Vol. 134, pp. 1–130.
Ezquerra, T. A. see *Baltá-Calleja, F. J.*: Vol. 108, pp. 1–48.

Fatkullin, N. see *Kimmich, R.*: Vol. 170, pp. 1–113.
Faust, R. see *Charleux, B.*: Vol. 142, pp. 1–70.
Faust, R. see *Kwon, Y.*: Vol. 167, pp. 107–135.
Fekete, E. see *Pukánszky, B.*: Vol. 139, pp. 109–154.
Fendler, J. H.: Membrane-Mimetic Approach to Advanced Materials. Vol. 113, pp. 1–209.
Fetters, L. J. see *Xu, Z.*: Vol. 120, pp. 1–50.
Fontenot, K. see *Schork, F. J.*: Vol. 175, pp. 129–255.
Förster, S., Abetz, V. and *Müller, A. H. E.*: Polyelectrolyte Block Copolymer Micelles. Vol. 166, pp. 173–210.
Förster, S. and *Schmidt, M.*: Polyelectrolytes in Solution. Vol. 120, pp. 51–134.
Freed, K. F. and *Dudowicz, J.*: Influence of Monomer Molecular Structure on the Miscibility of Polymer Blends. Vol. 183, pp. 63–126.
Freire, J. J.: Conformational Properties of Branched Polymers: Theory and Simulations. Vol. 143, pp. 35–112.
Frenkel, D. see *Hu, W.*: Vol. 191, pp. 1–35.
Frenkel, S. Y. see *Bronnikov, S. V.*: Vol. 125, pp. 103–146.
Frick, B. see *Baltá-Calleja, F. J.*: Vol. 108, pp. 1–48.
Fridman, M. L.: see *Terent'eva, J. P.*: Vol. 101, pp. 29–64.
Fuchs, G. see *Trimmel, G.*: Vol. 176, pp. 43–87.

Fukui, K. see Otaigbe, J. U.: Vol. 154, pp. 1–86.
Funke, W.: Microgels-Intramolecularly Crosslinked Macromolecules with a Globular Structure. Vol. 136, pp. 137–232.
Furusho, Y. see Takata, T.: Vol. 171, pp. 1–75.
Furuya, H. see Abe, A.: Vol. 181, pp. 121–152.

Galina, H.: Mean-Field Kinetic Modeling of Polymerization: The Smoluchowski Coagulation Equation. Vol. 137, pp. 135–172.
Gan, L. M. see Chow, P. Y.: Vol. 175, pp. 257–298.
Ganesh, K. see Kishore, K.: Vol. 121, pp. 81–122.
Gaw, K. O. and *Kakimoto, M.*: Polyimide-Epoxy Composites. Vol. 140, pp. 107–136.
Geckeler, K. E. see Rivas, B.: Vol. 102, pp. 171–188.
Geckeler, K. E.: Soluble Polymer Supports for Liquid-Phase Synthesis. Vol. 121, pp. 31–80.
Gedde, U. W. and *Mattozzi, A.*: Polyethylene Morphology. Vol. 169, pp. 29–73.
Gehrke, S. H.: Synthesis, Equilibrium Swelling, Kinetics Permeability and Applications of Environmentally Responsive Gels. Vol. 110, pp. 81–144.
Geil, P. H., Yang, J., Williams, R. A., Petersen, K. L., Long, T.-C. and *Xu, P.*: Effect of Molecular Weight and Melt Time and Temperature on the Morphology of Poly(tetrafluorethylene). Vol. 180, pp. 89–159.
de Gennes, P.-G.: Flexible Polymers in Nanopores. Vol. 138, pp. 91–106.
Georgiou, S.: Laser Cleaning Methodologies of Polymer Substrates. Vol. 168, pp. 1–49.
Geuss, M. see Munz, M.: Vol. 164, pp. 87–210.
Giannelis, E. P., Krishnamoorti, R. and *Manias, E.*: Polymer-Silicate Nanocomposites: Model Systems for Confined Polymers and Polymer Brushes. Vol. 138, pp. 107–148.
Van der Giessen, E. see Estevez, R.: Vol. 188, pp. 195–234.
Godovsky, D. Y.: Device Applications of Polymer-Nanocomposites. Vol. 153, pp. 163–205.
Godovsky, D. Y.: Electron Behavior and Magnetic Properties Polymer-Nanocomposites. Vol. 119, pp. 79–122.
Gohy, J.-F.: Block Copolymer Micelles. Vol. 190, pp. 65–136.
González Arche, A. see Baltá-Calleja, F. J.: Vol. 108, pp. 1–48.
Goranov, K. see Economy, J.: Vol. 117, pp. 221–256.
Gramain, P. see Améduri, B.: Vol. 127, pp. 87–142.
Grein, C.: Toughness of Neat, Rubber Modified and Filled β-Nucleated Polypropylene: From Fundamentals to Applications. Vol. 188, pp. 43–104.
Greish, K. see Maeda, H.: Vol. 193, pp. 103–121.
Grest, G. S.: Normal and Shear Forces Between Polymer Brushes. Vol. 138, pp. 149–184.
Grigorescu, G. and *Kulicke, W.-M.*: Prediction of Viscoelastic Properties and Shear Stability of Polymers in Solution. Vol. 152, p. 1–40.
Gröhn, F. see Rühe, J.: Vol. 165, pp. 79–150.
Grosberg, A. and *Nechaev, S.*: Polymer Topology. Vol. 106, pp. 1–30.
Grosche, O. see Engelhardt, H.: Vol. 150, pp. 189–217.
Grubbs, R., Risse, W. and *Novac, B.*: The Development of Well-defined Catalysts for Ring-Opening Olefin Metathesis. Vol. 102, pp. 47–72.
Gubler, U. and *Bosshard, C.*: Molecular Design for Third-Order Nonlinear Optics. Vol. 158, pp. 123–190.
Guida-Pietrasanta, F. and *Boutevin, B.*: Polysilalkylene or Silarylene Siloxanes Said Hybrid Silicones. Vol. 179, pp. 1–27.
van Gunsteren, W. F. see Gusev, A. A.: Vol. 116, pp. 207–248.
Gupta, B. and *Anjum, N.*: Plasma and Radiation-Induced Graft Modification of Polymers for Biomedical Applications. Vol. 162, pp. 37–63.

Gurtovenko, A. A. and *Blumen, A.*: Generalized Gaussian Structures: Models for Polymer Systems with Complex Topologies. Vol. 182, pp. 171–282.
Gusev, A. A., Müller-Plathe, F., van Gunsteren, W. F. and *Suter, U. W.*: Dynamics of Small Molecules in Bulk Polymers. Vol. 116, pp. 207–248.
Gusev, A. A. see Baschnagel, J.: Vol. 152, pp. 41–156.
Guillot, J. see Hunkeler, D.: Vol. 112, pp. 115–134.
Guyot, A. and *Tauer, K.*: Reactive Surfactants in Emulsion Polymerization. Vol. 111, pp. 43–66.

Hadjichristidis, N., Pispas, S., Pitsikalis, M., Iatrou, H. and *Vlahos, C.*: Asymmetric Star Polymers Synthesis and Properties. Vol. 142, pp. 71–128.
Hadjichristidis, N., Pitsikalis, M. and *Iatrou, H.*: Synthesis of Block Copolymers. Vol. 189, pp. 1–124.
Hadjichristidis, N. see Xu, Z.: Vol. 120, pp. 1–50.
Hadjichristidis, N. see Pitsikalis, M.: Vol. 135, pp. 1–138.
Hahn, O. see Baschnagel, J.: Vol. 152, pp. 41–156.
Hakkarainen, M.: Aliphatic Polyesters: Abiotic and Biotic Degradation and Degradation Products. Vol. 157, pp. 1–26.
Hakkarainen, M. and *Albertsson, A.-C.*: Environmental Degradation of Polyethylene. Vol. 169, pp. 177–199.
Halary, J. L. see Monnerie, L.: Vol. 187, pp. 35–213.
Halary, J. L. see Monnerie, L.: Vol. 187, pp. 215–364.
Hall, H. K. see Penelle, J.: Vol. 102, pp. 73–104.
Hamley, I. W.: Crystallization in Block Copolymers. Vol. 148, pp. 113–138.
Hammouda, B.: SANS from Homogeneous Polymer Mixtures: A Unified Overview. Vol. 106, pp. 87–134.
Han, M. J. and *Chang, J. Y.*: Polynucleotide Analogues. Vol. 153, pp. 1–36.
Harada, A.: Design and Construction of Supramolecular Architectures Consisting of Cyclodextrins and Polymers. Vol. 133, pp. 141–192.
Haralson, M. A. see Prokop, A.: Vol. 136, pp. 1–52.
Harding, S. E.: Analysis of Polysaccharides by Ultracentrifugation. Size, Conformation and Interactions in Solution. Vol. 186, pp. 211–254.
Hasegawa, N. see Usuki, A.: Vol. 179, pp. 135–195.
Hassan, C. M. and *Peppas, N. A.*: Structure and Applications of Poly(vinyl alcohol) Hydrogels Produced by Conventional Crosslinking or by Freezing/Thawing Methods. Vol. 153, pp. 37–65.
Hawker, C. J.: Dentritic and Hyperbranched Macromolecules Precisely Controlled Macromolecular Architectures. Vol. 147, pp. 113–160.
Hawker, C. J. see Hedrick, J. L.: Vol. 141, pp. 1–44.
He, G. S. see Lin, T.-C.: Vol. 161, pp. 157–193.
Hedrick, J. L., Carter, K. R., Labadie, J. W., Miller, R. D., Volksen, W., Hawker, C. J., Yoon, D. Y., Russell, T. P., McGrath, J. E. and *Briber, R. M.*: Nanoporous Polyimides. Vol. 141, pp. 1–44.
Hedrick, J. L., Labadie, J. W., Volksen, W. and *Hilborn, J. G.*: Nanoscopically Engineered Polyimides. Vol. 147, pp. 61–112.
Hedrick, J. L. see Hergenrother, P. M.: Vol. 117, pp. 67–110.
Hedrick, J. L. see Kiefer, J.: Vol. 147, pp. 161–247.
Hedrick, J. L. see McGrath, J. E.: Vol. 140, pp. 61–106.
Heine, D. R., Grest, G. S. and *Curro, J. G.*: Structure of Polymer Melts and Blends: Comparison of Integral Equation theory and Computer Sumulation. Vol. 173, pp. 209–249.
Heinrich, G. and *Klüppel, M.*: Recent Advances in the Theory of Filler Networking in Elastomers. Vol. 160, pp. 1–44.

Heinze, T. see Ebringerová, A.: Vol. 186, pp. 1–67.
Heinze, T. see El Seoud, O. A.: Vol. 186, pp. 103–149.
Heller, J.: Poly (Ortho Esters). Vol. 107, pp. 41–92.
Helm, C. A. see Möhwald, H.: Vol. 165, pp. 151–175.
Hemielec, A. A. see Hunkeler, D.: Vol. 112, pp. 115–134.
Hergenrother, P. M., Connell, J. W., Labadie, J. W. and *Hedrick, J. L.*: Poly(arylene ether)s Containing Heterocyclic Units. Vol. 117, pp. 67–110.
Hernández-Barajas, J. see Wandrey, C.: Vol. 145, pp. 123–182.
Hervet, H. see Léger, L.: Vol. 138, pp. 185–226.
Hiejima, T. see Abe, A.: Vol. 181, pp. 121–152.
Hikosaka, M., Watanabe, K., Okada, K. and *Yamazaki, S.*: Topological Mechanism of Polymer Nucleation and Growth – The Role of Chain Sliding Diffusion and Entanglement. Vol. 191, pp. 137–186.
Hilborn, J. G. see Hedrick, J. L.: Vol. 147, pp. 61–112.
Hilborn, J. G. see Kiefer, J.: Vol. 147, pp. 161–247.
Hillborg, H. see Vancso, G. J.: Vol. 182, pp. 55–129.
Hillmyer, M. A.: Nanoporous Materials from Block Copolymer Precursors. Vol. 190, pp. 137–181.
Hiramatsu, N. see Matsushige, M.: Vol. 125, pp. 147–186.
Hirasa, O. see Suzuki, M.: Vol. 110, pp. 241–262.
Hirotsu, S.: Coexistence of Phases and the Nature of First-Order Transition in Poly-N-isopropylacrylamide Gels. Vol. 110, pp. 1–26.
Höcker, H. see Klee, D.: Vol. 149, pp. 1–57.
Holm, C. see Arnold, A.: Vol. 185, pp. 59–109.
Holm, C., Hofmann, T., Joanny, J. F., Kremer, K., Netz, R. R., Reineker, P., Seidel, C., Vilgis, T. A. and *Winkler, R. G.*: Polyelectrolyte Theory. Vol. 166, pp. 67–111.
Holm, C., Rehahn, M., Oppermann, W. and *Ballauff, M.*: Stiff-Chain Polyelectrolytes. Vol. 166, pp. 1–27.
Holmes-Farley, S. R. see Dhal, P. K.: Vol. 192, pp. 9–58.
Hornsby, P.: Rheology, Compounding and Processing of Filled Thermoplastics. Vol. 139, pp. 155–216.
Houbenov, N. see Rühe, J.: Vol. 165, pp. 79–150.
Hromádková, Z. see Ebringerová, A.: Vol. 186, pp. 1–67.
Hu, W. and *Frenkel, D.*: Polymer Crystallization Driven by Anisotropic Interactions. Vol. 191, pp. 1–35.
Huber, K. see Volk, N.: Vol. 166, pp. 29–65.
Hugenberg, N. see Rühe, J.: Vol. 165, pp. 79–150.
Hui, C.-Y. see Creton, C.: Vol. 156, pp. 53–135.
Hult, A., Johansson, M. and *Malmström, E.*: Hyperbranched Polymers. Vol. 143, pp. 1–34.
Hünenberger, P. H.: Thermostat Algorithms for Molecular-Dynamics Simulations. Vol. 173, pp. 105–147.
Hunkeler, D., Candau, F., Pichot, C., Hemielec, A. E., Xie, T. Y., Barton, J., Vaskova, V., Guillot, J., Dimonie, M. V. and *Reichert, K. H.*: Heterophase Polymerization: A Physical and Kinetic Comparision and Categorization. Vol. 112, pp. 115–134.
Hunkeler, D. see Macko, T.: Vol. 163, pp. 61–136.
Hunkeler, D. see Prokop, A.: Vol. 136, pp. 1–52; 53–74.
Hunkeler, D. see Wandrey, C.: Vol. 145, pp. 123–182.
Huval, C. C. see Dhal, P. K.: Vol. 192, pp. 9–58.

Iatrou, H. see Hadjichristidis, N.: Vol. 142, pp. 71–128.

Iatrou, H. see Hadjichristidis, N.: Vol. 189, pp. 1–124.
Ichikawa, T. see Yoshida, H.: Vol. 105, pp. 3–36.
Ihara, E. see Yasuda, H.: Vol. 133, pp. 53–102.
Ikada, Y. see Uyama, Y.: Vol. 137, pp. 1–40.
Ikehara, T. see Jinnuai, H.: Vol. 170, pp. 115–167.
Ilavsky, M.: Effect on Phase Transition on Swelling and Mechanical Behavior of Synthetic Hydrogels. Vol. 109, pp. 173–206.
Imai, M. see Kaji, K.: Vol. 191, pp. 187–240.
Imai, Y.: Rapid Synthesis of Polyimides from Nylon-Salt Monomers. Vol. 140, pp. 1–23.
Inomata, H. see Saito, S.: Vol. 106, pp. 207–232.
Inoue, S. see Sugimoto, H.: Vol. 146, pp. 39–120.
Irie, M.: Stimuli-Responsive Poly(N-isopropylacrylamide), Photo- and Chemical-Induced Phase Transitions. Vol. 110, pp. 49–66.
Ise, N. see Matsuoka, H.: Vol. 114, pp. 187–232.
Ishikawa, T.: Advances in Inorganic Fibers. Vol. 178, pp. 109–144.
Ito, H.: Chemical Amplification Resists for Microlithography. Vol. 172, pp. 37–245.
Ito, K. and *Kawaguchi, S.*: Poly(macronomers), Homo- and Copolymerization. Vol. 142, pp. 129–178.
Ito, K. see Kawaguchi, S.: Vol. 175, pp. 299–328.
Ito, S. and *Aoki, H.*: Nano-Imaging of Polymers by Optical Microscopy. Vol. 182, pp. 131–170.
Ito, Y. see Suginome, M.: Vol. 171, pp. 77–136.
Ivanov, A. E. see Zubov, V. P.: Vol. 104, pp. 135–176.

Jacob, S. and *Kennedy, J.*: Synthesis, Characterization and Properties of OCTA-ARM Polyisobutylene-Based Star Polymers. Vol. 146, pp. 1–38.
Jacobson, K., Eriksson, P., Reitberger, T. and *Stenberg, B.*: Chemiluminescence as a Tool for Polyolefin. Vol. 169, pp. 151–176.
Jaeger, W. see Bohrisch, J.: Vol. 165, pp. 1–41.
Jaffe, M., Chen, P., Choe, E.-W., Chung, T.-S. and *Makhija, S.*: High Performance Polymer Blends. Vol. 117, pp. 297–328.
Jancar, J.: Structure-Property Relationships in Thermoplastic Matrices. Vol. 139, pp. 1–66.
Jen, A. K.-Y. see Kajzar, F.: Vol. 161, pp. 1–85.
Jerome, R. see Mecerreyes, D.: Vol. 147, pp. 1–60.
de Jeu, W. H. see Li, L.: Vol. 181, pp. 75–120.
Jiang, M., Li, M., Xiang, M. and *Zhou, H.*: Interpolymer Complexation and Miscibility and Enhancement by Hydrogen Bonding. Vol. 146, pp. 121–194.
Jin, J. see Shim, H.-K.: Vol. 158, pp. 191–241.
Jinnai, H., Nishikawa, Y., Ikehara, T. and *Nishi, T.*: Emerging Technologies for the 3D Analysis of Polymer Structures. Vol. 170, pp. 115–167.
Jo, W. H. and *Yang, J. S.*: Molecular Simulation Approaches for Multiphase Polymer Systems. Vol. 156, pp. 1–52.
Joanny, J.-F. see Holm, C.: Vol. 166, pp. 67–111.
Joanny, J.-F. see Thünemann, A. F.: Vol. 166, pp. 113–171.
Johannsmann, D. see Rühe, J.: Vol. 165, pp. 79–150.
Johansson, M. see Hult, A.: Vol. 143, pp. 1–34.
Joos-Müller, B. see Funke, W.: Vol. 136, pp. 137–232.
Jou, D., Casas-Vazquez, J. and *Criado-Sancho, M.*: Thermodynamics of Polymer Solutions under Flow: Phase Separation and Polymer Degradation. Vol. 120, pp. 207–266.
Jozefiak, T. H. see Dhal, P. K.: Vol. 192, pp. 9–58.

Kabanov, A. V., Batrakova, E. V., Sherman, S. and *Alakhov, V. Y.*: Polymer Genomics. Vol. 193, pp. 173–198.
Kaetsu, I.: Radiation Synthesis of Polymeric Materials for Biomedical and Biochemical Applications. Vol. 105, pp. 81–98.
Kaji, K., Nishida, K., Kanaya, T., Matsuba, G., Konishi, T. and *Imai, M.*: Spinodal Crystallization of Polymers: Crystallization from the Unstable Melt. Vol. 191, pp. 187–240.
Kaji, K. see Kanaya, T.: Vol. 154, pp. 87–141.
Kajzar, F., Lee, K.-S. and *Jen, A. K.-Y.*: Polymeric Materials and their Orientation Techniques for Second-Order Nonlinear Optics. Vol. 161, pp. 1–85.
Kakimoto, M. see Gaw, K. O.: Vol. 140, pp. 107–136.
Kaminski, W. and *Arndt, M.*: Metallocenes for Polymer Catalysis. Vol. 127, pp. 143–187.
Kammer, H. W., Kressler, H. and *Kummerloewe, C.*: Phase Behavior of Polymer Blends – Effects of Thermodynamics and Rheology. Vol. 106, pp. 31–86.
Kanaya, T. and *Kaji, K.*: Dynamcis in the Glassy State and Near the Glass Transition of Amorphous Polymers as Studied by Neutron Scattering. Vol. 154, pp. 87–141.
Kanaya, T. see Kaji, K.: Vol. 191, pp. 187–240.
Kandyrin, L. B. and *Kuleznev, V. N.*: The Dependence of Viscosity on the Composition of Concentrated Dispersions and the Free Volume Concept of Disperse Systems. Vol. 103, pp. 103–148.
Kaneko, M. see Ramaraj, R.: Vol. 123, pp. 215–242.
Kang, E. T., Neoh, K. G. and *Tan, K. L.*: X-Ray Photoelectron Spectroscopic Studies of Electroactive Polymers. Vol. 106, pp. 135–190.
Kaplan, D. L. see Singh, A.: Vol. 194, pp. 211–224.
Kaplan, D. L. see Xu, P.: Vol. 194, pp. 69–94.
Karlsson, S. see Söderqvist Lindblad, M.: Vol. 157, pp. 139–161.
Karlsson, S.: Recycled Polyolefins. Material Properties and Means for Quality Determination. Vol. 169, pp. 201–229.
Kataoka, K. see Nishiyama, N.: Vol. 193, pp. 67–101.
Kato, K. see Uyama, Y.: Vol. 137, pp. 1–40.
Kato, M. see Usuki, A.: Vol. 179, pp. 135–195.
Kausch, H.-H. and *Michler, G. H.*: The Effect of Time on Crazing and Fracture. Vol. 187, pp. 1–33.
Kausch, H.-H. see Monnerie, L. Vol. 187, pp. 215–364.
Kautek, W. see Krüger, J.: Vol. 168, pp. 247–290.
Kawaguchi, S. see Ito, K.: Vol. 142, pp. 129–178.
Kawaguchi, S. and *Ito, K.*: Dispersion Polymerization. Vol. 175, pp. 299–328.
Kawata, S. see Sun, H.-B.: Vol. 170, pp. 169–273.
Kazanskii, K. S. and *Dubrovskii, S. A.*: Chemistry and Physics of Agricultural Hydrogels. Vol. 104, pp. 97–134.
Kennedy, J. P. see Jacob, S.: Vol. 146, pp. 1–38.
Kennedy, J. P. see Majoros, I.: Vol. 112, pp. 1–113.
Kennedy, K. A., Roberts, G. W. and *DeSimone, J. M.*: Heterogeneous Polymerization of Fluoroolefins in Supercritical Carbon Dioxide. Vol. 175, pp. 329–346.
Khokhlov, A., Starodybtzev, S. and *Vasilevskaya, V.*: Conformational Transitions of Polymer Gels: Theory and Experiment. Vol. 109, pp. 121–172.
Kiefer, J., Hedrick, J. L. and *Hiborn, J. G.*: Macroporous Thermosets by Chemically Induced Phase Separation. Vol. 147, pp. 161–247.
Kihara, N. see Takata, T.: Vol. 171, pp. 1–75.

Kilian, H. G. and *Pieper, T.*: Packing of Chain Segments. A Method for Describing X-Ray Patterns of Crystalline, Liquid Crystalline and Non-Crystalline Polymers. Vol. 108, pp. 49–90.
Kim, J. see Quirk, R. P.: Vol. 153, pp. 67–162.
Kim, K.-S. see Lin, T.-C.: Vol. 161, pp. 157–193.
Kimmich, R. and *Fatkullin, N.*: Polymer Chain Dynamics and NMR. Vol. 170, pp. 1–113.
Kippelen, B. and *Peyghambarian, N.*: Photorefractive Polymers and their Applications. Vol. 161, pp. 87–156.
Kirchhoff, R. A. and *Bruza, K. J.*: Polymers from Benzocyclobutenes. Vol. 117, pp. 1–66.
Kishore, K. and *Ganesh, K.*: Polymers Containing Disulfide, Tetrasulfide, Diselenide and Ditelluride Linkages in the Main Chain. Vol. 121, pp. 81–122.
Kitamaru, R.: Phase Structure of Polyethylene and Other Crystalline Polymers by Solid-State 13C/MNR. Vol. 137, pp. 41–102.
Klapper, M. see Rusanov, A. L.: Vol. 179, pp. 83–134.
Klee, D. and *Höcker, H.*: Polymers for Biomedical Applications: Improvement of the Interface Compatibility. Vol. 149, pp. 1–57.
Klemm, E., Pautzsch, T. and *Blankenburg, L.*: Organometallic PAEs. Vol. 177, pp. 53–90.
Klier, J. see Scranton, A. B.: Vol. 122, pp. 1–54.
v. Klitzing, R. and *Tieke, B.*: Polyelectrolyte Membranes. Vol. 165, pp. 177–210.
Kloeckner, J. see Wagner, E.: Vol. 192, pp. 135–173.
Klüppel, M.: The Role of Disorder in Filler Reinforcement of Elastomers on Various Length Scales. Vol. 164, pp. 1–86.
Klüppel, M. see Heinrich, G.: Vol. 160, pp. 1–44.
Knuuttila, H., Lehtinen, A. and *Nummila-Pakarinen, A.*: Advanced Polyethylene Technologies – Controlled Material Properties. Vol. 169, pp. 13–27.
Kobayashi, S. and *Ohmae, M.*: Enzymatic Polymerization to Polysaccharides. Vol. 194, pp. 159–210.
Kobayashi, S. see Uyama, H.: Vol. 194, pp. 51–67.
Kobayashi, S. see Uyama, H.: Vol. 194, pp. 133–158.
Kobayashi, S., Shoda, S. and *Uyama, H.*: Enzymatic Polymerization and Oligomerization. Vol. 121, pp. 1–30.
Kobayashi, T. see Abe, A.: Vol. 181, pp. 121–152.
Köhler, W. and *Schäfer, R.*: Polymer Analysis by Thermal-Diffusion Forced Rayleigh Scattering. Vol. 151, pp. 1–59.
Koenig, J. L. see Bhargava, R.: Vol. 163, pp. 137–191.
Koenig, J. L. see Andreis, M.: Vol. 124, pp. 191–238.
Koike, T.: Viscoelastic Behavior of Epoxy Resins Before Crosslinking. Vol. 148, pp. 139–188.
Kokko, E. see Löfgren, B.: Vol. 169, pp. 1–12.
Kokufuta, E.: Novel Applications for Stimulus-Sensitive Polymer Gels in the Preparation of Functional Immobilized Biocatalysts. Vol. 110, pp. 157–178.
Konishi, T. see Kaji, K.: Vol. 191, pp. 187–240.
Konno, M. see Saito, S.: Vol. 109, pp. 207–232.
Konradi, R. see Rühe, J.: Vol. 165, pp. 79–150.
Kopecek, J. see Putnam, D.: Vol. 122, pp. 55–124.
Koßmehl, G. see Schopf, G.: Vol. 129, pp. 1–145.
Kostoglodov, P. V. see Rusanov, A. L.: Vol. 179, pp. 83–134.
Kozlov, E. see Prokop, A.: Vol. 160, pp. 119–174.
Kramer, E. J. see Creton, C.: Vol. 156, pp. 53–135.
Kremer, K. see Baschnagel, J.: Vol. 152, pp. 41–156.
Kremer, K. see Holm, C.: Vol. 166, pp. 67–111.

Kressler, J. see Kammer, H. W.: Vol. 106, pp. 31–86.
Kricheldorf, H. R.: Liquid-Cristalline Polyimides. Vol. 141, pp. 83–188.
Krishnamoorti, R. see Giannelis, E. P.: Vol. 138, pp. 107–148.
Krüger, J. and *Kautek, W.*: Ultrashort Pulse Laser Interaction with Dielectrics and Polymers, Vol. 168, pp. 247–290.
Kuchanov, S. I.: Modern Aspects of Quantitative Theory of Free-Radical Copolymerization. Vol. 103, pp. 1–102.
Kuchanov, S. I.: Principles of Quantitive Description of Chemical Structure of Synthetic Polymers. Vol. 152, pp. 157–202.
Kudaibergennow, S. E.: Recent Advances in Studying of Synthetic Polyampholytes in Solutions. Vol. 144, pp. 115–198.
Kuleznev, V. N. see Kandyrin, L. B.: Vol. 103, pp. 103–148.
Kulichkhin, S. G. see Malkin, A. Y.: Vol. 101, pp. 217–258.
Kulicke, W.-M. see Grigorescu, G.: Vol. 152, pp. 1–40.
Kumar, M. N. V. R., Kumar, N., Domb, A. J. and *Arora, M.*: Pharmaceutical Polymeric Controlled Drug Delivery Systems. Vol. 160, pp. 45–118.
Kumar, N. see Kumar, M. N. V. R.: Vol. 160, pp. 45–118.
Kummerloewe, C. see Kammer, H. W.: Vol. 106, pp. 31–86.
Kuznetsova, N. P. see Samsonov, G. V.: Vol. 104, pp. 1–50.
Kwon, Y. and *Faust, R.*: Synthesis of Polyisobutylene-Based Block Copolymers with Precisely Controlled Architecture by Living Cationic Polymerization. Vol. 167, pp. 107–135.

Labadie, J. W. see Hergenrother, P. M.: Vol. 117, pp. 67–110.
Labadie, J. W. see Hedrick, J. L.: Vol. 141, pp. 1–44.
Labadie, J. W. see Hedrick, J. L.: Vol. 147, pp. 61–112.
Lamparski, H. G. see O'Brien, D. F.: Vol. 126, pp. 53–84.
Laschewsky, A.: Molecular Concepts, Self-Organisation and Properties of Polysoaps. Vol. 124, pp. 1–86.
Laso, M. see Leontidis, E.: Vol. 116, pp. 283–318.
Lauprêtre, F. see Monnerie, L.: Vol. 187, pp. 35–213.
Lazár, M. and *Rychlý, R.*: Oxidation of Hydrocarbon Polymers. Vol. 102, pp. 189–222.
Lechowicz, J. see Galina, H.: Vol. 137, pp. 135–172.
Léger, L., Raphaël, E. and *Hervet, H.*: Surface-Anchored Polymer Chains: Their Role in Adhesion and Friction. Vol. 138, pp. 185–226.
Lenz, R. W.: Biodegradable Polymers. Vol. 107, pp. 1–40.
Leontidis, E., de Pablo, J. J., Laso, M. and *Suter, U. W.*: A Critical Evaluation of Novel Algorithms for the Off-Lattice Monte Carlo Simulation of Condensed Polymer Phases. Vol. 116, pp. 283–318.
Lee, B. see Quirk, R. P.: Vol. 153, pp. 67–162.
Lee, K.-S. see Kajzar, F.: Vol. 161, pp. 1–85.
Lee, Y. see Quirk, R. P.: Vol. 153, pp. 67–162.
Lehtinen, A. see Knuuttila, H.: Vol. 169, pp. 13–27.
Leónard, D. see Mathieu, H. J.: Vol. 162, pp. 1–35.
Lesec, J. see Viovy, J.-L.: Vol. 114, pp. 1–42.
Levesque, D. see Weis, J.-J.: Vol. 185, pp. 163–225.
Li, L. and *de Jeu, W. H.*: Flow-induced mesophases in crystallizable polymers. Vol. 181, pp. 75–120.
Li, L. see Chan, C.-M.: Vol. 188, pp. 1–41.
Li, M., Coenjarts, C. and *Ober, C. K.*: Patternable Block Copolymers. Vol. 190, pp. 183–226.
Li, M. see Jiang, M.: Vol. 146, pp. 121–194.

Liang, G. L. see *Sumpter, B. G.:* Vol. 116, pp. 27–72.
Lienert, K.-W.: Poly(ester-imide)s for Industrial Use. Vol. 141, pp. 45–82.
Likhatchev, D. see *Rusanov, A. L.:* Vol. 179, pp. 83–134.
Lin, J. and *Sherrington, D. C.:* Recent Developments in the Synthesis, Thermostability and Liquid Crystal Properties of Aromatic Polyamides. Vol. 111, pp. 177–220.
Lin, T.-C., Chung, S.-J., Kim, K.-S., Wang, X., He, G. S., Swiatkiewicz, J., Pudavar, H. E. and *Prasad, P. N.:* Organics and Polymers with High Two-Photon Activities and their Applications. Vol. 161, pp. 157–193.
Linse, P.: Simulation of Charged Colloids in Solution. Vol. 185, pp. 111–162.
Lippert, T.: Laser Application of Polymers. Vol. 168, pp. 51–246.
Liu, Y. see *Söderqvist Lindblad, M.:* Vol. 157, pp. 139–161.
Long, T.-C. see *Geil, P. H.:* Vol. 180, pp. 89–159.
López Cabarcos, E. see *Baltá-Calleja, F. J.:* Vol. 108, pp. 1–48.
Lotz, B.: Analysis and Observation of Polymer Crystal Structures at the Individual Stem Level. Vol. 180, pp. 17–44.
Löfgren, B., Kokko, E. and *Seppälä, J.:* Specific Structures Enabled by Metallocene Catalysis in Polyethenes. Vol. 169, pp. 1–12.
Löwen, H. see *Thünemann, A. F.:* Vol. 166, pp. 113–171.
Luo, Y. see *Schork, F. J.:* Vol. 175, pp. 129–255.

Macko, T. and *Hunkeler, D.:* Liquid Chromatography under Critical and Limiting Conditions: A Survey of Experimental Systems for Synthetic Polymers. Vol. 163, pp. 61–136.
Maeda, H., Greish, K. and *Fang, J.:* The EPR Effect and Polymeric Drugs: A Paradigm Shift for Cancer Chemotherapy in the 21st Century. Vol. 193, pp. 103–121.
Majoros, I., Nagy, A. and *Kennedy, J. P.:* Conventional and Living Carbocationic Polymerizations United. I. A Comprehensive Model and New Diagnostic Method to Probe the Mechanism of Homopolymerizations. Vol. 112, pp. 1–113.
Makhija, S. see *Jaffe, M.:* Vol. 117, pp. 297–328.
Malmström, E. see *Hult, A.:* Vol. 143, pp. 1–34.
Malkin, A. Y. and *Kulichkhin, S. G.:* Rheokinetics of Curing. Vol. 101, pp. 217–258.
Maniar, M. see *Domb, A. J.:* Vol. 107, pp. 93–142.
Manias, E. see *Giannelis, E. P.:* Vol. 138, pp. 107–148.
Martin, H. see *Engelhardt, H.:* Vol. 165, pp. 211–247.
Marty, J. D. and *Mauzac, M.:* Molecular Imprinting: State of the Art and Perspectives. Vol. 172, pp. 1–35.
Mashima, K., Nakayama, Y. and *Nakamura, A.:* Recent Trends in Polymerization of a-Olefins Catalyzed by Organometallic Complexes of Early Transition Metals. Vol. 133, pp. 1–52.
Mathew, D. see *Reghunadhan Nair, C. P.:* Vol. 155, pp. 1–99.
Mathieu, H. J., Chevolot, Y., Ruiz-Taylor, L. and *Leónard, D.:* Engineering and Characterization of Polymer Surfaces for Biomedical Applications. Vol. 162, pp. 1–35.
Matsuba, G. see *Kaji, K.:* Vol. 191, pp. 187–240.
Matsumura S.: Enzymatic Synthesis of Polyesters via Ring-Opening Polymerization. Vol. 194, pp. 95–132.
Matsumoto, A.: Free-Radical Crosslinking Polymerization and Copolymerization of Multivinyl Compounds. Vol. 123, pp. 41–80.
Matsumoto, A. see *Otsu, T.:* Vol. 136, pp. 75–138.
Matsuoka, H. and *Ise, N.:* Small-Angle and Ultra-Small Angle Scattering Study of the Ordered Structure in Polyelectrolyte Solutions and Colloidal Dispersions. Vol. 114, pp. 187–232.
Matsushige, K., Hiramatsu, N. and *Okabe, H.:* Ultrasonic Spectroscopy for Polymeric Materials. Vol. 125, pp. 147–186.

Mattice, W. L. see Rehahn, M.: Vol. 131/132, pp. 1–475.
Mattice, W. L. see Baschnagel, J.: Vol. 152, pp. 41–156.
Mattozzi, A. see Gedde, U. W.: Vol. 169, pp. 29–73.
Mauzac, M. see Marty, J. D.: Vol. 172, pp. 1–35.
Mays, W. see Xu, Z.: Vol. 120, pp. 1–50.
Mays, J. W. see Pitsikalis, M.: Vol. 135, pp. 1–138.
McGrath, J. E. see Hedrick, J. L.: Vol. 141, pp. 1–44.
McGrath, J. E., Dunson, D. L. and *Hedrick, J. L.*: Synthesis and Characterization of Segmented Polyimide-Polyorganosiloxane Copolymers. Vol. 140, pp. 61–106.
McLeish, T. C. B. and *Milner, S. T.*: Entangled Dynamics and Melt Flow of Branched Polymers. Vol. 143, pp. 195–256.
Mecerreyes, D., Dubois, P. and *Jerome, R.*: Novel Macromolecular Architectures Based on Aliphatic Polyesters: Relevance of the Coordination-Insertion Ring-Opening Polymerization. Vol. 147, pp. 1–60.
Mecham, S. J. see McGrath, J. E.: Vol. 140, pp. 61–106.
Meille, S. V. see Allegra, G.: Vol. 191, pp. 87–135.
Menzel, H. see Möhwald, H.: Vol. 165, pp. 151–175.
Meyer, T. see Spange, S.: Vol. 165, pp. 43–78.
Michler, G. H. see Kausch, H.-H.: Vol. 187, pp. 1–33.
Mikos, A. G. see Thomson, R. C.: Vol. 122, pp. 245–274.
Milner, S. T. see McLeish, T. C. B.: Vol. 143, pp. 195–256.
Mison, P. and *Sillion, B.*: Thermosetting Oligomers Containing Maleimides and Nadiimides End-Groups. Vol. 140, pp. 137–180.
Miyasaka, K.: PVA-Iodine Complexes: Formation, Structure and Properties. Vol. 108, pp. 91–130.
Miller, R. D. see Hedrick, J. L.: Vol. 141, pp. 1–44.
Minko, S. see Rühe, J.: Vol. 165, pp. 79–150.
Möhwald, H., Menzel, H., Helm, C. A. and *Stamm, M.*: Lipid and Polyampholyte Monolayers to Study Polyelectrolyte Interactions and Structure at Interfaces. Vol. 165, pp. 151–175.
Monkenbusch, M. see Richter, D.: Vol. 174, pp. 1–221.
Monnerie, L., Halary, J. L. and *Kausch, H.-H.*: Deformation, Yield and Fracture of Amorphous Polymers: Relation to the Secondary Transitions. Vol. 187, pp. 215–364.
Monnerie, L., Lauprêtre, F. and *Halary, J. L.*: Investigation of Solid-State Transitions in Linear and Crosslinked Amorphous Polymers. Vol. 187, pp. 35–213.
Monnerie, L. see Bahar, I.: Vol. 116, pp. 145–206.
Moore, J. S. see Ray, C. R.: Vol. 177, pp. 99–149.
Mori, H. see Bohrisch, J.: Vol. 165, pp. 1–41.
Morishima, Y.: Photoinduced Electron Transfer in Amphiphilic Polyelectrolyte Systems. Vol. 104, pp. 51–96.
Morton, M. see Quirk, R. P.: Vol. 153, pp. 67–162.
Motornov, M. see Rühe, J.: Vol. 165, pp. 79–150.
Mours, M. see Winter, H. H.: Vol. 134, pp. 165–234.
Müllen, K. see Scherf, U.: Vol. 123, pp. 1–40.
Müller, A. H. E. see Bohrisch, J.: Vol. 165, pp. 1–41.
Müller, A. H. E. see Förster, S.: Vol. 166, pp. 173–210.
Müller, A. J., Balsamo, V. and *Arnal, M. L.*: Nucleation and Crystallization in Diblock and Triblock Copolymers. Vol. 190, pp. 1–63.
Müller, M. and *Schmid, F.*: Incorporating Fluctuations and Dynamics in Self-Consistent Field Theories for Polymer Blends. Vol. 185, pp. 1–58.

Müller, M. see Thünemann, A. F.: Vol. 166, pp. 113–171.
Müller-Plathe, F. see Gusev, A. A.: Vol. 116, pp. 207–248.
Müller-Plathe, F. see Baschnagel, J.: Vol. 152, p. 41–156.
Mukerherjee, A. see Biswas, M.: Vol. 115, pp. 89–124.
Munz, M., Cappella, B., Sturm, H., Geuss, M. and *Schulz, E.*: Materials Contrasts and Nanolithography Techniques in Scanning Force Microscopy (SFM) and their Application to Polymers and Polymer Composites. Vol. 164, pp. 87–210.
Murat, M. see Baschnagel, J.: Vol. 152, p. 41–156.
Muthukumar, M.: Modeling Polymer Crystallization. Vol. 191, pp. 241–274.
Muzzarelli, C. see Muzzarelli, R. A. A.: Vol. 186, pp. 151–209.
Muzzarelli, R. A. A. and *Muzzarelli, C.*: Chitosan Chemistry: Relevance to the Biomedical Sciences. Vol. 186, pp. 151–209.
Mylnikov, V.: Photoconducting Polymers. Vol. 115, pp. 1–88.

Nagy, A. see Majoros, I.: Vol. 112, pp. 1–11.
Naka, K. see Uemura, T.: Vol. 167, pp. 81–106.
Nakamura, A. see Mashima, K.: Vol. 133, pp. 1–52.
Nakayama, Y. see Mashima, K.: Vol. 133, pp. 1–52.
Narasinham, B. and *Peppas, N. A.*: The Physics of Polymer Dissolution: Modeling Approaches and Experimental Behavior. Vol. 128, pp. 157–208.
Nechaev, S. see Grosberg, A.: Vol. 106, pp. 1–30.
Neoh, K. G. see Kang, E. T.: Vol. 106, pp. 135–190.
Netz, R. R. see Holm, C.: Vol. 166, pp. 67–111.
Netz, R. R. see Rühe, J.: Vol. 165, pp. 79–150.
Newman, S. M. see Anseth, K. S.: Vol. 122, pp. 177–218.
Nijenhuis, K. te: Thermoreversible Networks. Vol. 130, pp. 1–252.
Ninan, K. N. see Reghunadhan Nair, C. P.: Vol. 155, pp. 1–99.
Nishi, T. see Jinnai, H.: Vol. 170, pp. 115–167.
Nishida, K. see Kaji, K.: Vol. 191, pp. 187–240.
Nishikawa, Y. see Jinnai, H.: Vol. 170, pp. 115–167.
Nishiyama, N. and *Kataoka, K.*: Nanostructured Devices Based on Block Copolymer Assemblies for Drug Delivery: Designing Structures for Enhanced Drug Function. Vol. 193, pp. 67–101.
Noid, D. W. see Otaigbe, J. U.: Vol. 154, pp. 1–86.
Noid, D. W. see Sumpter, B. G.: Vol. 116, pp. 27–72.
Nomura, M., Tobita, H. and *Suzuki, K.*: Emulsion Polymerization: Kinetic and Mechanistic Aspects. Vol. 175, pp. 1–128.
Northolt, M. G., Picken, S. J., Den Decker, M. G., Baltussen, J. J. M. and *Schlatmann, R.*: The Tensile Strength of Polymer Fibres. Vol. 178, pp. 1–108.
Novac, B. see Grubbs, R.: Vol. 102, pp. 47–72.
Novikov, V. V. see Privalko, V. P.: Vol. 119, pp. 31–78.
Nummila-Pakarinen, A. see Knuuttila, H.: Vol. 169, pp. 13–27.

Ober, C. K. see Li, M.: Vol. 190, pp. 183–226.
O'Brien, D. F., Armitage, B. A., Bennett, D. E. and *Lamparski, H. G.*: Polymerization and Domain Formation in Lipid Assemblies. Vol. 126, pp. 53–84.
Ogasawara, M.: Application of Pulse Radiolysis to the Study of Polymers and Polymerizations. Vol. 105, pp. 37–80.
Ohmae, M. see Kobayashi, S.: Vol. 194, pp. 159–210.
Okabe, H. see Matsushige, K.: Vol. 125, pp. 147–186.

Okada, M.: Ring-Opening Polymerization of Bicyclic and Spiro Compounds. Reactivities and Polymerization Mechanisms. Vol. 102, pp. 1–46.
Okada, K. see Hikosaka, M.: Vol. 191, pp. 137–186.
Okano, T.: Molecular Design of Temperature-Responsive Polymers as Intelligent Materials. Vol. 110, pp. 179–198.
Okay, O. see Funke, W.: Vol. 136, pp. 137–232.
Onuki, A.: Theory of Phase Transition in Polymer Gels. Vol. 109, pp. 63–120.
Oppermann, W. see Holm, C.: Vol. 166, pp. 1–27.
Oppermann, W. see Volk, N.: Vol. 166, pp. 29–65.
Osad'ko, I. S.: Selective Spectroscopy of Chromophore Doped Polymers and Glasses. Vol. 114, pp. 123–186.
Osakada, K. and *Takeuchi, D.*: Coordination Polymerization of Dienes, Allenes, and Methylenecycloalkanes. Vol. 171, pp. 137–194.
Otaigbe, J. U., Barnes, M. D., Fukui, K., Sumpter, B. G. and *Noid, D. W.*: Generation, Characterization, and Modeling of Polymer Micro- and Nano-Particles. Vol. 154, pp. 1–86.
Otsu, T. and *Matsumoto, A.*: Controlled Synthesis of Polymers Using the Iniferter Technique: Developments in Living Radical Polymerization. Vol. 136, pp. 75–138.

de Pablo, J. J. see Leontidis, E.: Vol. 116, pp. 283–318.
Padias, A. B. see Penelle, J.: Vol. 102, pp. 73–104.
Pascault, J.-P. see Williams, R. J. J.: Vol. 128, pp. 95–156.
Pasch, H.: Analysis of Complex Polymers by Interaction Chromatography. Vol. 128, pp. 1–46.
Pasch, H.: Hyphenated Techniques in Liquid Chromatography of Polymers. Vol. 150, pp. 1–66.
Pasut, G. and *Veronese, F. M.*: PEGylation of Proteins as Tailored Chemistry for Optimized Bioconjugates. Vol. 192, pp. 95–134.
Paul, W. see Baschnagel, J.: Vol. 152, pp. 41–156.
Paulsen, S. B. and *Barsett, H.*: Bioactive Pectic Polysaccharides. Vol. 186, pp. 69–101.
Pautzsch, T. see Klemm, E.: Vol. 177, pp. 53–90.
Penczek, P., Czub, P. and *Pielichowski, J.*: Unsaturated Polyester Resins: Chemistry and Technology. Vol. 184, pp. 1–95.
Penczek, P. see Batog, A. E.: Vol. 144, pp. 49–114.
Penczek, P. see Bogdal, D.: Vol. 163, pp. 193–263.
Penelle, J., Hall, H. K., Padias, A. B. and *Tanaka, H.*: Captodative Olefins in Polymer Chemistry. Vol. 102, pp. 73–104.
Peppas, N. A. see Bell, C. L.: Vol. 122, pp. 125–176.
Peppas, N. A. see Hassan, C. M.: Vol. 153, pp. 37–65.
Peppas, N. A. see Narasimhan, B.: Vol. 128, pp. 157–208.
Petersen, K. L. see Geil, P. H.: Vol. 180, pp. 89–159.
Pet'ko, I. P. see Batog, A. E.: Vol. 144, pp. 49–114.
Pheyghambarian, N. see Kippelen, B.: Vol. 161, pp. 87–156.
Pichot, C. see Hunkeler, D.: Vol. 112, pp. 115–134.
Picken, S. J. see Northolt, M. G.: Vol. 178, pp. 1–108.
Pielichowski, J. see Bogdal, D.: Vol. 163, pp. 193–263.
Pielichowski, J. see Penczek, P.: Vol. 184, pp. 1–95.
Pieper, T. see Kilian, H. G.: Vol. 108, pp. 49–90.
Pispas, S. see Pitsikalis, M.: Vol. 135, pp. 1–138.
Pispas, S. see Hadjichristidis, N.: Vol. 142, pp. 71–128.
Pitsikalis, M., Pispas, S., Mays, J. W. and *Hadjichristidis, N.*: Nonlinear Block Copolymer Architectures. Vol. 135, pp. 1–138.

Pitsikalis, M. see Hadjichristidis, N.: Vol. 142, pp. 71–128.
Pitsikalis, M. see Hadjichristidis, N.: Vol. 189, pp. 1–124.
Pleul, D. see Spange, S.: Vol. 165, pp. 43–78.
Plummer, C. J. G.: Microdeformation and Fracture in Bulk Polyolefins. Vol. 169, pp. 75–119.
Pötschke, D. see Dingenouts, N.: Vol. 144, pp. 1–48.
Pokrovskii, V. N.: The Mesoscopic Theory of the Slow Relaxation of Linear Macromolecules. Vol. 154, pp. 143–219.
Pospíšil, J.: Functionalized Oligomers and Polymers as Stabilizers for Conventional Polymers. Vol. 101, pp. 65–168.
Pospíšil, J.: Aromatic and Heterocyclic Amines in Polymer Stabilization. Vol. 124, pp. 87–190.
Powers, A. C. see Prokop, A.: Vol. 136, pp. 53–74.
Prasad, P. N. see Lin, T.-C.: Vol. 161, pp. 157–193.
Priddy, D. B.: Recent Advances in Styrene Polymerization. Vol. 111, pp. 67–114.
Priddy, D. B.: Thermal Discoloration Chemistry of Styrene-co-Acrylonitrile. Vol. 121, pp. 123–154.
Privalko, V. P. and *Novikov, V. V.*: Model Treatments of the Heat Conductivity of Heterogeneous Polymers. Vol. 119, pp. 31–78.
Prociak, A. see Bogdal, D.: Vol. 163, pp. 193–263.
Prokop, A., Hunkeler, D., DiMari, S., Haralson, M. A. and *Wang, T. G.*: Water Soluble Polymers for Immunoisolation I: Complex Coacervation and Cytotoxicity. Vol. 136, pp. 1–52.
Prokop, A., Hunkeler, D., Powers, A. C., Whitesell, R. R. and *Wang, T. G.*: Water Soluble Polymers for Immunoisolation II: Evaluation of Multicomponent Microencapsulation Systems. Vol. 136, pp. 53–74.
Prokop, A., Kozlov, E., Carlesso, G. and *Davidsen, J. M.*: Hydrogel-Based Colloidal Polymeric System for Protein and Drug Delivery: Physical and Chemical Characterization, Permeability Control and Applications. Vol. 160, pp. 119–174.
Pruitt, L. A.: The Effects of Radiation on the Structural and Mechanical Properties of Medical Polymers. Vol. 162, pp. 65–95.
Pudavar, H. E. see Lin, T.-C.: Vol. 161, pp. 157–193.
Pukánszky, B. and *Fekete, E.*: Adhesion and Surface Modification. Vol. 139, pp. 109–154.
Putnam, D. and *Kopecek, J.*: Polymer Conjugates with Anticancer Acitivity. Vol. 122, pp. 55–124.
Putra, E. G. R. see Ungar, G.: Vol. 180, pp. 45–87.

Quirk, R. P., Yoo, T., Lee, Y., M., Kim, J. and *Lee, B.*: Applications of 1,1-Diphenylethylene Chemistry in Anionic Synthesis of Polymers with Controlled Structures. Vol. 153, pp. 67–162.

Ramaraj, R. and *Kaneko, M.*: Metal Complex in Polymer Membrane as a Model for Photosynthetic Oxygen Evolving Center. Vol. 123, pp. 215–242.
Rangarajan, B. see Scranton, A. B.: Vol. 122, pp. 1–54.
Ranucci, E. see Söderqvist Lindblad, M.: Vol. 157, pp. 139–161.
Raphaël, E. see Léger, L.: Vol. 138, pp. 185–226.
Rastogi, S. and *Terry, A. E.*: Morphological implications of the interphase bridging crystalline and amorphous regions in semi-crystalline polymers. Vol. 180, pp. 161–194.
Ray, C. R. and *Moore, J. S.*: Supramolecular Organization of Foldable Phenylene Ethynylene Oligomers. Vol. 177, pp. 99–149.
Reddinger, J. L. and *Reynolds, J. R.*: Molecular Engineering of p-Conjugated Polymers. Vol. 145, pp. 57–122.

Reghunadhan Nair, C. P., Mathew, D. and *Ninan, K. N.*: Cyanate Ester Resins, Recent Developments. Vol. 155, pp. 1–99.
Reichert, K. H. see Hunkeler, D.: Vol. 112, pp. 115–134.
Reihmann, M. and *Ritter, H.*: Synthesis of Phenol Polymers Using Peroxidases. Vol. 194, pp. 1–49.
Rehahn, M., Mattice, W. L. and *Suter, U. W.*: Rotational Isomeric State Models in Macromolecular Systems. Vol. 131/132, pp. 1–475.
Rehahn, M. see Bohrisch, J.: Vol. 165, pp. 1–41.
Rehahn, M. see Holm, C.: Vol. 166, pp. 1–27.
Reineker, P. see Holm, C.: Vol. 166, pp. 67–111.
Reitberger, T. see Jacobson, K.: Vol. 169, pp. 151–176.
Ritter, H. see Reihmann, M.: Vol. 194, pp. 1–49.
Reynolds, J. R. see Reddinger, J. L.: Vol. 145, pp. 57–122.
Richter, D. see Ewen, B.: Vol. 134, pp. 1–130.
Richter, D., Monkenbusch, M. and *Colmenero, J.*: Neutron Spin Echo in Polymer Systems. Vol. 174, pp. 1–221.
Riegler, S. see Trimmel, G.: Vol. 176, pp. 43–87.
Ringsdorf, H. see Duncan, R.: Vol. 192, pp. 1–8.
Risse, W. see Grubbs, R.: Vol. 102, pp. 47–72.
Rivas, B. L. and *Geckeler, K. E.*: Synthesis and Metal Complexation of Poly(ethyleneimine) and Derivatives. Vol. 102, pp. 171–188.
Roberts, C. J. see Ellis, J. S.: Vol. 193, pp. 123–172.
Roberts, G. W. see Kennedy, K. A.: Vol. 175, pp. 329–346.
Robin, J. J.: The Use of Ozone in the Synthesis of New Polymers and the Modification of Polymers. Vol. 167, pp. 35–79.
Robin, J. J. see Boutevin, B.: Vol. 102, pp. 105–132.
Rodríguez-Pérez, M. A.: Crosslinked Polyolefin Foams: Production, Structure, Properties, and Applications. Vol. 184, pp. 97–126.
Roe, R.-J.: MD Simulation Study of Glass Transition and Short Time Dynamics in Polymer Liquids. Vol. 116, pp. 111–114.
Roovers, J. and *Comanita, B.*: Dendrimers and Dendrimer-Polymer Hybrids. Vol. 142, pp. 179–228.
Rothon, R. N.: Mineral Fillers in Thermoplastics: Filler Manufacture and Characterisation. Vol. 139, pp. 67–108.
de Rosa, C. see Auriemma, F.: Vol. 181, pp. 1–74.
Rozenberg, B. A. see Williams, R. J. J.: Vol. 128, pp. 95–156.
Rühe, J., Ballauff, M., Biesalski, M., Dziezok, P., Gröhn, F., Johannsmann, D., Houbenov, N., Hugenberg, N., Konradi, R., Minko, S., Motornov, M., Netz, R. R., Schmidt, M., Seidel, C., Stamm, M., Stephan, T., Usov, D. and *Zhang, H.*: Polyelectrolyte Brushes. Vol. 165, pp. 79–150.
Ruckenstein, E.: Concentrated Emulsion Polymerization. Vol. 127, pp. 1–58.
Ruiz-Taylor, L. see Mathieu, H. J.: Vol. 162, pp. 1–35.
Rusanov, A. L.: Novel Bis (Naphtalic Anhydrides) and Their Polyheteroarylenes with Improved Processability. Vol. 111, pp. 115–176.
Rusanov, A. L., Likhatchev, D., Kostoglodov, P. V., Müllen, K. and *Klapper, M.*: Proton-Exchanging Electrolyte Membranes Based on Aromatic Condensation Polymers. Vol. 179, pp. 83–134.
Russel, T. P. see Hedrick, J. L.: Vol. 141, pp. 1–44.
Russum, J. P. see Schork, F. J.: Vol. 175, pp. 129–255.
Rychly, J. see Lazár, M.: Vol. 102, pp. 189–222.

Ryner, M. see Stridsberg, K. M.: Vol. 157, pp. 27–51.
Ryzhov, V. A. see Bershtein, V. A.: Vol. 114, pp. 43–122.

Sabsai, O. Y. see Barshtein, G. R.: Vol. 101, pp. 1–28.
Saburov, V. V. see Zubov, V. P.: Vol. 104, pp. 135–176.
Saito, S., Konno, M. and *Inomata, H.*: Volume Phase Transition of N-Alkylacrylamide Gels. Vol. 109, pp. 207–232.
Samsonov, G. V. and *Kuznetsova, N. P.*: Crosslinked Polyelectrolytes in Biology. Vol. 104, pp. 1–50.
Santa Cruz, C. see Baltá-Calleja, F. J.: Vol. 108, pp. 1–48.
Santos, S. see Baschnagel, J.: Vol. 152, p. 41–156.
Satchi-Fainaro, R., Duncan, R. and *Barnes, C. M.*: Polymer Therapeutics for Cancer: Current Status and Future Challenges. Vol. 193, pp. 1–65.
Satchi-Fainaro, R. see Duncan, R.: Vol. 192, pp. 1–8.
Sato, T. and *Teramoto, A.*: Concentrated Solutions of Liquid-Christalline Polymers. Vol. 126, pp. 85–162.
Schaller, C. see Bohrisch, J.: Vol. 165, pp. 1–41.
Schäfer, R. see Köhler, W.: Vol. 151, pp. 1–59.
Scherf, U. and *Müllen, K.*: The Synthesis of Ladder Polymers. Vol. 123, pp. 1–40.
Sherman, S. see Kabanov, A. V.: Vol. 193, pp. 173–198.
Schlatmann, R. see Northolt, M. G.: Vol. 178, pp. 1–108.
Schmid, F. see Müller, M.: Vol. 185, pp. 1–58.
Schmidt, M. see Förster, S.: Vol. 120, pp. 51–134.
Schmidt, M. see Rühe, J.: Vol. 165, pp. 79–150.
Schmidt, M. see Volk, N.: Vol. 166, pp. 29–65.
Scholz, M.: Effects of Ion Radiation on Cells and Tissues. Vol. 162, pp. 97–158.
Schönherr, H. see Vancso, G. J.: Vol. 182, pp. 55–129.
Schopf, G. and *Koßmehl, G.*: Polythiophenes – Electrically Conductive Polymers. Vol. 129, pp. 1–145.
Schork, F. J., Luo, Y., Smulders, W., Russum, J. P., Butté, A. and *Fontenot, K.*: Miniemulsion Polymerization. Vol. 175, pp. 127–255.
Schulz, E. see Munz, M.: Vol. 164, pp. 97–210.
Schwahn, D.: Critical to Mean Field Crossover in Polymer Blends. Vol. 183, pp. 1–61.
Seppälä, J. see Löfgren, B.: Vol. 169, pp. 1–12.
Sturm, H. see Munz, M.: Vol. 164, pp. 87–210.
Schweizer, K. S.: Prism Theory of the Structure, Thermodynamics, and Phase Transitions of Polymer Liquids and Alloys. Vol. 116, pp. 319–378.
Scranton, A. B., Rangarajan, B. and *Klier, J.*: Biomedical Applications of Polyelectrolytes. Vol. 122, pp. 1–54.
Sefton, M. V. and *Stevenson, W. T. K.*: Microencapsulation of Live Animal Cells Using Polycrylates. Vol. 107, pp. 143–198.
Seidel, C. see Holm, C.: Vol. 166, pp. 67–111.
Seidel, C. see Rühe, J.: Vol. 165, pp. 79–150.
El Seoud, O. A. and *Heinze, T.*: Organic Esters of Cellulose: New Perspectives for Old Polymers. Vol. 186, pp. 103–149.
Shabat, D. see Amir, R. J.: Vol. 192, pp. 59–94.
Shamanin, V. V.: Bases of the Axiomatic Theory of Addition Polymerization. Vol. 112, pp. 135–180.
Shcherbina, M. A. see Ungar, G.: Vol. 180, pp. 45–87.

Sheiko, S. S.: Imaging of Polymers Using Scanning Force Microscopy: From Superstructures to Individual Molecules. Vol. 151, pp. 61–174.
Sherrington, D. C. see Cameron, N. R.: Vol. 126, pp. 163–214.
Sherrington, D. C. see Lin, J.: Vol. 111, pp. 177–220.
Sherrington, D. C. see Steinke, J.: Vol. 123, pp. 81–126.
Shibayama, M. see Tanaka, T.: Vol. 109, pp. 1–62.
Shiga, T.: Deformation and Viscoelastic Behavior of Polymer Gels in Electric Fields. Vol. 134, pp. 131–164.
Shim, H.-K. and *Jin, J.*: Light-Emitting Characteristics of Conjugated Polymers. Vol. 158, pp. 191–241.
Shoda, S. see Kobayashi, S.: Vol. 121, pp. 1–30.
Siegel, R. A.: Hydrophobic Weak Polyelectrolyte Gels: Studies of Swelling Equilibria and Kinetics. Vol. 109, pp. 233–268.
de Silva, D. S. M. see Ungar, G.: Vol. 180, pp. 45–87.
Silvestre, F. see Calmon-Decriaud, A.: Vol. 207, pp. 207–226.
Sillion, B. see Mison, P.: Vol. 140, pp. 137–180.
Simon, F. see Spange, S.: Vol. 165, pp. 43–78.
Simon, G. P. see Becker, O.: Vol. 179, pp. 29–82.
Simon, P. F. W. see Abetz, V.: Vol. 189, pp. 125–212.
Simonutti, R. see Sozzani, P.: Vol. 181, pp. 153–177.
Singh, A. and *Kaplan, D. L.*: In Vitro Enzyme-Induced Vinyl Polymerization. Vol. 194, pp. 211–224.
Singh, A. see Xu, P.: Vol. 194, pp. 69–94.
Singh, R. P. see Sivaram, S.: Vol. 101, pp. 169–216.
Singh, R. P. see Desai, S. M.: Vol. 169, pp. 231–293.
Sinha Ray, S. see Biswas, M.: Vol. 155, pp. 167–221.
Sivaram, S. and *Singh, R. P.*: Degradation and Stabilization of Ethylene-Propylene Copolymers and Their Blends: A Critical Review. Vol. 101, pp. 169–216.
Slugovc, C. see Trimmel, G.: Vol. 176, pp. 43–87.
Smulders, W. see Schork, F. J.: Vol. 175, pp. 129–255.
Soares, J. B. P. see Anantawaraskul, S.: Vol. 182, pp. 1–54.
Sozzani, P., Bracco, S., Comotti, A. and *Simonutti, R.*: Motional Phase Disorder of Polymer Chains as Crystallized to Hexagonal Lattices. Vol. 181, pp. 153–177.
Söderqvist Lindblad, M., Liu, Y., Albertsson, A.-C., Ranucci, E. and *Karlsson, S.*: Polymer from Renewable Resources. Vol. 157, pp. 139–161.
Spange, S., Meyer, T., Voigt, I., Eschner, M., Estel, K., Pleul, D. and *Simon, F.*: Poly(Vinylformamide-co-Vinylamine)/Inorganic Oxid Hybrid Materials. Vol. 165, pp. 43–78.
Stamm, M. see Möhwald, H.: Vol. 165, pp. 151–175.
Stamm, M. see Rühe, J.: Vol. 165, pp. 79–150.
Starodybtzev, S. see Khokhlov, A.: Vol. 109, pp. 121–172.
Stegeman, G. I. see Canva, M.: Vol. 158, pp. 87–121.
Steinke, J., Sherrington, D. C. and *Dunkin, I. R.*: Imprinting of Synthetic Polymers Using Molecular Templates. Vol. 123, pp. 81–126.
Stelzer, F. see Trimmel, G.: Vol. 176, pp. 43–87.
Stenberg, B. see Jacobson, K.: Vol. 169, pp. 151–176.
Stenzenberger, H. D.: Addition Polyimides. Vol. 117, pp. 165–220.
Stephan, T. see Rühe, J.: Vol. 165, pp. 79–150.
Stevenson, W. T. K. see Sefton, M. V.: Vol. 107, pp. 143–198.
Stridsberg, K. M., Ryner, M. and *Albertsson, A.-C.*: Controlled Ring-Opening Polymerization: Polymers with Designed Macromoleculars Architecture. Vol. 157, pp. 27–51.

Sturm, H. see *Munz, M.*: Vol. 164, pp. 87–210.
Suematsu, K.: Recent Progress of Gel Theory: Ring, Excluded Volume, and Dimension. Vol. 156, pp. 136–214.
Sugimoto, H. and *Inoue, S.*: Polymerization by Metalloporphyrin and Related Complexes. Vol. 146, pp. 39–120.
Suginome, M. and *Ito, Y.*: Transition Metal-Mediated Polymerization of Isocyanides. Vol. 171, pp. 77–136.
Sumpter, B. G., Noid, D. W., Liang, G. L. and *Wunderlich, B.*: Atomistic Dynamics of Macromolecular Crystals. Vol. 116, pp. 27–72.
Sumpter, B. G. see *Otaigbe, J. U.*: Vol. 154, pp. 1–86.
Sun, H.-B. and *Kawata, S.*: Two-Photon Photopolymerization and 3D Lithographic Microfabrication. Vol. 170, pp. 169–273.
Suter, U. W. see *Gusev, A. A.*: Vol. 116, pp. 207–248.
Suter, U. W. see *Leontidis, E.*: Vol. 116, pp. 283–318.
Suter, U. W. see *Rehahn, M.*: Vol. 131/132, pp. 1–475.
Suter, U. W. see *Baschnagel, J.*: Vol. 152, pp. 41–156.
Suzuki, A.: Phase Transition in Gels of Sub-Millimeter Size Induced by Interaction with Stimuli. Vol. 110, pp. 199–240.
Suzuki, A. and *Hirasa, O.*: An Approach to Artifical Muscle by Polymer Gels due to Micro-Phase Separation. Vol. 110, pp. 241–262.
Suzuki, K. see *Nomura, M.*: Vol. 175, pp. 1–128.
Swiatkiewicz, J. see *Lin, T.-C.*: Vol. 161, pp. 157–193.

Tagawa, S.: Radiation Effects on Ion Beams on Polymers. Vol. 105, pp. 99–116.
Taguet, A., Ameduri, B. and *Boutevin, B.*: Crosslinking of Vinylidene Fluoride-Containing Fluoropolymers. Vol. 184, pp. 127–211.
Takata, T., Kihara, N. and *Furusho, Y.*: Polyrotaxanes and Polycatenanes: Recent Advances in Syntheses and Applications of Polymers Comprising of Interlocked Structures. Vol. 171, pp. 1–75.
Takeuchi, D. see *Osakada, K.*: Vol. 171, pp. 137–194.
Tan, K. L. see *Kang, E. T.*: Vol. 106, pp. 135–190.
Tanaka, H. and *Shibayama, M.*: Phase Transition and Related Phenomena of Polymer Gels. Vol. 109, pp. 1–62.
Tanaka, T. see *Penelle, J.*: Vol. 102, pp. 73–104.
Tauer, K. see *Guyot, A.*: Vol. 111, pp. 43–66.
Tendler, S. J. B. see *Ellis, J. S.*: Vol. 193, pp. 123–172.
Teramoto, A. see *Sato, T.*: Vol. 126, pp. 85–162.
Terent'eva, J. P. and *Fridman, M. L.*: Compositions Based on Aminoresins. Vol. 101, pp. 29–64.
Terry, A. E. see *Rastogi, S.*: Vol. 180, pp. 161–194.
Theodorou, D. N. see *Dodd, L. R.*: Vol. 116, pp. 249–282.
Thomson, R. C., Wake, M. C., Yaszemski, M. J. and *Mikos, A. G.*: Biodegradable Polymer Scaffolds to Regenerate Organs. Vol. 122, pp. 245–274.
Thünemann, A. F., Müller, M., Dautzenberg, H., Joanny, J.-F. and *Löwen, H.*: Polyelectrolyte complexes. Vol. 166, pp. 113–171.
Tieke, B. see *v. Klitzing, R.*: Vol. 165, pp. 177–210.
Tobita, H. see *Nomura, M.*: Vol. 175, pp. 1–128.
Tokita, M.: Friction Between Polymer Networks of Gels and Solvent. Vol. 110, pp. 27–48.
Traser, S. see *Bohrisch, J.*: Vol. 165, pp. 1–41.
Tries, V. see *Baschnagel, J.*: Vol. 152, p. 41–156.

Trimmel, G., Riegler, S., Fuchs, G., Slugovc, C. and *Stelzer, F.*: Liquid Crystalline Polymers by Metathesis Polymerization. Vol. 176, pp. 43–87.
Tsuruta, T.: Contemporary Topics in Polymeric Materials for Biomedical Applications. Vol. 126, pp. 1–52.

Uemura, T., Naka, K. and *Chujo, Y.*: Functional Macromolecules with Electron-Donating Dithiafulvene Unit. Vol. 167, pp. 81–106.
Ungar, G., Putra, E. G. R., de Silva, D. S. M., Shcherbina, M. A. and *Waddon, A. J.*: The Effect of Self-Poisoning on Crystal Morphology and Growth Rates. Vol. 180, pp. 45–87.
Usov, D. see Rühe, J.: Vol. 165, pp. 79–150.
Usuki, A., Hasegawa, N. and *Kato, M.*: Polymer-Clay Nanocomposites. Vol. 179, pp. 135–195.
Uyama, H. and *Kobayashi, S.*: Enzymatic Synthesis and Properties of Polymers from Polyphenols. Vol. 194, pp. 51–67.
Uyama, H. and *Kobayashi, S.*: Enzymatic Synthesis of Polyesters via Polycondensation. Vol. 194, pp. 133–158.
Uyama, H. see Kobayashi, S.: Vol. 121, pp. 1–30.
Uyama, Y.: Surface Modification of Polymers by Grafting. Vol. 137, pp. 1–40.

Vancso, G. J., Hillborg, H. and *Schönherr, H.*: Chemical Composition of Polymer Surfaces Imaged by Atomic Force Microscopy and Complementary Approaches. Vol. 182, pp. 55–129.
Varma, I. K. see Albertsson, A.-C.: Vol. 157, pp. 99–138.
Vasilevskaya, V. see Khokhlov, A.: Vol. 109, pp. 121–172.
Vaskova, V. see Hunkeler, D.: Vol. 112, pp. 115–134.
Verdugo, P.: Polymer Gel Phase Transition in Condensation-Decondensation of Secretory Products. Vol. 110, pp. 145–156.
Veronese, F. M. see Pasut, G.: Vol. 192, pp. 95–134.
Vettegren, V. I. see Bronnikov, S. V.: Vol. 125, pp. 103–146.
Vilgis, T. A. see Holm, C.: Vol. 166, pp. 67–111.
Viovy, J.-L. and *Lesec, J.*: Separation of Macromolecules in Gels: Permeation Chromatography and Electrophoresis. Vol. 114, pp. 1–42.
Vlahos, C. see Hadjichristidis, N.: Vol. 142, pp. 71–128.
Voigt, I. see Spange, S.: Vol. 165, pp. 43–78.
Volk, N., Vollmer, D., Schmidt, M., Oppermann, W. and *Huber, K.*: Conformation and Phase Diagrams of Flexible Polyelectrolytes. Vol. 166, pp. 29–65.
Volksen, W.: Condensation Polyimides: Synthesis, Solution Behavior, and Imidization Characteristics. Vol. 117, pp. 111–164.
Volksen, W. see Hedrick, J. L.: Vol. 141, pp. 1–44.
Volksen, W. see Hedrick, J. L.: Vol. 147, pp. 61–112.
Vollmer, D. see Volk, N.: Vol. 166, pp. 29–65.
Voskerician, G. and *Weder, C.*: Electronic Properties of PAEs. Vol. 177, pp. 209–248.

Waddon, A. J. see Ungar, G.: Vol. 180, pp. 45–87.
Wagener, K. B. see Baughman, T. W.: Vol. 176, pp. 1–42.
Wagner, E. and *Kloeckner, J.*: Gene Delivery Using Polymer Therapeutics. Vol. 192, pp. 135–173.
Wake, M. C. see Thomson, R. C.: Vol. 122, pp. 245–274.
Wandrey, C., Hernández-Barajas, J. and *Hunkeler, D.*: Diallyldimethylammonium Chloride and its Polymers. Vol. 145, pp. 123–182.
Wang, K. L. see Cussler, E. L.: Vol. 110, pp. 67–80.

Wang, S.-Q.: Molecular Transitions and Dynamics at Polymer/Wall Interfaces: Origins of Flow Instabilities and Wall Slip. Vol. 138, pp. 227–276.
Wang, S.-Q. see Bhargava, R.: Vol. 163, pp. 137–191.
Wang, T. G. see Prokop, A.: Vol. 136, pp. 1–52; 53–74.
Wang, X. see Lin, T.-C.: Vol. 161, pp. 157–193.
Watanabe, K. see Hikosaka, M.: Vol. 191, pp. 137–186.
Webster, O. W.: Group Transfer Polymerization: Mechanism and Comparison with Other Methods of Controlled Polymerization of Acrylic Monomers. Vol. 167, pp. 1–34.
Weder, C. see Voskerician, G.: Vol. 177, pp. 209–248.
Weis, J.-J. and *Levesque, D.*: Simple Dipolar Fluids as Generic Models for Soft Matter. Vol. 185, pp. 163–225.
Whitesell, R. R. see Prokop, A.: Vol. 136, pp. 53–74.
Williams, R. A. see Geil, P. H.: Vol. 180, pp. 89–159.
Williams, R. J. J., Rozenberg, B. A. and *Pascault, J.-P.*: Reaction Induced Phase Separation in Modified Thermosetting Polymers. Vol. 128, pp. 95–156.
Winkler, R. G. see Holm, C.: Vol. 166, pp. 67–111.
Winter, H. H. and *Mours, M.*: Rheology of Polymers Near Liquid-Solid Transitions. Vol. 134, pp. 165–234.
Wittmeyer, P. see Bohrisch, J.: Vol. 165, pp. 1–41.
Wood-Adams, P. M. see Anantawaraskul, S.: Vol. 182, pp. 1–54.
Wu, C.: Laser Light Scattering Characterization of Special Intractable Macromolecules in Solution. Vol. 137, pp. 103–134.
Wunderlich, B. see Sumpter, B. G.: Vol. 116, pp. 27–72.

Xiang, M. see Jiang, M.: Vol. 146, pp. 121–194.
Xie, T. Y. see Hunkeler, D.: Vol. 112, pp. 115–134.
Xu, P., Singh, A. and *Kaplan, D. L.*: Enzymatic Catalysis in the Synthesis of Polyanilines and Derivatives of Polyanilines. Vol. 194, pp. 69–94.
Xu, P. see Geil, P. H.: Vol. 180, pp. 89–159.
Xu, Z., Hadjichristidis, N., Fetters, L. J. and *Mays, J. W.*: Structure/Chain-Flexibility Relationships of Polymers. Vol. 120, pp. 1–50.

Yagci, Y. and *Endo, T.*: N-Benzyl and N-Alkoxy Pyridium Salts as Thermal and Photochemical Initiators for Cationic Polymerization. Vol. 127, pp. 59–86.
Yamaguchi, I. see Yamamoto, T.: Vol. 177, pp. 181–208.
Yamamoto, T.: Molecular Dynamics Modeling of the Crystal-Melt Interfaces and the Growth of Chain Folded Lamellae. Vol. 191, pp. 37–85.
Yamamoto, T., Yamaguchi, I. and *Yasuda, T.*: PAEs with Heteroaromatic Rings. Vol. 177, pp. 181–208.
Yamaoka, H.: Polymer Materials for Fusion Reactors. Vol. 105, pp. 117–144.
Yamazaki, S. see Hikosaka, M.: Vol. 191, pp. 137–186.
Yannas, I. V.: Tissue Regeneration Templates Based on Collagen-Glycosaminoglycan Copolymers. Vol. 122, pp. 219–244.
Yang, J. see Geil, P. H.: Vol. 180, pp. 89–159.
Yang, J. S. see Jo, W. H.: Vol. 156, pp. 1–52.
Yasuda, H. and *Ihara, E.*: Rare Earth Metal-Initiated Living Polymerizations of Polar and Nonpolar Monomers. Vol. 133, pp. 53–102.
Yasuda, T. see Yamamoto, T.: Vol. 177, pp. 181–208.
Yaszemski, M. J. see Thomson, R. C.: Vol. 122, pp. 245–274.
Yoo, T. see Quirk, R. P.: Vol. 153, pp. 67–162.

Yoon, D. Y. see Hedrick, J. L.: Vol. 141, pp. 1–44.
Yoshida, H. and *Ichikawa, T.*: Electron Spin Studies of Free Radicals in Irradiated Polymers. Vol. 105, pp. 3–36.

Zhang, H. see Rühe, J.: Vol. 165, pp. 79–150.
Zhang, Y.: Synchrotron Radiation Direct Photo Etching of Polymers. Vol. 168, pp. 291–340.
Zheng, J. and *Swager, T. M.*: Poly(arylene ethynylene)s in Chemosensing and Biosensing. Vol. 177, pp. 151–177.
Zhou, H. see Jiang, M.: Vol. 146, pp. 121–194.
Zhou, Z. see Abe, A.: Vol. 181, pp. 121–152.
Zubov, V. P., Ivanov, A. E. and *Saburov, V. V.*: Polymer-Coated Adsorbents for the Separation of Biopolymers and Particles. Vol. 104, pp. 135–176.

Subject Index

Acrylic acid based monomers 212–215
Alkyl peroxides 5
Amylose 182–183
Antigenase 206

Benzyl peroxide 5
Bilirubin oxidase 77
Biocatalysis 211–224
Biocompatible templates 81–82
Biodegradable polymers 122–124, 133–158
Biopolymer-polyphenol conjugates 62–65
– chitosan conjugates 62
– poly(amino acid) conjugates 62–64
– synthetic polymers 64–65
Bisphenols 36–39

Cardanol 40, 42
Catalytic amino acid residues 107–108
Catechins 56–58
Catechol derivatives 52–55
– polymerization 52–53
– urushiol 53–55
Cellulose 167–182
– high-order molecular structures 174–177
– history 167–168
– reaction mechanism 171–174
– synthesis via enzymatic polymerization 168–170
Cellulose-related saccharides 177–182
Chemical recycling 122–127
– continuous degradation of polyesters 124–125
– lipase-catalyzed transformation 122–124
– poly(ester/carbonate urethane) 125–127
Chiral polyesters 145–146

Chitin 183–192
– enzymatic polymerization 183–185
– high-order molecular structures 190–192
– mechanistic aspects 185–190
Chitin oligomer derivatives 192
Chitosan conjugates 62
Chondroitin 200–202
– enzymatic polymerization 200–201
Chondroitin derivatives 201–202
Circular dichroism 91
Compound I 8–10
Compound II 10
Conjugates 51–67
Copolymerization 104–105
m-Cresol 33–35
p-Cresol 22, 23
Curing 51–67
Cyclic carbonate oligomers 120
Cyclic diacid anhydrides, polymerization 115–116
Cyclic oligomers 103–104

Degradative transformation 122–127
Diaminoazobenzene 72
Dicarboxylic acid 135–137
Dicarboxylic-acid-activated esters 138–142
Dicarboxylic acid alkyl esters 137–138
Dicarboxylic acid derivatives, lipase-catalyzed polycondensation 135–143
Diols 115–116
Dip-pen nanolithography 86–88

Electroactive polyaniline films 77
Enantioselective ring-opening polymerization 112–115
End-functionalized polyesters 109–110

Enzymatic oxidative polymerization 51–67
– catechol derivatives 52–55
– flavonoids 55–62
Enzymatic polymerization 159–210
– acrylic acid based monomers 212–215
– cellulose 168–170
– characteristics of 161–164
– chitin 183–185
– chondroitin 200–201
– hyaluronan 196–199
– kinetics and mechanism 220–222
– polysaccharides 164–205
– styrene based monomers 215–217
– vinyl polymers 211–224
Enzyme catalysis 1–49
– organic reactions 2
– polyanilines 69–94
– polymer chemistry 2–4
Enzyme immobilization 105
– horseradish peroxidase 74–75
Enzyme switch 74
Epigallocatechin gallate 60–62
Ethylphenol 42
Extracellular matrices 193

Flavonoids 55–62
– catechins 56–58
– polymeric 59–62
Fluorescent polymers 38–39
Functional polyesters, polycondensation 145–152
– chiral polyesters 145–146
– reactive polyesters 148–151
– sugar-containing polyesters 146–148

Gelatin 63–64
Glycosaminoglycans 193–203
Glycosylation 164–167
Green chemistry 1–49
Green fluorescent protein 62
Green solvents 127

Hammett equation 75
Hematin 77–78
– modified 82
HEPES 14
Horseradish peroxidase 1–49, 5, 71
– bisphenols 36–39
– catalytic cycle 7–10

– immobilization 74–75
– phenol polymerization 18–39
– m-substituted phenols 32–35
– o-substituted phenols 35–39
– p-substituted phenols 20–32
– unsubstituted phenols 18–20
Hyaluronan 194–200
– enzymatic polymerization 196–199
Hyaluronan derivatives 199–200
Hyaluronidase 63
Hydrogen peroxide 5
Hydrolases 162
Hydroquinone 30

Interfacial polymerization 86
Irreversible ping-pong mechanism 7
Isomerases 162

Laccase 40, 53
Lactones 95–132
– copolymerization 104–105
– enantioselective ring-opening polymerization 112–115
– five-membered 100
– four-membered 98–100
– naturally derived 110–111
– polymerization by PHB depolymerase 108–109
– regioselective ring-opening polymerization 109–112
– ring-opening polymerization 98–109
– seven-membered 102
– six-membered 100–102
Ligases 162
Lipase 133–158
Lipase-catalyzed polycondensation
– cyclic and polymeric anhydrides 142–143
– oxyacid derivatives 143–145
Lipase-catalyzed transformation 122–124
– green solvent 127
Lyases 162

Macrolides 102–103
Matrix metalloproteinases 63
Micellar polymerization 85
Michaelis-Menten kinetics 107
Model complexes 41–44

Subject Index

Oxidoreductases 162
Oxiranes 115–116
Oxyacid derivatives, lipase-catalyzed polycondensation 143–145

Peroxidases 4–18, 39–41
– biological functions 6
– catalyzed reactions 6
– definitions 4
– in nature 5–7
– phenol polymer formation 13–14
– side reactions 10–12
– solvent composition 14–18
PHB depolymerase 108–109
Phenol oxidation 10
Phenol polymerases, model complexes 41–44
Phenol polymerization
– horseradish peroxidase 18–39
– peroxidases 4–18, 39–41
Phenol polymers 1–49
– formation of 13–14
– thermal cross-linking 27
Phenols
– m-substituted 32–35
– o-substituted 35–39
– p-substituted 20–32
– unsubstituted 18–20
4-Phenoxyphenol 43
Poly(amino acid) conjugates 62–64
Polyaniline 71
– analytic studies 88–92
Polyaniline synthesis
– aqueous solutions 73
– enzymatic approach 73–74
– hematin 77–78
– horseradish peroxidase immobilization 74–75
– organic solvents 71–73
– substrate behavior and dubstituent sites 75–76
Polycarbonates
– enzymatic synthesis 154
– ring-opening polymerization 116–120
Polycondensation 133–158
– dicarboxylic acids 135–143
– functional polyesters 145–152
– oxyacid derivatives 143–145
Poly(ester/carbonate urethane) 125–127
Polyesters 95–132, 133–158

– continuous degradation 124–125
Polyhedral oligomeric silsesquioxane 64–65
Polymer chemistry, enzyme catalysis 2–4
Polyphenols 1–49
Polyphosphate, ring-opening polymerization 121–122
Polypyrroles 76
Poly[(R)-3-hydroxybutylate] polymerase 152–154
Polysaccharides
– enzymatic polymerization 164–205
– glycosylation 164–167
– unnatural 203–205
Polythioester, ring-opening polymerization 121–122
Polythiophenes 76
Poly(vinylphosphonic acid) 81
Pummerer's ketone 13, 22

Quercetin 58

Reactive oxygen species 59
Reactive polyesters 148–151
Regioselective ring-opening polymerization 109–112
– end-functionalized polyesters 109–110
– naturally derived lactones 110–111
– structurally characteristic polymers 111–112
Ring-opening polymerization 95–132
– enantioselective 112–115
– lipase-catalyzed 105–107
– polycarbonates 116–120
– polyphosphate 121–122
– polythioester 120–121
– regioselective 109–112
Rutin 58–59

Screen-printed electrode 74
Solid-state NMR studies 89–90
Solvent/buffer ratios 15
Soybean peroxidase 39
Structurally characteristic polymers 111–112
Styrene based monomers 215–217
m-Substituted phenols 32–35
o-Substituted phenols 35–39

p-Substituted phenols   20–32
Substituted trimethylene carbonate
   118–119
Sugar-containing polyesters   146–148
Sulfonated polystyrene template   79–81
Synthetic polymers   64–65

Template-assisted polymerization   79–82
– application of fast kinetics   85
– biocompatible templates   81–82
– modified hematin   82
– role of template   82–84
– SPS template   79–81
– substituent position   84
Thermal cross-linking   27
Transferases   162

Transition-state analog substrate   169
Trimethylene carbonate   117–118
– substituted   118–119
Tyrosine polymers   31

Unnatural polysaccharides   203–205
Unsubstituted phenols   18–20
Urushi   53–55

Vinyl polymers
– enzymatic polymerization   211–224
– vitamin C functionalized   218–220
Vitamin C   211–224

Xanthine oxidase   59
Xylan   183

Printing: Krips bv, Meppel
Binding: Stürtz, Würzburg